International
REVIEW OF
Neurobiology
Volume 66

International
REVIEW OF
Neurobiology

Volume 66

Neuroimaging

Part A

EDITED BY

MICHAEL F. GLABUS

Department of Psychiatry, School of Medicine
Louisiana State University Health Sciences Center
Shreveport, Louisiana

ELSEVIER

AMSTERDAM • BOSTON • HEIDELBERG • LONDON
NEW YORK • OXFORD • PARIS • SAN DIEGO
SAN FRANCISCO • SINGAPORE • SYDNEY • TOKYO
Academic Press is an imprint of Elsevier

Elsevier Academic Press
525 B Street, Suite 1900, San Diego, California 92101-4495, USA
84 Theobald's Road, London WC1X 8RR, UK

This book is printed on acid-free paper. ∞

For all information on all Elsevier Academic Press publications visit our Web site at www.books.elsevier.com

ISBN-13: 978-0-12-366867-7
ISBN-10: 0-12-366867-0

PRINTED IN THE UNITED STATES OF AMERICA
06 07 08 09 9 8 7 6 5 4 3 2

CONTENTS

Modeling Brain Responses

KARL J. FRISTON, WILLIAM PENNY, AND OLIVIER DAVID

Voxel-Based Morphometric Analysis Using Shape Transformations

CHRISTOS DAVATZIKOS

The Cutting Edge of fMRI and High-Field fMRI

DAE-SHIK KIM

Quantification of White Matter Using Diffusion-Tensor Imaging

HAE-JEONG PARK

Perfusion fMRI for Functional Neuroimaging

GEOFFREY K. AGUIRRE, JOHN A. DETRE, AND JIONGJIONG WANG

Functional Near-Infrared Spectroscopy: Potential and Limitations in Neuroimaging Studies

YOKO HOSHI

Neural Modeling and Functional Brain Imaging: The Interplay Between the Data-Fitting and Simulation Approaches

BARRY HORWITZ AND MICHAEL F. GLABUS

Combined EEG and fMRI Studies of Human Brain Function

V. MENON AND S. CROTTAZ-HERBETTE

CONTRIBUTORS

Numbers in parentheses indicate the pages on which the authors' contributions begin.

Autumn Agnoli (55), The fMRI Data Center, Dartmouth College, Hanover, New Hampshire 03755

Geoffrey K. Aguirre (213), Department of Neurology, University of Pennsylvania, Philadelphia, Pennsylvania 19104

S. Crottaz-Herbette (291), Department of Psychiatry & Behavioral Sciences, Stanford University School of Medicine Stanford, California 94305

Christos Davatzikos (125), Section of Biomedical Image Analysis, Department of Radiology, University of Pennsylvania, Philadelphia, Pennsylvania 19104

Olivier David (89), The Wellcome Department of Cognitive Neurology, University College London, London WC1N 3BG, United Kingdom

John A. Detre (213), Department of Neurology, University of Pennsylvania, Philadelphia, Pennsylvania 19104

James Dobson (55), The fMRI Data Center, Dartmouth College, Hanover, New Hampshire 03755

Karl J. Friston (89), The Wellcome Department of Cognitive Neurology, University College London, London WC1N 3BG, United Kingdom

Michael F. Glabus (267), Laboratory of Biological Psychiatry, Department of Psychiatry, Louisiana State University Health Sciences Center, Shreveport, Louisiana 71130

Barry Horwitz (267), Section on Brain Imaging and Modeling, National Institute on Deafness and Other Communication Disorders, National Institutes of Health, Bethesda, Maryland 20892

Yoko Hoshi (237), Department of Integrated Neuroscience, Tokyo Institute of Psychiatry, Tokyo 156-8585, Japan

Dae-Shik Kim (147), Center for Biomedical Imaging, Boston University School of Medicine, Boston, Massachusetts 02118

V. Menon (291), Department of Psychiatry & Behavioral Sciences, Program in Neuroscience, and Neuroscience Institute at Stanford, Stanford University School of Medicine, Stanford, California 94305

Hae-Jeong Park (167), Department of Diagnostic Radiology, Yonsei University, College of Medicine, Seoul 120-749, Korea

William Penny (89), The Wellcome Department of Cognitive Neurology, University College London, London WC1N 3BG, United Kingdom

Michael Schmitt (55), The fMRI Data Center, Dartmouth College, Hanover, New Hampshire 03755

Sarene Schumacher (55), The fMRI Data Center, Dartmouth College, Hanover, New Hampshire 03755

Paul M. Thompson (1), Laboratory of Neuro Imaging, Department of Neurology, UCLA School of Medicine, Los Angeles, California 90095

Arthur W. Toga (1), Laboratory of Neuro Imaging, Department of Neurology, UCLA School of Medicine, Los Angeles, California 90095

Bennet Vance (55), The fMRI Data Center, Dartmouth College, Hanover, New Hampshire 03755

John Darrell Van Horn (55), The fMRI Data Center, Dartmouth College, Hanover, New Hampshire 03755

Jens Voeckler (55), Distributed Systems Laboratory, Department of Computer Science, University of Chicago, Chicago, Illinois 60637

Jiongjiong Wang (213), Department of Radiology, University of Pennsylvania, Philadelphia, Pennsylvania 19104

John Wolfe (55), The fMRI Data Center, Dartmouth College, Hanover, New Hampshire 03755

Jeffrey Woodward (55), The fMRI Data Center, Dartmouth College, Hanover, New Hampshire 03755

Yong Zhao (55), Distributed Systems Laboratory, Department of Computer Science, University of Chicago, Chicago, Illinois 60637

PREFACE

The Decade of the Brain was the subject of Presidential Proclamation 6158 by President George Bush, and ended in the year 2000. The aim was "*to enhance public awareness of the benefits to be derived from brain research.*" During that period, there began a concerted and unprecedented effort in neuroscience research that has revolutionized our understanding of human brain function and the mind-brain interface. The revolution owed much of its success to the rapid development of sophisticated neuroimaging techniques. Concurrently, there has been a gradual merging of information and expertise from exponents of advanced methodologies deployed to other areas of neuroscience. The combination of these approaches offers arguably the most exciting prospects yet for the study of human behavior, embracing cognitive neurology, biological psychiatry, neuropsychology, and cognitive neuroscience.

Perhaps the most important developments in neuroimaging are those that will lead to direct clinical applications. Diffusion tensor imaging has huge potential for opening up our understanding of the nature of neurodevelopment and the course of neurological disorders. Perfusion fMRI has the capacity to reveal new information about cerebral perfusion dynamics at a time-scale and spatial resolution not possible with radioisotope imaging. High-field fMRI has progressed to the point where it is now possible to view the individual layers of the cerebral cortex, while structural MRI research has facilitated a detailed picture of normal and abnormal neurodevelopmental trajectories not seen before.

In March 2000, in his opening address to the 10th Annual Conference of the Rotman Research Institute, "*The Frontal Lobes*", Dr Donald T. Stuss stated that the next decade would be the *Decade of The Mind* and emphasized the central role that neuroimaging would have in it. This book provides a timely review of what is state-of-the-art in the field of neuroimaging and details techniques that have evolved from being purely experimental in nature to becoming standard instruments of research in the behavioral neurosciences.

This volume is divided into two parts with Part A focusing on neuroimaging methods and Part B on neuroimaging applications in research and clinical environments. In Part A, the first chapter by Drs. Toga and Thompson details the use of brain atlases for the study of normal and diseased populations and their role in future research. The concept of the neuroimaging database as a resource for data sharing and meta-analysis was recognized in 1994. However, Dr. Van Horn and colleagues, with particular reference to the fMRI Data Center at Dartmouth College, NH, elucidate its role as a resource for scientific discovery. The next chapter by Dr. Friston and colleagues describes methods for modeling brain responses with particular reference to fMRI and EEG models. These models are used to draw inference on brain function, and answer questions about how interactions among brain regions are mediated. The evolution of automated voxel-based morphometric analysis methods has revolutionized the analysis of structural MRI data analysis. Dr. Davatzikos reviews this approach and focuses on the hypothesis-free, unbiased, shape transformation methods in Chapter 4. This chapter presents data on the application of these methods to the study of dementia, schizophrenia, and genetic influence on brain development. In Chapter 5, Dr. Kim reviews the rapid evolution in the last decade of fMRI BOLD imaging through to the cutting edge of fMRI and high field fMRI. Dr. Kim describes how inclusion of the "initial dip" of the BOLD response can significantly improve spatial specificity and how results from optical imaging confirm the validity of this method. Chapter 6 is a review by Dr. Park of the relatively new neuroimaging method for studying white matter tractography. The theoretical construct for diffusion weighted and diffusion tensor imaging is outlined with the concepts for visual presentation of white matter tractography, which are beautifully illustrated. The clinical application of diffusion tensor imaging based on white matter pathologies includes brain ischemia, trauma, tumor, focal epilepsy, multiple sclerosis, Alzheimer's disease, and schizophrenia. The recent advances in fiber tractography are likely to prove to be an invaluable tool in modeling studies of cognition, where anatomical connectivity is guided by microelectrode recording and post-mortem tracer studies in primate brain. One of the most interesting recent developments in fMRI has been the evolution of the perfusion fMRI method. Dr. Aguirre and colleagues review the technique of arterial spin labeling and describe the advantages of this method over conventional BOLD imaging. The principal advantage is that ASL is well suited to studies where changes in cognitive processes occur over relatively long time-periods, which is problematic with conventional BOLD where low frequency noise is dominant. ASL also demonstrates equivalent spatial resolution to "initial dip" BOLD imaging. Near-infrared spectroscopy has been used to measure changes in Hb in brain tissue since 1977. In Chapter 8, Dr. Hoshi reviews the method of functional NIRS, which has been widely applied in functional neuroimaging studies since the early 1990s. Time-resolved spectroscopy and phase-resolved spectroscopy are now established techniques for overcoming some of the quantification problems with this method of neuroimaging. However, despite a number of theoretical and experimental NIR investigations, the mechanism for light propagation in the human head is not fully understood. Nevertheless, relatively inexpensive multiple-sensor systems are appearing in research environments and fNIRS will open new dimensions in brain research and clinical applications in the future. Chapter 9, by Drs. Horwitz and Glabus, complements the work of Dr. Friston and colleagues in Chapter 3 with a review of methods for systems-level and large-scale neural models. This work reviews techniques for data-fitting and simulation and shows how these two types of modeling can be combined to obtain insight into interregional neural interactions. Chapter 10 formally introduces the technique for combining two different neuroimaging methods: fMRI and EEG. Dr. Menon describes the neurophysiological basis of the emergent EEG and evoked potential signals and their relationship with the BOLD response. A clear description of the problems encountered when recording EEG in the MRI environment and proposed solutions to them are given. The essentials for integrating ERP and fMRI data are described, including a review of dipole modeling methods.

Part B of this volume is a collection of chapters that review the application of neuroimaging methods in clinical diagnostic and research settings. Chapter 1 reviews the results from a number of studies of anxiety disorders. Drs. Nitschke and Heller give balanced reviews of cognitive and neuroimaging findings for each of six illnesses: Obsessive-compulsive disorder, posttraumatic stress disorder, panic disorder, specific phobia, social phobia, and generalized anxiety disorder. The neuroimaging findings reviewed are from structural (MRI) and functional (EEG, SPECT, PET, fMRI) neuroimaging studies. Dr. Ebmeier and colleagues outline the diagnostic criteria and review the neuroimaging findings for the dementias in diagnostic and research settings in Chapter 2. Neuroimaging has consistently played an important role in clinical diagnostic and research studies, and a review of cognitive findings in parallel with brain structural and functional neuroimaging studies (MRI, PET, SPECT), is presented. The role of functional neuroimaging in diagnosis of the dementias, and the use of neuroimaging in differential diagnosis is described with clarity. Avolition is a key feature in depression and is probably linked to symptom heterogeneity, disease course, and differential treatment response in mood disorders. In Chapter 3, Drs. Nitschke and Mackiewicz focus on morphometric and functional neuroimaging research in depression, with particular reference to the role of the dorsolateral prefrontal cortex and anterior cingulate in volition. The authors present an in-depth review of findings that reveal detailed information on the neurobiological substrate, and its manifestation in the cognitive domain. Schizophrenia is the most debilitating and probably the most intensely studied mental illness. Neuroimaging studies in the past decade in particular have shed new light on the pathophysiology of this disorder. Chapter 4 by Dr. Tost and colleagues is a review of functional neuroimaging research in schizophrenia and focuses on voluntary motor control, the visual system, auditory system, selective attention, working memory, medication effects, and molecular imaging. One of the most puzzling and challenging areas of diagnostic medicine is the group of conditions known as the functional somatic syndromes. These include chronic fatigue syndrome, fibromyalgia syndrome, irritable bowel syndrome and other miscellaneous disorders. Chapter 5 by Dr. Wood gives an enlightening exposition on the possible neurobiological substrate for this group of disorders by reviewing and dissecting the body of emerging evidence. This chapter reviews many functional and structural neuroimaging studies and shows that, despite an enduring lack of peripheral pathology, a combination of methods offers the best possibility for characterizing these conditions, thus providing hope for new diagnostic criteria and treatment options. Dr. Minagar and colleagues describe the role of neuroimaging in the diagnosis and treatment of multiple sclerosis in Chapter 6. This chapter outlines the clinical manifestations and the pathology as it relates to neuroimaging. Sections 4 and 5 are devoted to reviewing findings in MS using T1- and T2-weighted MRI, MRS, MTI, fMRI, DTI, PET and SPECT and is probably the broadest and most current review of neuroimaging MS available. The last section reviews neuroimaging and neuropsychological findings that provide important complementary diagnostic information. This chapter will have broad appeal to diagnosticians and subsections on advanced neuroimaging techniques and psychosocial manifestations will be of particular interest to researchers. Drs. Kelley and Gonzlaez-Toledo review neuroimaging of stroke in Chapter 7. The chapter reviews what is current in stroke neuroimaging research including the use of fMRI, MRS, and susceptibility weighted imaging. This chapter will appeal in particular to clinicians involved in the diagnosis and treatment of stroke. The neuroimaging environment is particularly challenging for the study of young children and, until recently, such studies have mainly been concerned with diagnosis. The evolution of new technology and fast MRI sequences has made it possible to acquire neuroimaging data in a more compact period and the incidence of pediatric neuroimaging studies, including *in utero* MRI, has increased dramatically in recent years. Drs. Seyffert and Castellanos devote Chapter 8 to review functional MRI in pediatric neurobehavioral disorders. This chapter covers a large range of fMRI studies of typically developing children and adolescents that have examined executive function, language, and mood along with studies of Autism and Autism Spectrum Disorders, Dyslexia, Attention Deficit Hyperactivity Disorder, Conduct Disorder, Tourette's Disorder, Anxiety and Depressive Disorders, Anorexia, and juvenile Bipolar Disorder. The authors outline some of the problems associated with pediatric neuroimaging and offer some thought provoking ideas for overcoming them, including the use of combinations of imaging modalities. Chapter 9 superbly complements Chapter 8 and covers structural MRI and brain development. Dr. Thompson and colleagues have been at the forefront of this field for a number of years and their work focuses on neurodevelopmental trajectories. The use of atlas-parcellation and anatomical mapping methods are described, along with the results of applying them to healthy brain maturation, and visualizing how it is derailed in neuropsychiatric disorders such as schizophrenia, autism, fetal alcohol syndrome, and Williams syndrome. The illustrative work of Dr. Thompson is, as usual, a visual feast. One of the most important scientific developments in recent years has been the application of neuroimaging to the investigation of the genetics of brain structure and function. Neuroimaging has already proven to be a powerful tool for characterizing candidate endophenotypes in a range of mental illnesses. In Chapter 10, Dr Winterer and colleagues present a review of some groundbreaking studies: EEG and ERP studies that quantify electrophysiological phenotypes; structural brain studies that reflect chromosomal abnormalities; PET studies that show correlation between neuroreceptor binding and phenotypic severity; fMRI-phenotype studies of cognition. These approaches span a series of studies in dementia, mental instability, schizophrenia, and mood and anxiety disorders. A large part of this chapter is devoted to the pioneering work of this group in studying the role of the 5-HTTLPR s allele in accounting for individual differences in emergent emotional behavior, with particular focus on the amygdala. The final chapter in Part B is a review of the methods and applications of neuroreceptor imaging in psychiatry using PET and SPECT, by Dr. Frankle and colleagues. This chapter provides an excellent theoretical construct and describes the model-based methods for quantitating the outcome measure derived in neuroreceptor imaging studies, the binding potential. The distribution volume and rate constant equations are rigorously derived and the authors provide the theoretical foundation for deriving outcome measures using equilibrium analysis, kinetic analysis, and the graphical method. The final part of the chapter focuses on a large range of mental illnesses (schizophrenia, affective disorders, major depressive disorders, bipolar disorder, anxiety, panic disorder, social phobia, OCD, PTSD, and substance abuse) and comprehensively reviews the extant literature on each with specific reference to each relevant neuroreceptor system, including dopamine, serotonin, GABA and the effect of different classes of medication. This chapter will prove equally valuable to students and those involved in research, due to the depth and breadth of material provided.

I would like to thank all the authors, who contributed so much to this book and made it what it is. It is their work and they deserve great credit for their efforts. I would also like to thank Cindy Minor, Senior Development Editor at Elsevier, for her constant support and assistance in bringing this large body of work to press. On behalf of the contributing authors, I would like to express our collective appreciation to the readers for their interest in this area. I would also like to apologize to those readers that feel their particular area of interest has been overlooked in this work. To an extent this was inevitable, given the magnitude of the neuroimaging field. Finally, I hope that all readers will refer to this work as a source of stimulation in their future scientific endeavours.

Michael F. Glabus

BRAIN ATLASES OF NORMAL AND DISEASED POPULATIONS

Arthur W. Toga and Paul M. Thompson

Laboratory of Neuro Imaging, Department of Neurology, UCLA School of Medicine
Los Angeles, California 90095

INTERNATIONAL REVIEW OF
NEUROBIOLOGY, VOL. 66
DOI: 10.1016/S0074-7742(05)66001-1

Human beings are highly diverse. The ability to statistically and visually compare brain-image data from multiple individuals is essential for the understanding of normal variability within a particular population as well as for the differentiation of normal populations from diseased. This chapter introduces the use of brain atlases to describe specific subpopulations, measure their variability, and characterize the structural differences between them. We describe extensions to the atlas concept to enable comparisons with other data, to act as a framework for other studies, and to provide database-like features. Utilizing data from structural MRIs, we have built atlases with defined coordinate systems, creating a framework to map data from functional, histological, and other studies of the same population. This paper describes the basic approach to the calculation of atlases and the underlying mathematical constructs that enable it. We include examples of atlas results from various normal and diseased populations.

I. Introduction

The concept of atlases in neuroscience is an old one (see Toga and Mazziotta, 1996 for a historical survey). Cartographic approaches have been used for centuries to identify and target specific regions in the brain and to establish relationships between a coordinate and a structure. In a brain atlas, spatial correspondence between a point in a coordinate system and a neuroanatomic feature was traditionally (and is sometimes still) achieved through definition of a set of rules governing pitch, roll, and yaw between a given brain and the atlas. Originally, the brain atlas was built around a rigid Cartesian coordinate system, but modern brain atlases incorporate flexible, computable systems that accommodate the sometimes considerable differences between a brain and the atlas. The transition from a static atlas representation to a computational one has resulted in dramatic extensions to the atlas concept. The brain atlas is now equivalent to a database: it incorporates a multitude of data points, all of which are organized, relational, extendable, and queriable. Originally, the brain atlas was purely neuroanatomical; now it can include functional information like descriptions of gene expression, receptor patterns, or connectivity. Originally, the brain atlas was built from a single, supposedly representative, example of the species at a single age (or weight); now it represents whole populations of individuals and statistically incorporates their distribution. Originally, the brain atlas was three dimensional (3D) (although most published book-form atlases were not accurate in the z axis). Now it incorporates time (making it four dimensional [4D]), from a scale of milliseconds (functional activity) to years (development and aging). Originally, the brain atlas depicted only the normal brain; now it can be used to describe a particular disease, and it may soon become a routinely applied biomarker for the detection of early stages in a pathological process or for assessing drug effects in clinical trials. This chapter reviews the background, evolution, and application of brain atlases in health and disease.

II. Background

There are numerous depictions of brain structure, and there are various methods to elucidate them. Comprehensive maps of brain structure have been created at a variety of spatial scales from each of the following media: 3D tomographic images (Damasio, 1995), anatomical specimens (Duvernoy, 1991; Ono *et al.*, 1990; Talairach and Szikla, 1967; Talairach and Tournoux, 1988), and a variety of histological preparations that reveal regional cytoarchitecture (Brodmann, 1909), myelination patterns (Smith, 1907), protein densities, and mRNA distributions. Most early atlases of the human brain were derived from one, or at best a few, individual postmortem specimens (Brodmann, 1909; Matsui and Hirano, 1978; Ono *et al.*, 1990; Schaltenbrand and Bailey, 1959; Schaltenbrand and Wahren, 1977; Talairach and Szikla, 1967; Talairach and Tournoux, 1988). Such atlases take the form of anatomical references or represent a particular feature of the brain (Van Buren and Borke, 1972; Van Buren and Maccubin, 1962), such as a specific neurochemical distribution (Mansour *et al.*, 1995) or the cellular architecture of the cerebral cortex (Brodmann, 1909).

A. Cryosection Imaging

Several digital atlases have been developed using photographic images of cryoplaned frozen specimens (Bohm *et al.*, 1983; Greitz *et al.*, 1991). Photographed material, though it provides superior anatomical detail, has limitations. For accurate correlations, data must be placed in a plane equivalent to that of the image of interest. Digital imaging, however, overcomes many limitations of conventional film photography. With digital color cameras that have 1024^2 pixel-imaging planes and 24 bits per pixel, spatial resolution can be as high as 100 microns per voxel for whole-human-head cadaver preparations, or higher for isolated brain regions (Toga *et al.*, 1994). Cryosectioning in micron increments permits data collection with high spatial resolution in the axis orthogonal to the sectioning plane. Acquisition of images in series directly from the consistently positioned cryoplaned blockface also eliminates the need for serial image registration prior to reconstruction. Serial images can be reconstructed as a 3D anatomical volume that is amenable to various resampling and positioning schemes.

Beyond the anatomical atlases based on postmortem and histological material mentioned above, the use of magnetic resonance (MR) to acquire detailed descriptions of anatomy *in vivo* is a driving force in brain-mapping research. MR imaging (MRI) data have the advantage of (largely) retaining the spatial integrity of the data by maintaining the intrinsic three-axis registration and simple volumetric coordinates (Damasio, 1995). Furthermore, there is an increasing ability to image a variety of structural (functional and chemical) signatures, including tissue type and white-matter tracts. Unfortunately, even high-resolution MR atlases, with up to 150 slices, a section thickness of 1 mm, and 256^2 pixel-imaging planes (Evans *et al.*, 1991; Lehmann *et al.*, 1991) still result in resolutions (and contrast levels) that inhibit accurate and precise depictions of the complexity of many neuroanatomical structures. However, technological advances continue to push improvements in spatial and contrast resolution. A simple innovation in the collection of atlas-quality MRIs involves the averaging of multiple coregistered scans ($N = 27$) from a single subject to overcome the lack of contrast and relatively poor signal to noise (Holmes *et al.*, 1998). Some atlases provide coregistered data from multiple MRI pulse sequences (e.g., proton density, T1- and T2-weighted images) to provide additional features. This allows multispectral characterization of the anatomy, in the sense that the contrast depends on different physical mechanisms, reflecting lipid content, for example, or water diffusivity in different tissues (Van Leemput *et al.*, 1999).

Other brain atlases describe function, quantified by positron emission tomography (PET) (Minoshima *et al.*, 1994), functional MRI (Le Bihan, 1996), or electrophysiology (Avoli *et al.*, 1991; Palovcik *et al.*, 1992). Maps also have been developed to represent neuronal connectivity and circuitry (Annese *et al.*, 2004; Van Essen and Maunsell, 1983) based on compilations of empirical evidence (Berger, 1929; Brodmann, 1909; Penfield and Boldrey, 1937). Recently, chemoarchitectural maps that describe receptor densities relative to functional attributes have been created (Zilles *et al.*, 2002), and these help bridge functional and anatomical observations.

B. Multimodality Atlases

Characterizing a single subject with multiple imaging devices clearly combines the strengths of each imaging modality. In the Visible Human Project (Spitzer, 1996), two cadavers (one male and one female) were cryoplaned and imaged at 1.0-mm intervals, and their entire bodies were also reconstructed via 5,000 postmortem CT and MRI images. The resulting digital data sets consist of over 15 gigabytes (GB) of image data. While not an atlas *per se*, the Visible Human imagery has sufficient quality and accessibility to make it a test platform for developing methods and standards (Spitzer, 1996). The data have served as a foundation for the development of related atlases of regions of the cerebral cortex (Drury and Van Essen, 1997), and for high-quality brain models and visualizations (Schiemann *et al.*, 1996; Stewart *et al.*, 1996). Using multimodality data from a patient with a localized pathology, and more recently the Visible Human data, Höhne and coworkers developed a commercially available brain atlas called VOXEL-MAN designed for teaching neuroanatomy (Höhne *et al.*, 1990, 1992; Pommert *et al.*, 1994; Tiede *et al.*, 1993). Data from single subjects, premortem and postmortem, provide a unique view into the relationship between *in vivo* imaging and histological assessment. Mega *et al.* (1997) scanned Alzheimer's patients in the terminal stages of their disease using both MRI and PET. These data were combined with postmortem 3D histological images showing the gross anatomy (Toga *et al.*, 1994) and a Gallyas stain of neurofibrillary tangles. This multimodal, but single-subject, atlas of Alzheimer's disease relates the anatomical and histopathological underpinnings of *in vivo* metabolic and perfusion maps of this disease.

However, none of these atlases was developed with consideration for any of the others. Thus, disappointingly, each of the brain maps contained in these atlases (postmortem and *in vivo*) has a different spatial scale and resolution, utilizes different data structures, emphasizes different structural or functional characteristics, and is inherently incompatible with the others. Though each mapping strategy clearly has its place in a collective effort to map the brain, unless certain integrative approaches are implemented (including spatial normalization), these brain maps can only remain as individual and independent efforts. An integrated, comprehensive whole, combining the many diverse mapping approaches, will never be achieved.

Population-based brain atlases (the focus of this chapter) offer a powerful framework with which to synthesize the results of disparate imaging studies. These atlases use novel analytical tools to fuse data across subjects, modalities, and times. They detect group-specific features not apparent in individual scans. Population-based atlases can be stratified into subpopulations to reflect a particular (clinical or demographic) subgroup (Toga and Mazziotta, 1996). The effort to design appropriate reference systems for brain data presents considerable challenges, since these systems must capture how brain structure and function vary in large populations, across age and gender, in different disease states, across imaging modalities, and even across species.

Imaging algorithms are also significantly improving the flexibility of digital brain atlases. *Deformable brain atlases* are adaptable in that they can be individualized to reflect the anatomy of new subjects, and *probabilistic atlases* retain information on cross-subject variations in brain structure and function. These atlases are powerful new tools with broad clinical and research applications (Kikinis *et al.*, 1996; Mangin *et al.*, 2004; Roland and Zilles, 1994; Toga and Thompson, 2003). Despite the significant challenges in expanding the atlas concept to include the time dimension, dynamic brain atlases include probabilistic information on growth rates that may assist research into pediatric disorders (Gogtay *et al.*, 2004; Sowell *et al.*, 2004; Thompson *et al.*, 2000), lifelong normal changes (Sowell *et al.*, 2003), or rates of degenerative tissue loss in aging and dementing diseases (Leow *et al.*, 2005; Sowell *et al.*, 2003; Thompson *et al.*, 2003).

Anatomical variations severely hamper the integration and comparison of data across subjects and groups (Meltzer and Frost, 1994; Woods, 1996). Motivated by the need to standardize data across subjects, researchers developed analytic methods to remove size and shape differences that distinguish one brain from another (Talairach and Tournoux, 1988). Spatial transformation of individual brain maps onto a 3D digital brain atlas removes subject-specific shape variations and allows subsequent comparison of brain structure or function between individuals (Ashburner and Friston, 2000; Collins *et al.*, 1994; Thompson *et al.*, 2004). Conversely, deformable brain atlases are based on the idea that a digital brain atlas can be elastically deformed to fit a new subject's anatomy (Christensen *et al.*, 1996; Evans *et al.*, 1991; Haller *et al.*, 1997; Toga, 1998). High-dimensional brain *warping* algorithms (Avants and Gee, 2004; Bro-Nielsen and Gramkow, 1996; Christensen *et al.*, 1996; Collins *et al.*, 1994; Davatzikos, 1996; Miller, 2004; Thirion, 1995; Thompson and Toga, 1996) affect the transfer of 3D maps of structure, maps of function, and other descriptions, such as information on cytoarchitecture, histological content, and neurochemical content (Mega *et al.*, 1997) onto the scan of any subject.

The coordinate system used to equate brain topology with an index must include carefully selected features common to all brains. Further, these features must be readily identifiable and sufficiently distributed anatomically to avoid bias. Once parameters are defined, rigorous systems for matching or spatially normalizing a brain to this coordinate system must be utilized. This allows individual data to be transformed to match the space occupied by the atlas. In the Talairach stereotaxic system (Talairach and Tournoux, 1988; Talairach *et al.*, 1967), piecewise affine transformations are applied to 12 rectangular regions of the brain, defined by vectors from the anterior and posterior commissures to the extrema of the cortex. These transformations reposition the anterior commissure of the subject's scan at the origin of the 3D coordinate space, vertically align the interhemispheric plane, and horizontally orient the line connecting the two commissures. Each point in the incoming brain image, after it is registered into the atlas space, is labeled by an (x,y,z) address indexed to the atlas brain. Although originally developed to help interpret brain-stem and ventricular studies acquired using pneumoencephalography (Talairach *et al.*, 1967), the Talairach stereotaxic system rapidly became an international standard for reporting functional activation sites in PET studies, allowing researchers to compare and contrast results from different laboratories (Fox *et al.*, 1985, 1988; Friston *et al.*, 1995).

Perhaps surprisingly, few atlases of neuropathology use a standardized three-dimensional coordinate system to integrate data across patients, techniques, and acquisitions. Atlases with a well-defined coordinate space (Evans *et al.*, 1992; Friston *et al.*, 1995; Van Essen, 2004), together with algorithms to align data with them (Toga, 1998), have enabled the pooling of brain-mapping data from multiple subjects and sources, including large patient populations. Automated algorithms can then capitalize on atlas descriptions of anatomical variance to guide image segmentation (Fischl *et al.*, 2004; Pitiot *et al.*, 2004), tissue classification (Duncan *et al.*, 2004; Shattuck *et al.*, 2001), functional image analysis (Dinov *et al.*, 2000; Rasser *et al.*, 2005), and pathology detection (Thompson *et al.*, 2004).

Without methods to overcome the problems of anatomical variability, the statistical power to resolve disease and treatment effects is seriously undermined. First, normal anatomical variation results in an overlap between diseased and normal subjects on most anatomical measures. Second, these difficulties are exacerbated in disease-related changes such as atrophy (Leow *et al.*, 2005; Mega *et al.*, 1998; Meltzer and Frost, 1994; Thompson *et al.*, 1998; Woods, 1996) or other progressive and dynamic anatomical changes. In the case of the cortex, profiles of gray-matter loss are difficult to calibrate against a reference population due to the lack of statistics on expected changes in these populations. To fully capitalize on neuroimaging data in disease, researchers will need an appropriately complex mathematical framework to address these challenges. Once resolved, brain maps can then be compared across patients and across time (Janke *et al.*, 2001; Mazziotta *et al.*, 1995; Miller, 2004; Studholme *et al.*, 2004; Thompson *et al.*, 2004).

We have constructed and applied this atlas strategy in normal populations as well as in descriptions of diseased individuals, which can be designed to reflect the unique anatomy and physiology of a particular clinical subpopulation. The resulting statistical representations of anatomy have revealed the profile of structural brain deficits in studies of Alzheimer's disease (Thompson *et al.*, 2003), epilepsy (Lin *et al.*, 2003), unipolar depression (Ballmeier *et al.*, 2004), childhood and adult-onset schizophrenia (Cannon *et al.*, 2002; Narr *et al.*, 2004; Thompson *et al.*, 2001), attention-deficit/hyperactivity disorder (Sowell *et al.*, 2003), fetal alcohol syndrome (Sowell *et al.*, 2002), Tourette's syndrome (Sowell *et al.*, 2004), bipolar disorder (Gogtay *et al.*, 2004), autism (Leow *et al.*, 2005; Levitt *et al.*, 2003), Williams syndrome (Thompson *et al.*, 2005), and methamphetamine abuse (Thompson *et al.*, 2004). Based on well-characterized patient groups, these atlases contain composite maps and visualizations of structural variability, asymmetry, and group-specific differences. This quantitative framework can be used to recognize anomalies or patterns and label structures in new patients. Because they retain information on group anatomical variability, disease-specific atlases are a type of probabilistic atlas specialized to represent a particular clinical group. The resulting atlases can identify patterns of altered structure or function and can guide algorithms for knowledge-based image analysis (Collins *et al.*, 1994; Dinov *et al.*, 2000; Pitiot *et al.*, 2004). Pathological change can be tracked over time and disease-specific features resolved. Rather than simply fusing information from multiple subjects and sources, these strategies resolve group-specific features not apparent in individual scans. In essence, these atlases can serve as biomarker reference systems for a variety of pathological conditions.

III. Strategy and Methodology

To ensure the fidelity of the atlas, we rely on anatomical features contained within the data to define structures. We have developed model-driven algorithms that deform them to match the anatomy of new subjects (Thompson and Toga, 1996; Thompson *et al.*, 2000, 2004; Toga and Thompson, 1997). Anatomical models provide an explicit geometry for individual structures in each scan, such as landmark points, curves, or surfaces. Because the digital models reside in the same stereotaxic space as the atlas data, surface and volume models stored as lists of vector coordinates are amenable to digital transformation as well as to geometric and statistical measurement (Gerig *et al.*, 2001; Narr *et al.*, 2000; Thompson *et al.*, 1996; Zhou *et al.*, 1999). The underlying 3D coordinate system is central to all atlas systems, since it supports the linkage of structure models and associated image data with spatially indexed neuroanatomical labels, preserving spatial information and adding anatomical knowledge. Furthermore, any other ancillary data that can be indexed with the coordinate system can be related to the anatomical framework.

A. Registration and Tissue Mapping

Three-dimensional volumetric MRI scans are first rotated and scaled to match a standardized brain template in stereotaxic space. This template can be either an average-intensity brain dataset constructed from a population of young normal subjects (Mazziotta *et al.*, 2001) or a brain dataset specially constructed to reflect the average anatomy of a specific subpopulation, such as young children (Wilke *et al.*, 2003) or elderly subjects (e.g., Janke *et al.*, 2001; Mega *et al.*, 2000; Narr *et al.*, 2004; Thompson *et al.*, 2000, 2004). Once aligned, a measure of the brain scaling imposed is retained as a covariate for statistical analysis. If a given subject has been scanned repeatedly, the same scaling is applied to both baseline and follow-up scans, to ensure that observed differences reflect true brain changes. A tissue-classification algorithm then splits up the scan into regions representing gray matter, white matter, cerebrospinal fluid (CSF), and nonbrain tissues. Stereotaxic maps of each tissue type are retained.

B. Cortical Pattern Matching

Depending on the specific pulse sequence and the characteristics of the protocol, MRI scans have sufficient resolution and tissue contrast to track cortical gray-matter loss in individual patients. Nonetheless, extreme variability in gyral patterns confounds efforts to compare this loss against a normative population and to determine the average profile of tissue loss in a group. *Cortical pattern matching* methods (Thompson *et al.*, 2004) are designed to overcome these challenges. They encode both gyral patterning and gray-matter variation. This can substantially improve the statistical power to localize deficits. These cortical analyses tease apart the effects of group gyral-shape variation from gray-matter change, and they can also be used to measure cortical asymmetries (Luders *et al.*, 2005a; Narr *et al.*, 2001; Sowell *et al.*, 2001; Thompson *et al.*, 2001).

Briefly, a 3D geometric model of the cortical surface is extracted from the MRI scan (Fig. 1) and flattened to a 2D planar format (to avoid cuts, a spherical topology can be retained [Fischl *et al.*, 2001; Lerch and Evans, 2005; Thompson *et al.*, 1997; van Essen, 2004]). A complex deformation, or warping transform, is then applied that aligns the sulcal anatomy of each subject with an average sulcal pattern derived for the group. The same strategy can be applied to any feature set, in addition to or instead of sulcal anatomy. To improve feature alignment across subjects, all sulci that can be reliably delineated and that occur consistently can be digitized (Narr *et al.*, 2004; Sowell *et al.*, 2004) and used to constrain this transformation. As far as possible, this procedure adjusts for differences across subjects in cortical patterning and shape. Cortical measures can then be compared across subjects and groups. Sulcal landmarks are used as anchors, as homologous cortical regions are better aligned after sulci are matched than after data are just averaged at each point in stereotaxic space (see, e.g., functional MRI studies by Rasser *et al.*, 2005; Rex *et al.*, 2001; Zeineh *et al.*, 2003). Given that the deformation maps associate cortical locations with the same relation to the primary folding pattern across subjects, a local measurement of any number of structural, functional, or other characteristics can be performed in each subject and *averaged across equivalent cortical locations*. One such characteristic is a measure of local gray matter. To quantify local gray matter, one can use a measure termed '*gray-matter density*' to compare the spatial distribution of gray matter across subjects. This measures the proportion of gray matter in a small region of fixed radius (usually 15 mm) around each cortical point (Ashburner and Friston, 2000; Baron *et al.*, 2001; Bullmore *et al.*, 1999; Good *et al.*, 2001; Mummery *et al.*, 2000; Rombouts *et al.*, 2000; Sowell *et al.*, 1999, 2003; Thompson *et al.*, 2004; Wright *et al.*, 1995). Given the large anatomical variability in some cortical regions, high-dimensional elastic matching of cortical patterns (Thompson *et al.*, 2000, 2001) is used to associate measures of gray-matter density from

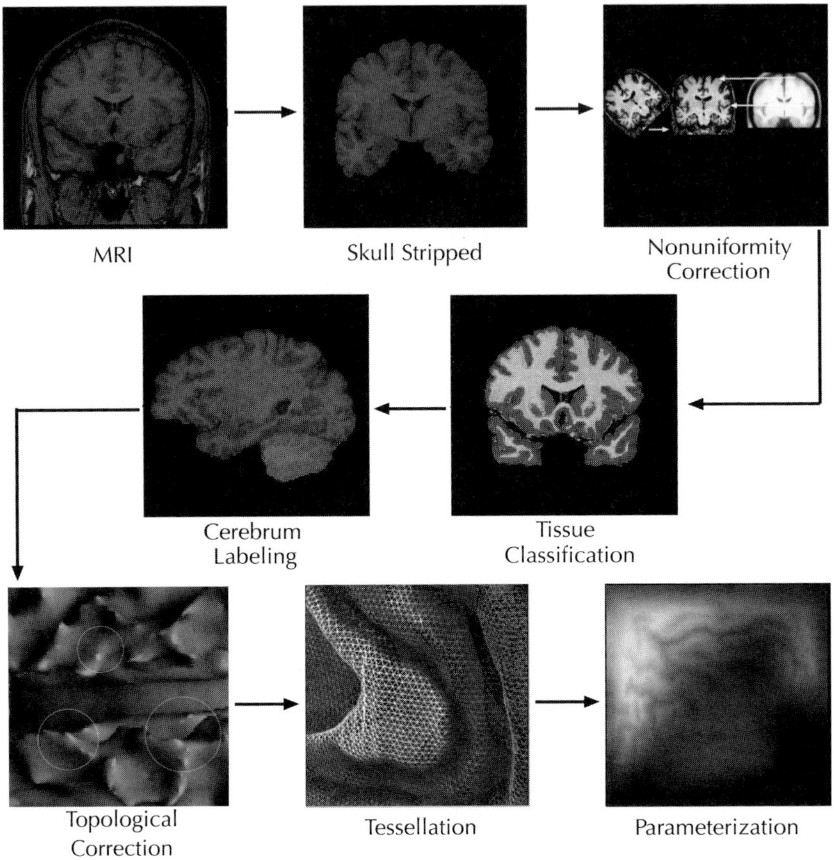

FIG. 1. Analyzing cortical data. This schematic shows a sequence of image-processing steps that can be applied to MRI scans to map how development and disease affect the cortex. The steps include manual or automated removal of the scalp and other nonbrain tissues from the image (to create 'skull stripped' data), alignment of the data to a standard coordinate space, correction for intensity nonuniformity, tissue classification, parcellation, and creation of cortical surface models (These procedures are detailed in the main text). To help in the comparison of cortical anatomy across subjects, a three-dimensional (3D) cortical surface model can be reconstructed from the scan, defining the interface between gray and white matter. If these interfaces are defined using intensity isosurfaces in the images, additional steps may be needed in order to ensure that the reconstructed surface is topologically equivalent to a sphere (a process called topology correction, which removes small bridges and connections). After this step, the surface can be tessellated using a triangular or rectilinear mesh structure (in tessellation, a small region including the gneu of the corpus callosum is shown). Finally, the surface can be flattened into a square, disc, or sphere, which enables points on the 3D surface to be referenced using 2D coordinates (a process called parameterization). When surfaces are flattened, information on the 3D locations of cortical points can be retained through a color-coding scheme that stores x,y,z locations using the color channels red, green, and blue, respectively. Flat maps can then be compared across subjects, as shown in Fig. 2. (See Color Insert.)

homologous cortical regions, first across time and then also across subjects (as shown in Figs. 2 and 3).

More recently, a number of algorithms have been developed that map the profile of cortical gray-matter thickness in three dimensions (see Fig. 3 for an overview) (Fischl *et al.*, 2001; Lerch and Evans, 2005; Lohmann *et al.*, 2003; Miller *et al.*, 2003; Ratnanather *et al.*, 2001; Thompson *et al.*, 2004, 2005; Xu *et al.*, 2005). Regional alterations in cortical thickness have been found in

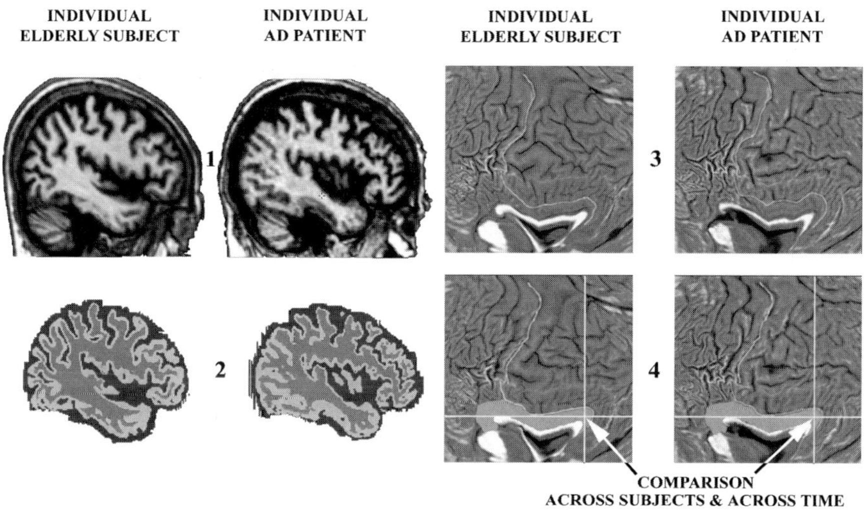

FIG. 2. Comparing gray matter across subjects. Gray matter is easier to compare across subjects if adjustments are first made for the gyral patterning differences across subjects. This adjustment can be made using cortical pattern matching , which is illustrated here on example brain MRI datasets from a healthy control subject (left column) and from a patient with Alzheimer's disease (right column). First, the MRI images have extracerebral tissues deleted from the scans (stage 1), and individual voxels are classified as gray matter, white matter, or cerebrospinal fluid (CSF) (shown here in green, red, and blue colors) (stage 2). After a 3D geometric model of the cortex is flattened (stage 3), features such as the central sulcus (light blue curve) and cingulate sulcus (green curve) can be reidentified (*http://www.loni.ucla.edu/~esowell/edevel/new_sulcvar.html*). An elastic warp is applied (stage 4) that moves these features, and entire gyral regions (pink colors), into the same reference position in "flat space." After alignment of sulcal patterns from all individual subjects, group comparisons can be made at each 2D pixel (yellow cross-hairs) that effectively compare gray-matter measures across corresponding cortical regions. The cortical measure that is compared, across groups or over time, may be cortical thickness (Fig. 3) or gray-matter density, which is the amount of gray matter (stage 2) that lies within a fixed radius (typically 10–15 mm) of each cortical point. The results of these statistical comparisons can then be plotted back onto an average three-dimensional (3D) cortical model made for the group, and significant findings can be visualized as color-coded maps. Such algorithms bring gray-matter maps from different subjects into a common anatomical reference space, overcoming individual differences in gyral patterns and shape by matching locations point by point throughout the cortex. This enhances the precision of intersubject statistical procedures that detect localized changes in gray matter. (See Color Insert.)

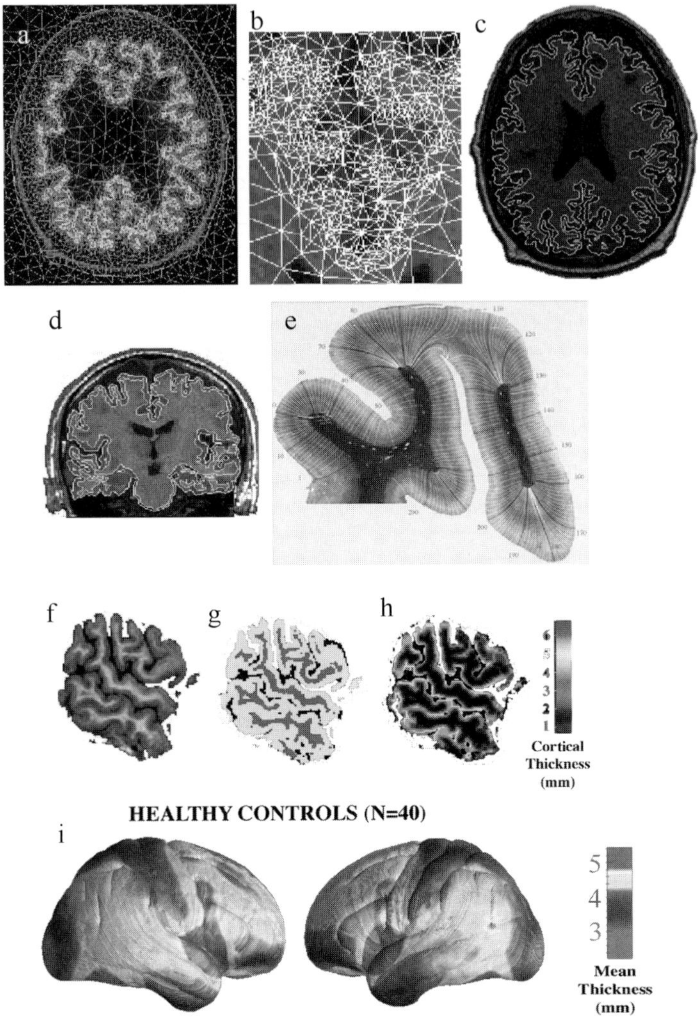

FIG. 3. Mapping cortical thickness. Several algorithms have been developed to measure the thickness of the cortical gray matter, which is affected in a variety of diseases and changes throughout development. In one approach (Xu *et al.*, 2003, 2005), we used adaptive finite element meshes (a) (shown close up in b), with refinement in a *narrow band* around the cortex (colors, a), to segment the inner and outer cortical surfaces from MRI data (c, d). A parallel line of work measured cortical thickness in two-dimensional (2D) histology (Annese *et al.*, 2002) by solving the heat equation and integrating the gradient of the solution (e). In related work (Thompson *et al.*, 2004, 2005), we computed average maps of cortical thickness for several populations. MRI scans (f) are classified as gray matter, white matter, CSF, or a background class (represented by green, red, black, and white, respectively, in g). To quantify cortical gray-matter thickness, we used the 3D distance measured from the cortical white matter–gray matter boundary in the tissue-classified brain volumes to the cortical

Williams syndrome (Thompson *et al.*, 2004), in Alzheimer's disease (Fischl *et al.*, 2004; Lerch and Evans, 2005), in HIV/AIDS (Thompson *et al.*, 2005), and in schizophrenia patients taking different medications (Narr *et al.*, 2004; Szeszko *et al.*, 2004). These localized thickness changes have also been related to cognitive changes that occur as children develop (Sowell *et al.*, 2004), as well as to changes in language function (Lu *et al.*, 2005). There are also sex differences and hemispheric asymmetries in the profiles of average cortical thickness (Luders *et al.*, 2005b).

Accurate modeling of the 3D gray/white matter and gray matter/CSF interfaces (Xu *et al.*, 2005), as well as high-dimensional normalization of the anatomy across subjects (Davatzikos *et al.*, 2001; Thompson *et al.*, 2004), are both critical for obtaining accurate group maps of mean cortical thickness with sufficient signal-to-noise ratio to detect regional group differences (Fig. 3). The accuracy of an average thickness map is degraded by any errors in alignment of the anatomy across subjects, but these can be overcome to some degree by cortical pattern matching, which imposes a high-dimensional normalization of anatomy across subjects.

One advantage of cortical matching is that it localizes deficits relative to gyral landmarks or any other feature of interest; it also averages data from corresponding gyri, for example, which would be impossible if data were only linearly mapped into stereotaxic space. Application of this approach over time yields annualized 4D maps of gray-matter loss rates within each subject, which are then elastically realigned for averaging and comparison across diagnostic groups (Fig. 4). The effects of age, gender, medication, disease, and other measures on gray matter can be assessed at each cortical point.

C. Statistical Maps

To determine whether there are group differences in cortical features, an algorithm is used to fit a statistical model, such as the *general linear model* (GLM

---

surface (gray matter–CSF boundary) in each subject (h). We first resampled tissue-classified brain volumes to 0.33-mm isotropic voxels to obtain distance measures indexing gray-matter thickness at subvoxel spatial resolution. Gray-matter thickness, measured at thousands of homologous cortical locations in each subject, is then compared across subjects and averaged at each cortical surface location, which provides spatially detailed maps of local thickness differences within or between groups. The thickness data may also be smoothed by a surface-based kernel to enhance the signal-to-noise ratio, before cross-subject comparisons are made. Frame i shows the mean cortical thickness in a group of 40 healthy young adults, ranging from low values in primary sensorimotor and visual cortices (2–3 mm; yellow colors) to the highest values on the medial wall in cingulate areas (up to 6 mm; purple colors). The regional variations in these maps agree with those found in the classical cortical thickness maps derived postmortem by von Economo (Sowell *et al.*, 2004). (See Color Insert.)

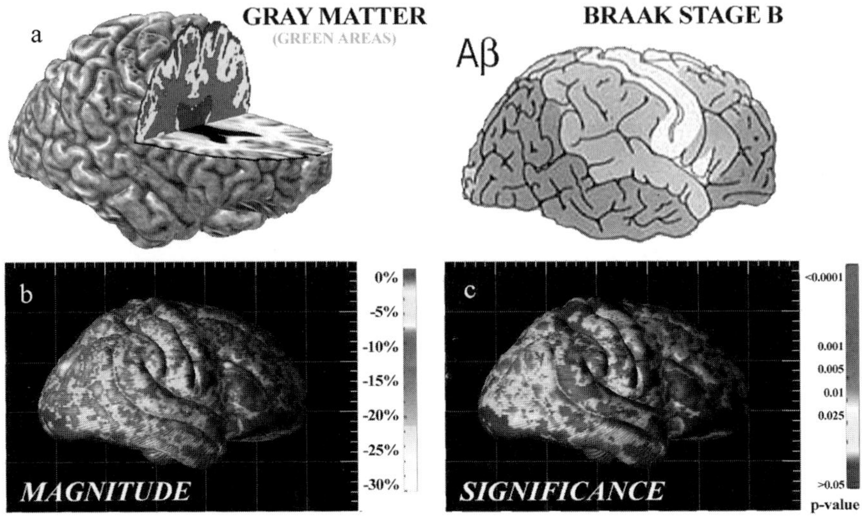

FIG. 4. Gray-matter deficits in early Alzheimer's disease. Here, the local amount of cortical gray matter (shown in green colors; a) is compared across 26 patients with mild-to-moderate Alzheimer's (age: 75.8 ± 1.7 years; MMSE score: 20.0 ± 0.9) and 20 matched elderly controls (72.4 ± 1.3 years). At this stage of Alzheimer's disease, there is a reduction in gray matter that reaches 30% in the temporal–parietal regions (b). Frame c shows a map of the statistical significance of these deficits (Thompson *et al.*, 2003). Intriguingly, the pattern of temporal-lobe gray-matter loss, seen with MRI, spatially matches the pattern of beta-amyloid (Aβ) deposition seen postmortem (Braak and Braak, 1991, 1997; Braak *et al.*, 2000). The inset panel (Braak Stage B) is adapted from data reported by Braak and Braak (1997). It shows regions with minimal (white), moderate (orange), and severe (red) beta-amyloid deposition. Because amyloid deposition and gray-matter loss may not be synchronized, these maps can represent different stages of Alzheimer's disease; however, there is a clear spatial agreement between MRI and beta-amyloid maps in the severity of the deficits. Intriguingly, both maps indicated that *primary sensorimotor regions* (white in the amyloid map) and the *superior temporal gyrus* (blue colors in c) showed little damage relative to other temporal-lobe gyri. These overall MRI patterns are similar to those in independent studies by Baron *et al.* (2001), O'Brien *et al.* (2001), and Burton *et al.* (2002). (See Color Insert.)

[Friston *et al.*, 1995]) to the data at each cortical location. This results in a variety of parameters that characterize how gray-matter variation is linked with other variables. The significance of these links can be plotted as a significance map. In addition, it is possible to plot certain estimated parameters, such as the local rates of gray-matter loss at each cortical location (e.g., as a percentage change per year), regression parameters that identify disease effects, and even nonlinearities in the rates of brain change over time (e.g., quadratic regression coefficients) (Sowell *et al.*, 2003). In principle, any statistical model can be fitted, including genetic models that estimate genetic or allelic influences on brain structure

(Cannon *et al.*, 2005; Thompson *et al.*, 2001, 2003). Finally, permutation testing is typically used to ascribe an overall significance value for the observed map. This can be adjusted for the fact that multiple statistical tests are made when a whole map of statistics is visualized. Patients and controls are randomly assigned to groups; on a supercomputer this often happens many millions of times (see, e.g., Thompson *et al.*, 2001). A null distribution is built to estimate the probability that the observed effects could have occurred by chance, and this is reported as a significance value for the overall map.

When deforming an atlas to match an individual's anatomy, a researcher can use mesh-based models of anatomical systems to help guide the mapping of one brain to another. These anatomically driven algorithms guarantee biological as well as computational validity, generating meaningful object-to-object correspondences, especially at the cortex. In this model-based approach (Davatzikos, 1996; Gerig *et al.*, 2001; Thompson and Toga, 1996, 2004), the first step is to extract systems of surfaces from each dataset, to guide the volumetric mapping. The model surfaces can include (m)any functional, cytoarchitectonic, and lobar boundaries in three dimensions. Both the surfaces and the landmark curves within them are reconfigured to match their counterparts in the target data sets exactly (Leow *et al.*, 2005).

D. Anatomical Models

Since much of the functional territory of the human cortex is buried in sulci, it is useful to build a generic structure to model them (Le Goualher *et al.*, 1999; Mangin *et al.*, 2004; Thompson and Toga, 1996; Zhou *et al.*, 1999). The underlying data structure is a connected system of surface meshes in which the individual meshes are parametric. These surfaces are 3D sheets that divide and join at curved junctions to form a connected network of models. With the help of these meshes, each patient's anatomy is modeled in sufficient detail to be sensitive to subtle differences in disease. Separate surfaces model the deep internal trajectories of features such as the parieto-occipital sulcus, the anterior and posterior calcarine sulcus, the Sylvian fissure, and the cingulate, marginal, and supracallosal sulci in both hemispheres. Additional gyral boundaries are represented by parameterized curves lying in the cortical surface. The ventricular system is modeled as a closed system of 14 connected surface elements whose junctions reflect cytoarchitectonic boundaries of the adjacent tissue (Thompson and Toga, 1998). Any anatomical feature in the brain can be modeled this way, as can functional features, histological features, and even genetic features that include spatial information, such as *in situ* hybridization. Information on the meshes' spatial relations, including their surface topology (closed or open), anatomical names, mutual connections, directions of parameterization, and common

3D junctions and boundaries, is stored in a hierarchical graph structure. This ensures the continuity of displacement-vector fields defined at mesh junctions.

E. Surface Parameterization

After imposition of an identical regular grid structure on anatomical surfaces from different subjects, the explicit geometry can be exploited to drive and constrain correspondence maps that associate anatomical points in different subjects. Structures that can be extracted automatically in parametric form include the external cortical surface, ventricular surfaces, and several deep sulcal surfaces. Recent success of sulcal extraction approaches based on deformable surfaces (Vaillant *et al.*, 1999) led us to combine a 3D skeletonization algorithm with deformable-curve and surface-governing equations to automatically produce parameterized models of cingulate, parieto-occipital, and calcarine sulci, without manual initialization (Zhou *et al.*, 1999). Other segmentation approaches that utilize template methods (Fischl *et al.*, 2004) are often amenable to subsequent parameterization (Gerig *et al.*, 2001). Additional, manually segmented surfaces can also be given a uniform rectilinear parameterization using algorithms described by Thompson *et al.* (1996), and they can be used to drive the warping algorithm. Each resultant surface mesh is analogous in form to a uniform rectangular grid drawn on a rubber sheet that is subsequently stretched to match all data points. Association of points on each surface with the same mesh coordinate produces a dense correspondence vector field between surface points in different subjects. This procedure is carried out under stringent conditions to ensure that landmark curves and points known to the anatomist appear in corresponding locations in each parametric grid.

F. Maps of the Cortical Parameter Space

Detailed models of cortical anatomy are also created when a tiled, spherical mesh is driven into the configuration of each subject's cortex (MacDonald *et al.*, 2000; Thompson and Toga, 1996). Because these cortical models are obtained by deformation of a spherical mesh, any point on the cortical surface must map to exactly one point on the sphere and *vice versa*. Each cortical surface is parameterized with an invertible mapping, so sulcal curves and landmarks in the folded brain surface can be reidentified in the spherical map (see Fischl *et al.*, 1999; Hurdal and Stephenson, 2004; Sereno *et al.*, 1996; Wang *et al.*, 2004, for similar approaches). To retain relevant 3D information, in our approach cortical surface-point position vectors in 3D stereotaxic space are color coded, so a unique color is placed at each position on the spherical map indicating the 3D position of the cortical point that maps to it (Thompson *et al.*, 2004). In other words, the colors represent 3D locations and the entire set of colors forms an image on the sphere in a color image format. To match between cortical regions in different subjects, we first derive a spherical map for each respective cortical surface model and then perform a matching process in the spherical parametric space. A flow field is calculated on the sphere that brings corresponding gyral and sulcal regions into the same spherical locations across subjects (Bakircioglu *et al.*, 1999; Davatzikos, 1996; Drury *et al.*, 1996; Fischl *et al.*, 1999; Leow *et al.*, 2005; Van Essen *et al.*, 1997; Wang *et al.*, 2004). This warp can be set up in a variety of ways. Spherical harmonic functions are an orthonormal basis on the sphere, which means that any smooth flow on the sphere can be represented with high accuracy using a linear combination of these functions, so long as a sufficient number of functions is used (Gerig *et al.*, 2001; Thompson and Toga, 1996). Alternatively, a flow field in the cortical parameter space can be computed to align maps of surface convexity, intrinsic curvatures, or surface features such as conformal structures (Fischl *et al.*, 1999; Wang *et al.*, 2005). The resulting flow field can align structural or functional information across subjects, bringing curve and surface interfaces into exact registers in the process (Leow *et al.*, 2005). Alternatively, an approach based on covariant partial differential equations can be used for matching cortical surfaces (Joshi *et al.*, 2005; Memoli *et al.*, 2004; Thompson *et al.*, 2000). This precisely matches cortical landmarks across subjects, and it creates maps that are independent of the surface metrics. Related approaches, based on level sets (Osher and Sethian, 1988), allow feature-based and intensity-based registration techniques to be combined in the same algorithm, making it easier to match corresponding anatomy across subjects (Leow *et al.*, 2005). The approach ensures that the way cortical structures are matched in three dimensions is independent of the way the surfaces are computationally represented (irrespective of their tile density and parameterizations).

G. Tensor Maps of Directional Variation

Structures do not vary to the same degree in every coordinate, and even these directional biases vary by cortical system. The principal directions of anatomical variability in a group can be shown in a *tensor map* (Fig. 5) (Kindlmann *et al.*, 2004; Thompson *et al.*, 2000). The maps have two uses. First, they make it easier to detect anomalies, which may be small in magnitude but headed in unusual directions. Second, they significantly increase the information content of Bayesian priors used for automated structure extraction and identification and for nonlinear image registration (Gee *et al.*, 1995; Mangin *et al.*, 2004; Pitiot *et al.*, 2004).

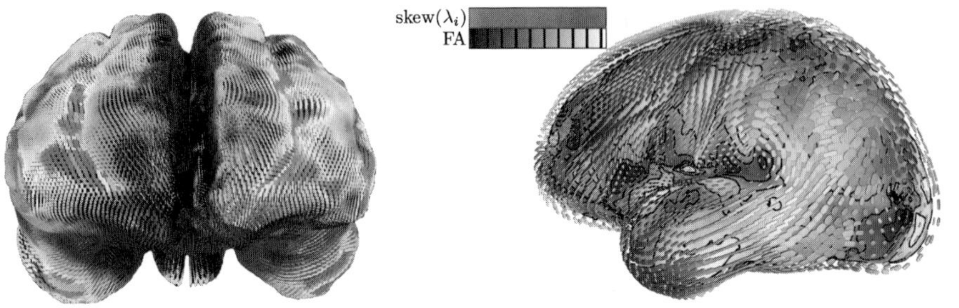

FIG. 5. Mapping directional patterns of brain variation. This image shows a tensor map of variability for normal subjects, created after 20 elderly subjects' data were mapped into Talairach space (all right-handed; 10 males, 10 females). Ellipsoidal glyphs indicate the principal directions of variation—they are most elongated along directions where anatomic variation is greatest across subjects. Each glyph represents the covariance tensor of the vector fields that map individual subjects onto their group average. Because gyral patterns constrain the mappings, the fields reflect variations in cortical organization at a more local level than can be achieved by matching only global cortical geometry. Note the elongated glyphs in the anterior temporal cortex and the very low variability (in any direction) in the entorhinal and inferior frontal areas. By better defining the parameters of allowable normal variations, we can use the resulting information to distinguish normal anatomical variants from abnormal. Frame b shows a rendering of the same data using superquadric glyphs, which help to visualize the principal axes of variation (Kindlmann *et al.*, 2004). More subtle aspects of anatomical variation can be visualized, including the skew of the eigenvalues of variation (shown in color in b and the fractional anisotropy (FA) of the variation (shown in grayscale). The anisotropy indicates whether there is a strong directional preference in the population variability. (See Color Insert.)

H. Mathematical Details

Deformation-based methods to track brain change have often been based on continuum mechanics, which describes physical models of elastic or fluid bodies (Chung *et al.*, 2001; Freeborough and Fox, 1997; Haney *et al.*, 2001; Thompson and Toga, 2000; Toga and Thompson, 1998). The 3D shape of one brain, imaged at one time-point, is imagined to be embedded in a physical medium, such as an elastic block or a fluid. This deformable template is reconfigured to match its shape in a later image (intensity changes over time may also be modeled [Joshi *et al.*, 1997]). As described above, in some approaches, a complex 3D deformation field is computed that matches large numbers of surface, curve, and point landmarks in the two brains. When features are added to constrain the deformation, key anatomical and functional interfaces can be matched up when one scan is deformed into the shape of the other. In one approach (Thompson and Toga, 1996, 2002), parametric mesh models of brain structures are used to drive a 3D deformation vector map $U:\mathbf{x}\rightarrow\mathbf{u}(\mathbf{x})$, which is derived from the Navier equilibrium equations for linear elasticity:

$$\mu\nabla^2\mathbf{u} + (\lambda + \mu)\nabla(\nabla \cdot \mathbf{u}(\mathbf{x})) + \mathbf{F}(\mathbf{x}\text{-}\mathbf{u}(\mathbf{x})) = \mathbf{0}, \forall\mathbf{x} \in \mathrm{R} \tag{1}$$

All the terms in this equation just describe forces and distortions in a 3D material in which the image is considered to be embedded. R is a discrete lattice representation of the scan to be transformed, $\nabla \cdot \mathbf{u}(\mathbf{x}) = \Sigma\partial \mathrm{u_j}/\partial \mathrm{x_j}$ is the divergence or cubical dilation of the medium, ∇^2 is the Laplacian operator that measures the irregularity of the deformation, $\mathbf{F}(\mathbf{x})$ is the internal force vector, and Lamé's coefficients λ and μ refer to the elastic properties of the medium. Matching of cortical surfaces, across time and subsequently across subjects (for data averaging), can also be enforced. Mappings based on high-dimensional elastic and fluid models can recover extremely complex patterns of change (see Fig. 5 for examples of these warping fields) (see also Freeborough and Fox, 1997; Rey *et al.*, 2002).

I. Fluid Modeling of Brain Change

Building the work on elastic image registration, Christensen *et al.* (1996) developed a compressible fluid-deformation model for image registration that forces the deformation matching the scans to be smooth and preserve the topology, even under large deformations. The top two rows of Fig. 6 show examples of our implementation of this fluid image-matching algorithm (Chiang *et al.*, 2005). Strictly speaking, the Navier equations (Eq. 1) are derived under a small deformation assumption (which is valid for growth and atrophic processes), but the fluid model uses a regridding approach when necessary to ensure a smooth final solution. The forces that drive one image to match the other were also designed to match regions in each dataset with high intensity similarity. Transformation parameters were determined by gradient descent on a cost functional (Eq. 2) that penalizes squared intensity mismatch between the deforming template $\mathrm{T}(\mathbf{x}\text{-}\mathbf{u}(\mathbf{x},t))$ and target $\mathrm{S}(\mathbf{x})$ while guaranteeing the smoothness of the transformation:

$$\mathrm{C}(\mathrm{T}(\mathbf{x}), \mathrm{S}(\mathbf{x}), \mathbf{u}) = (1/2)\int_\Omega |\mathrm{T}(\mathbf{x}\text{-}\mathbf{u}(\mathbf{x}, t)) - \mathrm{S}(\mathbf{x})|^2 \mathbf{dx} \tag{2}$$

The driving force, which deforms the anatomical template, is defined as the variation of the cost functional with respect to the displacement field:

$$\mathbf{F}((\mathbf{x}, \mathbf{u}(\mathbf{x}, t)) = -(\mathrm{T}(\mathbf{x}\text{-}\mathbf{u}(\mathbf{x}, t)) - \mathrm{S}(\mathbf{x}))\nabla\mathrm{T}|_{\mathrm{x\text{-}u(x,}t)} \tag{3}$$

$$\mu\nabla^2\mathbf{v}(\mathbf{x}, t) + (\lambda + \mu)\nabla(\nabla^{\mathrm{T}} \cdot \mathbf{v}(\mathbf{x}, t)) + \mathbf{F}(\mathbf{x}, \mathbf{u}(\mathbf{x}, t)) = \mathbf{0} \tag{4}$$

$$\partial\mathbf{u}(\mathbf{x}, t)/\partial t = \mathbf{v}(\mathbf{x}, t) - \nabla\mathbf{u}(\mathbf{x}, t)\mathbf{v}(\mathbf{x}, t) \tag{5}$$

The deformation velocity is governed by the creeping flow momentum equation for a Newtonian fluid and the conventional displacement field in a

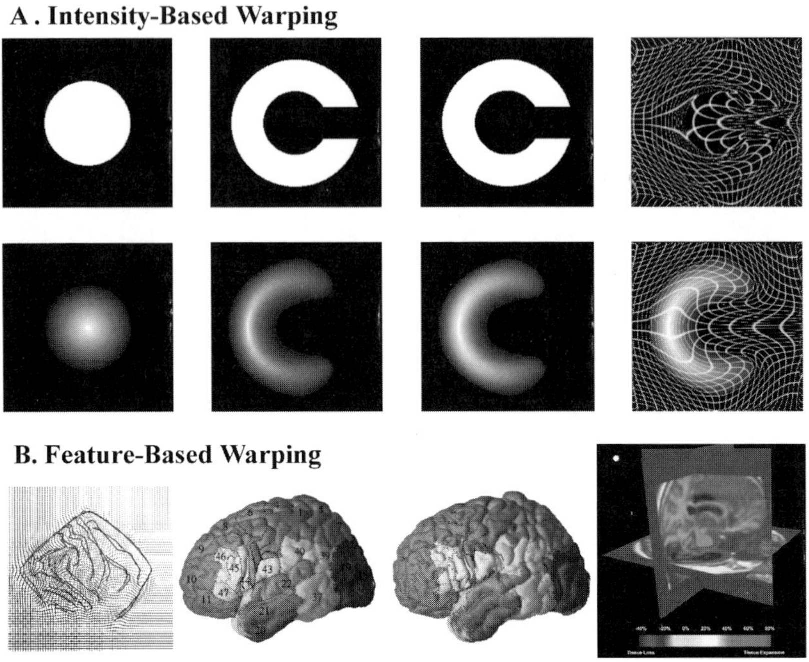

FIG. 6. Warping algorithms used in brain imaging. As illustrated here, brain image–warping algorithms (Toga, 1998) fall into two major categories: intensity-based algorithms (A, first two rows), and feature-based algorithms (B, bottom row), depending on the information used to align one image with the other. The top row shows a circle and a C shape, and then shows the result of deformation of the circle to the C shape using an intensity-based fluid warping algorithm (data from Chiang *et al.*, 2005). The final column shows the complex pattern of fluid deformation required to transform the circle to the C. Note that this algorithm does not require any user intervention; it automatically finds the deformation field, or "flow," that best aligns one image-intensity distribution with the other. The second row shows a similar experiment in which the circle and the C shape have a range of intensities; the circle is then deformed so that its intensity profile matches that of the C shape. The final row shows examples of feature-based warping. In the first panel, the flattened sulcal pattern of an individual is aligned with an average template of the cortical sulci; in the next two panels, one cortical surface (labeled in color with Brodmann areas) has its labels transferred to another subject's cortical surface with surface-based warping (Rasser *et al.*, 2004; Thompson *et al.*, 2000). In this mapping, three-dimensional (3D) sulcal curves lying in the cortex drive the warped mapping of one surface onto the other, and the Brodmann labels are carried along by the mapping. The final panel shows an application of warping to measure rates of annual tissue loss in semantic dementia patients scanned serially with MRI (data from Leow *et al.*, 2005). A combination of surface-based and intensity-based approaches (i.e., a hybrid approach) is used to derive a 3D elastic deformation mapping in which blue colors denote tissue loss and red colors denote regional expansion (observed here in the lateral ventricles). (See Color Insert.)

Lagrangian reference system is connected to a Eulerian velocity field by the relation of material differentiation. Experimental results were excellent (Chiang *et al.*, 2005; Christensen *et al.*, 1996). Because fluid matching is computationally intensive, subsequent work focused on deriving separable (and therefore computationally faster) filters to approximate the continuum-mechanical filters derived above (Cachier and Ayache, 2003; Chiang *et al.*, 2005; Gramkow, 1996; Miller *et al.*, 2002; Nielsen *et al.*, 1994). Some elastic matching algorithms are now fast enough to track brain change in real-time, in surgical applications that use intraoperative scanning (Warfield *et al.*, 1998). Other work has focused on the intensity-matching term, using measures from information theory to help guide the automatic mapping of one dataset onto another (see Figure 6, top two rows; Chiang *et al.*, 2005).

J. Population-Based Atlasing of Brain Change

In different individuals, growth processes or tissue losses occur in anatomies that are geometrically different. Additional warping techniques are needed to compare growth profiles across subjects. This additional warping is needed to compute average profiles of growth in a group and to define statistical differences in rates of growth or loss. Figure 6 (last panel) shows an example in which warping is applied to map rates of tissue loss in a semantic dementia patient (Leow *et al.*, 2005). Mathematically, if $U^i(\mathbf{x},t_i)$ is the 3D displacement vector required to deform the anatomy at position $\mathbf{x}$ in subject i at reference time 0 to its corresponding homologous position at time t_i, then a linear approximation of the local rate of volumetric growth (Chung *et al.*, 2001) can be written in terms of the identity tensor and displacement gradient tensor as follows:

$$\Lambda^i(\mathbf{x}) = \partial \mathrm{J}^i/\partial \mathrm{t} = det(I + \nabla U^i)/t_i, \qquad (6)$$

If A^i is the secondary deformation mapping transforming the baseline anatomy of individual i onto the atlas, then the set of registered growth maps $\Lambda^i(A^i(\mathbf{x}))$ can be treated as observations from a spatially parameterized random field whose mean and variance can be estimated. Statistical effects of age, gender, genotype, or medication can then be detected using random field theory to produce statistical maps.

K. Improved Dynamic Models

In the development of dynamic atlases for clinical applications, there is a particular interest in modeling developmental processes that speed up or slow down. Diseases can accelerate, or their rate of progression can be slowed down by therapy. If individuals are scanned more than twice over large timespans, there is an opportunity to more accurately detect brain change, as well as to encode these changes in a group atlas. To compare growth patterns in different groups of subjects, the GLM (Friston *et al.*, 1995) can be used to analyze the registered growth profiles (or degenerative profiles). For the *i*th individual's *j*th measure, we have

$$Y_{ij} = f(\mathrm{Age}_{ij}, \beta) + \varepsilon_{ij} \tag{7}$$

Here, Y_{ij} signifies the outcome measure at a voxel or surface point, such as growth or tissue loss; f denotes a constant, linear, quadratic, cubic, or other function of the individual's age for that scan; and β denotes the regression/ANOVA coefficients to be estimated. Age (Age_{ij}) can be replaced by time from the onset of disease, from the start of medication, or from the onset of puberty (e.g., Giedd *et al.*, 1999). This flexibility in the parameters of the time axis allows one to temporally register dynamic patterns using criteria that are expected to bring into line temporal features of interest that appear systematically in a group (Janke *et al.*, 2001). For example, the independent variable could be a cognitive score such as mini-mental status (Thompson *et al.*, 2002), which declines over time in disease. In a developmental study, this independent variable could be a measure of physical or psychological maturity, which might reflect the developmental stage of the subject better than would age alone. Parameterization of dynamic effects using measures other than time (e.g., clinical status) also provides a mechanism by which to align new patients' time series with a dynamic atlas.

L. Random-Effects Modeling

In the statistical model of brain change (Eq. 7), the coefficient vector, β, is assumed to be constant, i.e., a fixed effect. The ε_{ij} are assumed normally distributed and uncorrelated, both between and within individuals. If multiple scans are available over time, a random-effects model can also model brain changes in a population:

$$Y_{ij} = \alpha_i + f(\mathrm{Age}_{ij}, \beta) + \varepsilon_{ij} \tag{8}$$

Here, the model is the same as the GLM except for the α_i term, which is called a *random effect.* It describes the correlation between an individual's multiple scans. Random-effects models may also be fitted with *correlated* errors (Davidian and Giltinan, 1995; Verbeke and Molenberghs, 2000). If this is done, ε_{ij} and ε_{ik} (where k is not equal to j) are assumed correlated, with the correlation a function of the time elapsed between the two measurements (Giedd *et al.*, 1999). In models whose fit is confirmed as significant, e.g., by permutation, loadings on nonlinear

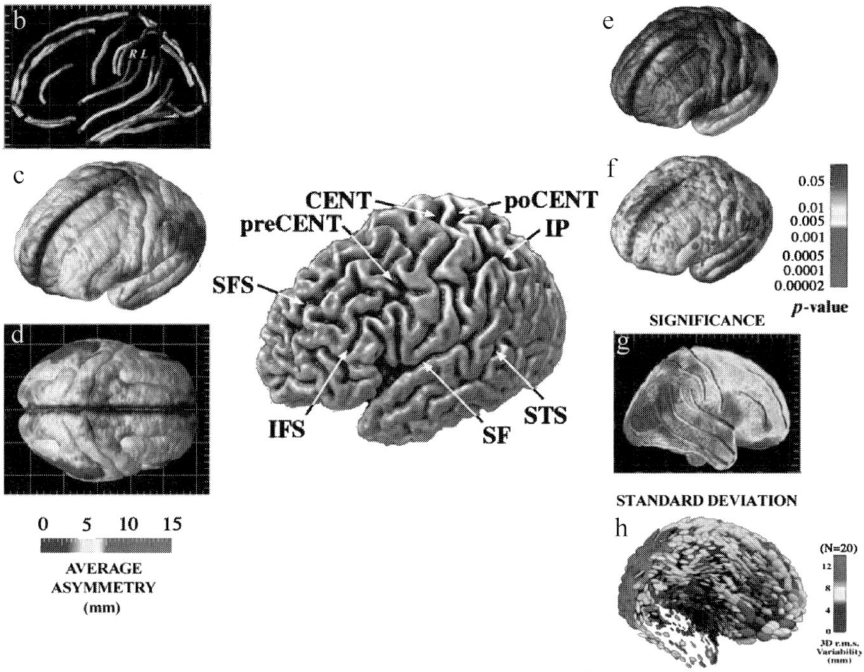

FIG. 7. Multisubject maps of brain asymmetry. Image-analysis techniques make it possible to distinguish systematic asymmetries in a population or a specific group of subjects from random fluctuations in anatomy. After individual MRI scans are aligned and scaled into a standard three-dimensional (3D) space, 3D curves representing the primary sulcal pattern are digitized (center image). (Sulci include central [CENT], precentral [preCENT], postcentral [poCENT], intraparietal [IP], superior frontal [SFS], inferior frontal [IFS], superior temporal, and Sylvian fissures [SF]). The magnitude of asymmetry in the average anatomy, from the average of these curves across 20 normal subjects (b), is shown in color (red colors denote greater asymmetry). Extension of these methods to surfaces (c, d) reveals prominent asymmetries in Broca's anterior speech area and in language regions surrounding the Sylvian fissure. By comparison of the average magnitude of these asymmetries to their standard error (e), regions of significant asymmetry are identified (f). Asymmetries are greatest in brain regions with greatest gyral pattern variability across subjects (g, h). The tensor map (h; see also Fig. 5) shows that the preferred directions of intersubject anatomical variability are also approximately aligned with the direction of interhemispheric asymmetry. (See Color Insert.)

parameters can be visualized as attribute maps $\beta(\mathbf{x})$. This reveals the topography of accelerated or decelerated brain change. The result is a formal approach to assess whether, and where, brain change is speeding up or slowing down. This is a key feature in developmental and medication studies, and it is a key element of developmental atlases that are currently being built.

M. Mapping Gyral-Pattern Differences in a Population

Once cortical models are available for a large number of subjects, in a common 3D coordinate space, patterns of cortical variability and cortical change over time can be calculated. Cortical anatomy can be compared, between any pair of subjects, by computation of the warped mapping that elastically transforms one cortex into the shape of the other. Due to variations in gyral patterning, cortical differences among subjects will be severely underestimated unless elements of the gyral pattern are matched from one subject to another. This matching is also required for cortical averaging; otherwise, corresponding gyral features will not be averaged together. Fortunately, the major gyri and sulci of the cortical surface have similar spatial layouts across subjects (Ono *et al.*, 1990; Regis, 1994; see Thompson *et al.*, 2002 for some caveats), even though their geometry varies substantially. Transformations can therefore be developed that match large networks of gyral and sulcal features with their counterparts in the target brain (Davatzikos, 1996; Fischl and Dale, 2001; Van Essen *et al.*, 1997).

In one approach, a maximal set, or template, is specified that contains all primary sulci that consistently occur in normal subjects (Fig. 7, center panel, shows some of these).

To find good matches among cortical regions, many groups perform the matching process in the cortical surface's parametric space, which permits of more tractable mathematics (Fig. 3; Fig. 6, bottom row). This vector flow field in the parametric space indirectly specifies a correspondence field in 3D, which drives one cortical surface into the shape of another. This mapping matches not only the overall cortical geometry but the entire network of the 38 landmark curves to their counterparts in the target brain, and thus it is a valid encoding of cortical variation. The flow in parameter space (Fig. 6, bottom row) can be represented by spherical harmonics (Gerig *et al.*, 2001), which are eigenfunctions of the spherical Laplacian, or by the solution of an elastic or fluid partial differential equation (PDE) that aligns sulcal and gyral landmarks (Barkircioglu *et al.*, 1999; Miller *et al.*, 2002) or curvature maps (Fischl and Dale, 2001). In one approach based on covariant PDEs, these flows are made invariant to the way the cortical surfaces are parameterized (Thompson *et al.*, 2000). When the self-adjoint differential operator governing the PDE is discretized, fields of Christoffel symbols are derived from the metric tensor of the surface domain and added as correction terms. The matching fields are then independent of the surface metrics and can be used to associate signals from corresponding cortical regions across subjects.

On the sphere, the parameter shift function $\mathbf{u(r)}:\Omega \rightarrow \Omega$ is given by the solution $F:\mathbf{r}\rightarrow\mathbf{r}-\mathbf{u}(\mathbf{r})$ to a curve-driven warp in the spherical parametric space $\Omega=[0,2\pi)\times$

$[0,\pi)$. For points $\mathbf{r}$=(r,s) in the parameter space (Fig. 6, bottom row), a system of simultaneous partial differential equations is written for the flow field $\mathbf{u}(\mathbf{r})$:

$$L^{\ddagger}(\mathbf{u}(\mathbf{r})) + \mathbf{F}(\mathbf{r} - \mathbf{u}(\mathbf{r})) = \mathbf{0}, \forall \mathbf{r} \in \Omega, \text{with } \mathbf{u}(\mathbf{r}) = \mathbf{u}_0(\mathbf{r}), \forall \mathbf{r} \in \mathrm{M}_0 \cup \mathrm{M}_1 \quad (9)$$

Here M_0, M_1 are sets of points and sulcal or gyral curves where displacement vectors $\mathbf{u}(\mathbf{r}) = \mathbf{u_0}(\mathbf{r})$ matching corresponding anatomy across subjects are known. The flow behavior is modeled using continuum mechanical equations. L can be any second-order self-adjoint differential operator; a common example is the Cauchy–Navier differential operator $L = \mu\nabla^2 + (\lambda+\mu)\nabla(\nabla^T\bullet)$ with body force $\mathbf{F}$ (Chiang *et al.*, 2005; Christiansen *et al.*, 1996; Gee and Bajcsy, 1998; Leow *et al.*, 2005; Thompson *et al.*, 2000). To create mappings that are independent of the surface metrics (parameterizations), we use $L^{\ddagger}$, the *covariant* form of the differential operator L. $L^{\ddagger}$, all of L's partial derivatives, are replaced with *covariant* derivatives with respect to the metric tensor of the surface domain where calculations are performed. The covariant derivative of a (contravariant) vector field, $u^i(\mathbf{x})$, is $u^i_{,k} = \partial u^j/\partial x^k + \mathbf{\Gamma}^{\mathbf{j}}_{\mathbf{ik}}\mathbf{u}^{\mathbf{i}}$ where the *Christoffel symbols of the second kind*, $\mathbf{\Gamma_{ik}}^{\mathbf{j}}$, are computed from derivatives of the metric tensor components $g_{jk}(\mathbf{x})$:

$$\mathbf{\Gamma}^{\mathbf{i}}_{\mathbf{jk}} = (1/2)g^{il}(\partial g_{lj}/\partial X^k + \partial g_{lk}/\partial X^j - \partial g_{jk}/\partial X^i) \quad (10)$$

These correction terms are then used in the solution of PDEs (Thompson *et al.*, 2000), producing a family of 3D deformation maps, $\mathbf{U}_i(\mathbf{r})$, matching each individual cortex in 3D to the average cortex for a group. Here, $\mathbf{U}_i$ is a 3D location on the *i*th subject's cortex, and $\mathbf{r}$ is the location it maps to, after warping, in the cortical parameter space.

N. Mapping Gray-Matter Deficits

To help understand the approach, first we describe a cross-sectional study of gray-matter deficits in dementia, in which each subject is imaged once (longitudinal data are described next). Even in a study with a single scan from each subject, gyral-pattern variation across subjects makes it difficult to infer precisely where gray matter is lost in a group. If gray-matter maps are directly averaged together in stereotaxic space, it is difficult to localize results to specific cortical regions. To address this, cortical-pattern matching can help in the computation of group averages and statistics. As a first step, all MRIs are radiofrequency (RF)-corrected and segmented with a Gaussian mixture classifier, producing binary maps of gray matter. Let $g_{i,r}(\mathbf{x})$ be the gray-matter density, i.e., the proportion of voxels classified as gray matter that fall within a sphere (center $\mathbf{x}$, radius $\mathbf{r}$) in the *i*th subject's scan. Then, for a point at parameter location $\mathbf{r}$ on the group average cortex, $g_{i,r}(\mathbf{U}_i(\mathbf{r}))$ is the gray-matter density at the corresponding cortical point in subject *i*.

After the aligned maps of gray-matter density are averaged across groups of patients with Alzheimer's disease and healthy controls, the spatial profile of gray-matter deficits in disease is revealed (Fig. 4). By averaging the aligned maps, and then texturing them back onto a group-average model of the cortex, researchers can compute the average magnitude of gray-matter loss for the Alzheimer's-disease population (Fig. 4). Regions with 10–20% reduction in the measure are demarcated from adjacent regions with little detectable loss. The group effect size can also be measured through attachment of a field of t statistics, $t(\mathbf{r})$, to the cortical parameter space, and computation of the area of the t field on the group-average cortex above a fixed threshold ($p < 0.01$, uncorrected). For groups that are not demographically matched, more sophisticated regression models could be applied, resulting in F fields (or other nonparametric fields) that indicate the significance of the overall fit and of how individual model parameters help explain the loss. If whole surfaces of statistics are surveyed, there are several approaches that are routinely used to make a multiple-comparisons correction, which is required to confirm the significance of the overall effect. In permutation approaches, we can confirm the significance of the deficits by permuting the assignment of subjects to groups repeatedly, and then estimating the null distribution of statistics on the surface. Under stronger assumptions, Gaussian field methods can also be used, which analyze the topology and smoothness of the statistical fields and their level sets.

O. Mapping Surface-Area Changes

A similar cortical-matching approach has been used to map localized changes in cortical surface *area* over time. Chung *et al.* (2001) noted that if $X = X(v_1, v_2, t)$ is a parameterization of the cortical surface S_t, its surface metric tensor is $g_{ij} = X_i' X_j$, where the prime indicates the matrix transpose and $X_i = dX/dv_i$ denotes the partial derivative vector. The rate of local surface-area change per unit surface area, or *area-dilatation rate*, is then approximated by

$$d(\ln\ L)/dt = tr[g^{-1}(DX)'(d(DU)/dt)DX] \tag{11}$$

where the *local surface area element*, $L(t)$, is given by

$$L(t) = det^{1/2}(g) = (g_{11}g_{22} - g_{12}g_{21})^{1/2} \tag{12}$$

Here, $DX = (X_1, X_2)$ is a 3 2 matrix and DU is a 3 3 displacement gradient matrix. By approximation of the resulting surface-based parameters as t-distributed random fields, or chi-squared and Hotelling's T^2-distributed random fields, then null distributions and statistical criteria can be developed to tell where significant brain change has occurred. Laplace–Beltrami smoothing (Chung *et al.*, 2001) and statistical flattening (Worsley *et al.*, 1999) can also help to optimize signal detection in the resulting surface-based fields.

IV. Applications

Using a variety of populations with imaging data collected in the same fashion, we have created a series of probabilistic atlases that retain information on anatomical and functional variability (Mazziotta *et al.*, 1995; Thompson *et al.*, 1997). Descriptions of several of these follow. As the subject database increases in size and content, the digital form of these atlases allows efficient statistical comparisons of individuals or groups. In addition, the population that an atlas represents can be divided into subpopulations to represent specific disease types, and these subpopulations can subsequently be separated again according to age, gender, handedness (right or left), or genetic factors.

A. Pathology Detection

Normal anatomical complexity makes it difficult to design automated strategies that detect abnormal brain structure. At the same time, brain structure is so variable that group-specific patterns of anatomy and function are often obscured. Reports of structural differences in the brain linked to gender, IQ, and handedness are a topic of intense controversy (Toga and Thompson, 2003, 2005), and it is even less clear how these factors affect disease-specific abnormalities. The importance of these linkages has propelled computational anatomy to the forefront of brain-imaging investigations. To distinguish abnormalities from normal variants, a realistically complex mathematical framework is required to encode information on anatomical variability in homogeneous populations (Grenander and Miller, 1998). We employed elastic registration or *warping* algorithms to achieve distinct advantages for the encoding of patterns of anatomical variation and the detection of pathology. Cortical patterns are altered in schizophrenia (Narr *et al.*, 2000), Alzheimer's disease (Thompson *et al.*, 1998, 2000), and a wide variety of developmental disorders, including fetal alcohol syndrome (Sowell *et al.*, 2003), autism (Levitt *et al.*, 2003), and Williams syndrome (Thompson *et al.*, 2005). Through the use of specialized strategies for group averaging of anatomy, specific features of anatomy emerge that are not observed in individual representations due to their considerable variability. Group-specific patterns of cortical organization or asymmetry can then be mapped out and visualized (Narr *et al.*, 1999; Thompson *et al.*, 2004).

B. Deformable Probabilistic Atlases

Warping algorithms create deformation maps that indicate 3D patterns of anatomical differences between any pair of subjects. By defining probability distributions on the space of deformation transformations that drive the anatomy of different subjects into correspondence (Thompson and Toga, 1997), we can estimate statistical parameters of these distributions from databased anatomical data to determine the magnitude and directional biases of anatomical variation. Encoding of local variation can then be used to assess the severity of structural variants outside of the normal range, which, in brain data, may be a sign of disease (Thompson *et al.*, 1997).

C. Encoding Brain Variation

To see if disease-specific features could be detected in individual patients, we developed a *random vector field* approach to construct a population-based brain atlas (Thompson and Toga, 1997). Briefly, given a 3D MR image of a new subject, a warping algorithm calculates a set of high-dimensional volumetric maps that elastically match this image with other scans from an anatomical image database. Target scans are selected from subjects matched for age, handedness, gender, and other demographic factors (Thompson *et al.*, 1997, 1998). The resulting family of volumetric warps provides empirical information on local variability patterns. A probability space of random transformations with statistical flattening corrections, based on the theory of anisotropic Gaussian random fields (Thompson *et al.*, 2000; Worsley *et al.*, 1999), is then used to encode the variations. For the cortex, specialized approaches are needed to represent variations in gyral patterns (Thompson and Toga, 1998; Thompson *et al.*, 2004). Confidence limits in stereotaxic space are determined, for points in the new subject's brain, and probability maps can be created to highlight and quantify regional patterns of deformity (Thompson *et al.*, 1997).

D. Brain Asymmetry

One feature observable from the construction of average anatomical models is that consistent patterns of brain asymmetry can be mapped despite wide variations in asymmetry in individual subjects (Fig. 7). In dementia, the increased cortical asymmetry probably indicates an asymmetric progression by the disease. Figure 8 shows average maps of the lateral ventricles, once again from Alzheimer's disease and matched elderly normal populations. As expected, the ventricles are significantly enlarged in dementia. Notice, however, that a pronounced asymmetry is observed in both groups (the left volume is larger than the right; $p < 0.05$). This is an example of an effect that becomes clear after group averaging of anatomy and is not universally apparent in individual subjects. It is, however, consistent with prior volumetric measurements (Aso *et al.*, 1995; Shenton *et al.*,

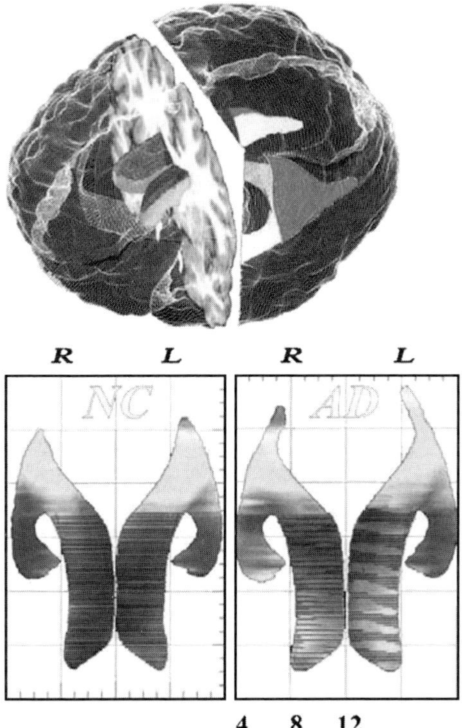

FIG. 8. Population-based maps of average ventricular anatomy in normal aging and in Alzheimer's disease. In patients and controls, three-dimensional (3D) parametric surface meshes were used to model 14 ventricular elements, and meshes representing each surface element were averaged by hemisphere in each group. An average model for Alzheimer's patients (red) is superimposed on an average model for matched normal controls (blue). Mesh averaging reveals enlarged occipital horns in the Alzheimer's patients and high stereotaxic variability in both groups. Extreme variability at the occipital horn tips also contrasts sharply with the stability of septal and temporal ventricular regions. A top view of these averaged surface meshes reveals localized asymmetry, variability, and displacement within and between groups. These subcortical asymmetries emerge only after averaging of anatomical maps in large groups of subjects. (See Color Insert.)

1992). Anatomical averaging can also be cross-validated with a traditional volumetric approach. Occipital horns are, on average, 17.1% larger on the left in the normal group (4070.1 ± 479.9 mm^3) than on the right (3475.3 ± 334.0 mm^3; $p < 0.05$), but no significant asymmetry is found for the superior horns (left: 8658.0 ± 976.7 mm^3; right: 8086.4 ± 1068.2 mm^3; $p > 0.19$) or for the inferior horns (left: 620.6 ± 102.6 mm^3; right: 573.7 ± 85.2 mm^3; $p > 0.37$). The asymmetry is clearly localized in the 3D group-average anatomical representations. In particular, the occipital horn extends (on average) 5.1 mm farther posteriorly on the left than the right. Similar approaches for ventricular-shape mapping have been applied to detect longitudinal progression of Alzheimer's disease (Thompson *et al.*, 2004), as well as ventricular expansion in schizophrenia (Narr *et al.*, 2003) and HIV/AIDS (Thompson *et al.*, 2005). The capacity to resolve asymmetries in a group atlas can assist in studies of disease-specific cortical organization (Narr *et al.*, 2004; Thompson *et al.*, 2004).

E. Asymmetry in Disease

To see whether cortical asymmetries were disturbed in schizophrenia, we made average cortical representations for schizophrenic patients (N = 25; 15 males, 10 females; all right-handed) and matched controls (N = 28; 15 males, 13 females; all right-handed). Thirty-six major sulcal curves were used to drive each subject's gyral pattern into a group mean configuration (Fig. 7). The magnitude of anatomical variation in each brain region was also computed from the deformation vector fields and was shown in color as a variability map. Perhaps surprisingly, asymmetry was not lower in the patient group. Marked asymmetries were observed in the sagittal projections of average anatomy for each group. Significant asymmetries were confirmed by calculations of the curvature and extent measures from the parametric mesh models (Narr *et al.*, 1999). In the frontal cortex, the patients also displayed greater variability than the controls. Developmental studies using these techniques have shown that these asymmetries are also age dependent, and the magnitude of the perisylvian asymmetry increases throughout childhood and adolescence, even after adjustment for age-related increases in overall cerebral volume (Sowell *et al.*, 2003).

F. Corpus Callosum Differences

We also attempted to identify regionally selective patterns of callosal deficits in patient groups with Alzheimer's disease (Thompson *et al.*, 1998), schizophrenia (Narr *et al.*, 1999, 2001), autism (Leow *et al.*, 2005), and Williams syndrome (Chiang *et al.*, 2005). In one Alzheimer's disease study (Thompson *et al.*, 1998), the mid-sagittal callosum was first partitioned into 5 sectors (Duara *et al.*, 1991; Larsen *et al.*, 1992). This roughly segregates callosal fibers from distinct cortical regions. In Alzheimer's disease, focal fiber loss was expected at the callosal isthmus (sector 2), whose fibers selectively innervate the temporal–parietal regions with early neuronal loss and perfusion deficits (Brun and Englund, 1981). Consistent with this hypothesis, a significant area reduction at the isthmus was found, reflecting a dramatic 24.5% decrease from 98.0 ± 8.6 mm^2 in controls to 74.0 ± 5.3 mm^2 in Alzheimer's disease patients ($p < 0.025$). Terminal sectors (1 and 5) were not significantly atrophied, and the central midbody sector showed only a trend toward significance (16.6% mean area loss; $p < 0.1$), due to substantial intergroup overlap.

G. Sex Differences in Brain Structure

Sex differences are observed in normal brain morphology (Davatzikos and Resnick, 1998; DeLacoste-Utamsing and Holloway, 1982; Harasty *et al.*, 1997). Despite an overall sex difference in brain volume (men have cerebral volumes around 9% higher than those of women), it is worth noting that it would be hard to distinguish the brain MRI scan of a man or a woman on the basis of brain volume or any other MRI parameter, given the large intersubject variability. Some additional sex differences in anatomy are found consistently, but these are found on average, and there is some overlap between the sexes on all brain measures that have been assessed. On average, adult women have higher cortical-surface complexity (Luders *et al.*, 2005b), as measured by a fractal dimension measure computed from computational models of the cortex (Thompson *et al.*, 1996, 2005d). This difference is attributable partly to gender differences in cortical curvature and partly to the differences in spatial frequency of gyrification (Luders *et al.*, 2004, 2005). Adult women also have different profiles of cortical thickness, including greater cortical thickness than men in some brain regions even if no correction is made for differences in brain scale (Luders *et al.*, 2005c). Hemispheric asymmetries in cortical thickness and gray-matter density are also found, and these are modulated by gender in normal subjects (Luders *et al.*, 2005a) and in subjects with schizophrenia (Narr *et al.*, 2004). Sex differences in brain structure do not necessarily imply that there are corresponding differences in functional organization of the brain in the same regions, but they are important to take into account, as they may interact with group differences in development, genotype, or disease progression. For example, there is a complex (nonlinear) relationship between the size of brain substructures and the overall size of the brain. This allometric relation follows a log–linear power law, i.e., in larger brains, some substructures will necessarily occupy a larger proportion of the total brain volume, and this will be true both in women and in men. If only linear adjustments are made for brain size in a comparison of structures across genders, or if brains are scaled into a standard brain space, this may result in misleading assertions about the remaining differences, because brain-scale effects have not been entirely factored out (see Thompson *et al.*, 2000, for a discussion).

H. Sex Differences in Schizophrenia

Clear sex differences are also found in the structural brain abnormalities that are observed in schizophrenia (see Fig. 9), and this finding accords with gender differences in the clinical manifestation of the illness (DeLisi *et al.*, 1989; Gur *et al.*,

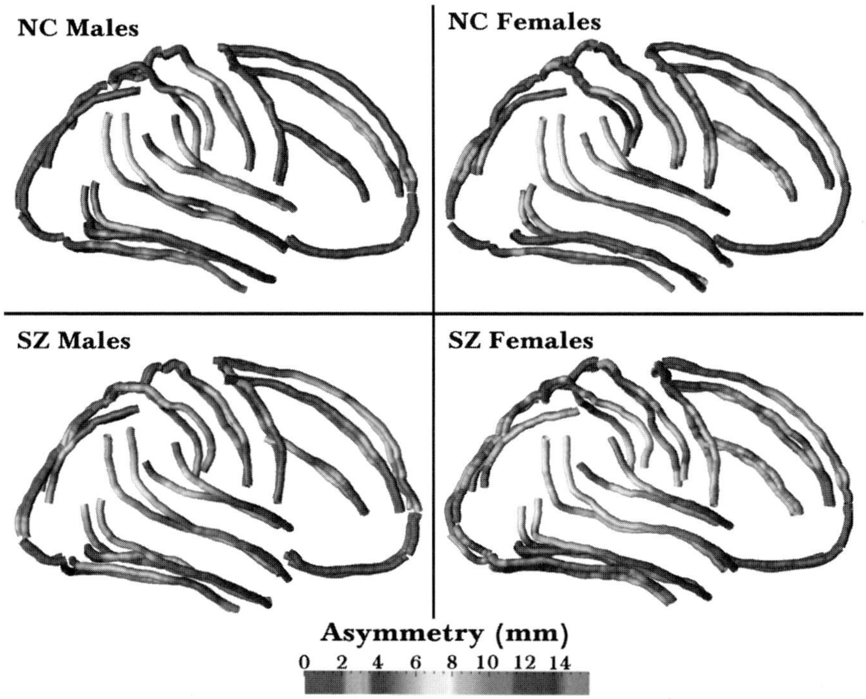

FIG. 9. Brain asymmetry in schizophrenia. To map group variability and asymmetry in sulci and surrounding anatomy, we obtain three-dimensional (3D) cortical extractions from magnetic resonance (MR) scans and bring them into register using surface-warping algorithms that enable the averaging of equivalent cortical regions that can be delineated through the tracing of major hemispheric sulci across subjects. The atlas data can be stratified by gender and disease group (SZ, schizophrenic patients; NC, normal controls) in order to determine whether asymmetries are altered, on average, in each group. Here the color code denotes the magnitude of the asymmetry (in millimeters) in the average sulcal pattern for each group. (See Color Insert.)

1996; Leung and Chue, 2000). For example, in Fig. 10, displacements that represent an upward bowing of the corpus callosum are observed in patients with schizophrenia but not in controls, and they are more pronounced between male patients and male controls than between female patients and female controls (Narr *et al.*, 2000, 2002). These sex differences in callosal displacements may relate to the sexual dimorphism in the etiology and course of schizophrenia, where male patients appear to show increased negative symptoms, an earlier age of onset, and a worse course of illness than do female patients (e.g., DeLisi *et al.*, 1989; Gur *et al.*, 1996). Interestingly, genetic rather than shared environmental or disease-specific influences appear to contribute to displacements of the corpus

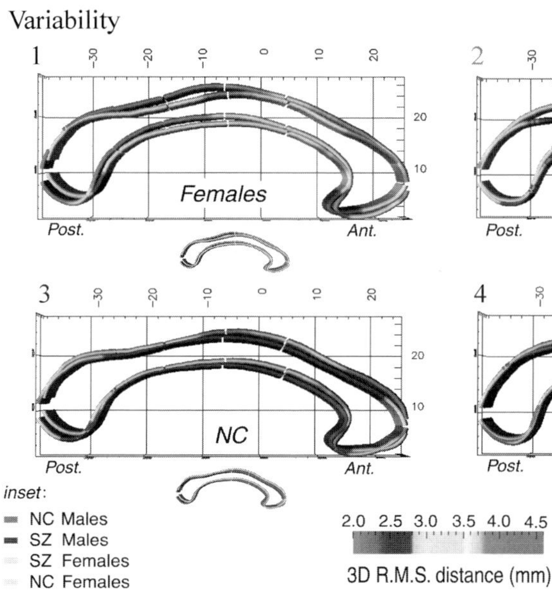

FIG. 10. Average corpus callosum shapes in schizophrenia. Midsagittal corpus callosum boundaries were averaged from 25 patients with chronic schizophrenia (DSM-III-R criteria; 15 males, 10 females; age: 31.1 ± 5.6 years) and from 28 control subjects matched for age (30.5 ± 8.7 years), gender (15 males, 13 females), and handedness (one left-handed subject per group). Profiles of anatomic variability around the group averages are also shown (in color) as a root-mean-squared (RMS) deviation from the mean. Anatomical averaging reveals a much more pronounced and significant bowing effect in the schizophrenic patients than in the nonschizophrenic controls. Male patients show a significant increase in curvature for superior and inferior callosal boundaries ($p < 0.001$), with a highly significant sex-by-diagnosis interaction ($p < 0.004$). The sample was stratified by sex and diagnosis, and separate group averages show that the disease induces less bowing in females (panel 1) than in males (panel 2). While gender differences are not apparent in controls (panel 3), a clear gender difference is seen in the schizophrenic patients (panel 4). Abnormalities localized in a disease-specific atlas can therefore be analyzed to reveal interactions between disease and demographic parameters. Data from Narr *et al.*, 2000. (See Color Insert.)

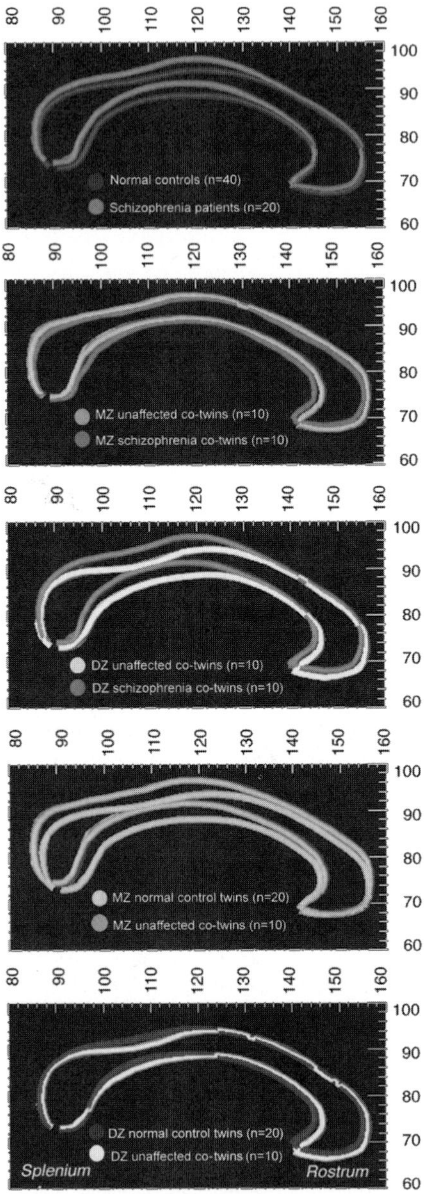

FIG. 11. Callosal surface averages mapped in groups defined by a biological risk for schizophrenia. Average anatomical mesh models of the corpus callosum are shown in different colors to illustrate differences between (1) monozygotic (MZ) and dizygotic (DZ) co-twins discordant for schizophrenia and (2) healthy monozygotic and dizygotic twin pairs. From the top, midsagittal

callosum, where effects are related to both lateral and third-ventricle enlargements (Fig. 11) (Narr *et al.*, 2002). Stratification of probabilistic atlases by gender or other genetic factors provides a computationally fast way to visualize these effects and relate them to epidemiologic or genetic data (Cannon *et al.*, 2005; Narr *et al.*, 2004; Roybal *et al.*, 2005).

I. Comparing a Subject to an Atlas

In one validation experiment (Thompson *et al.*, 1997), probability maps were created to highlight abnormal deviations in the callosal and midline anatomy of a tumor patient. The two regions of metastatic tissue induced marked distortions in the normal architecture of the brain. Variations in deep surface anatomy were stored as a spatially adaptive covariance tensor field, and then probability maps were generated for the tumor patient. In the tumor patient, the herniation effects apparent in the blockface imagery were detected in the probability maps of structures near the lesion sites.

In one experiment, mappings that deform one cortex into gyral correspondence with another were used to create an *average* cortex for patients with mild-to-moderate Alzheimer's disease (Fig. 12 [second row]). Thirty-six gyral curves for 9 Alzheimer's disease patients were transferred to the cortical parameter space uniformly reparameterized, and a set of 36 average gyral curves for the group was created by vector averaging of point locations on each curve. Each individual cortical pattern was then aligned with the average curve set by use of a spherical flow field. The 9 flow fields (one for each patient) were then used to create an average cortex in 3D space in the following way. When a code (that indexes 3D locations) is carried along with the flow that aligns each individual with the average folding pattern, information can then be recovered at a particular location in the average folding pattern, specifying the 3D cortical points that map to it in each subject. A regular grid is then ruled over the warped coded map, and 3D position values for each subject are read off, which allows the recovery of cortical positions in any subject's *original* 3D anatomy. This produces a new coordinate grid on a given subject's cortex, in which particular grid-points appear in the same location relative to the primary gyral pattern across all subjects (see Fischl *et al.*, 1999, for a similar approach). By averaging these 3D positions across subjects, we constructed an average 3D cortical model for the group (Fig. 12). The resulting mapping is guaranteed to average together all

---

callosal averages are mapped in (1) schizophrenia patients and controls; (2) unaffected and affected MZ co-twins; (3) unaffected and affected DZ co-twins; (4) unaffected MZ co-twins of the schizophrenia probands and MZ control twin pairs; and (5) unaffected DZ co-twins of the schizophrenia probands and DZ control twin pairs. Data from Narr *et al.*, 2003. (See Color Insert.)

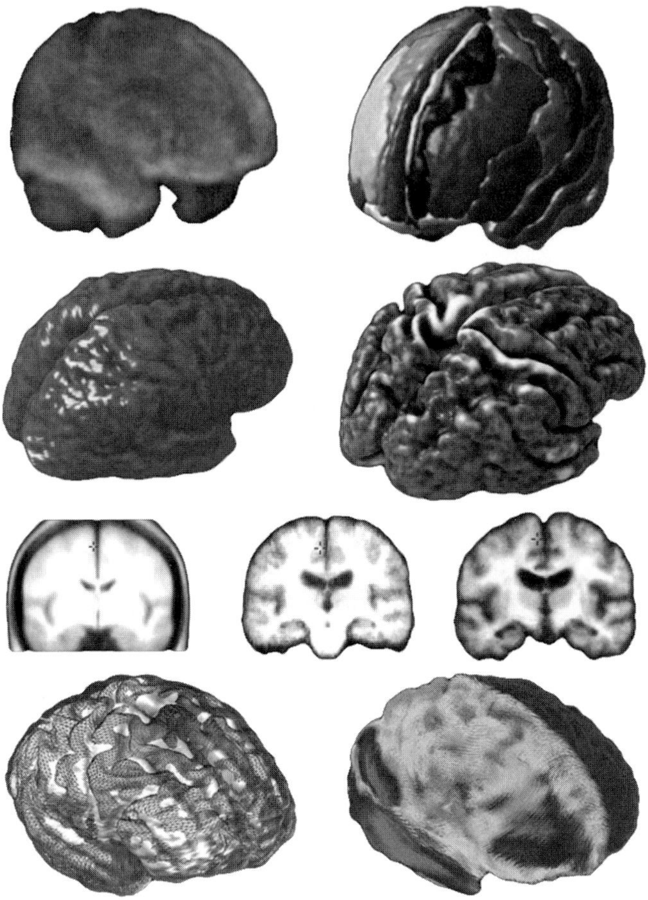

FIG. 12. Methods for averaging brains. (a) In a widely used average brain image template (ICBM305; row 3, panel 1) based on voxel-wise intensity averaging of 305 young normal subjects' scans (Evans *et al.*, 1994), anatomical features are not well resolved at the cortex. Cortical variability is represented by probability clouds (top left) that describe the frequency of incidence for each gyrus at each stereotaxic voxel, after linear registration and voxel-by-voxel comparison. These statistical/probabilistic maps of anatomy, or SPAMs, can be used to parcellate anatomy in new subjects, in order to help derive automated, but approximate, measures of volumes for brain substructures. In an affine brain template (row 2, first panel, and row 3, second panel), similarly constructed from Alzheimer's disease patients' scans, the cortical average is also poorly resolved. By contrast, anatomical features are highly resolved, even at the cortex, in the *Continuum–Mechanical Brain Template* (row 3, final panel), which applies a continuum–mechanical transformation to each brain before intensity averaging is done. Scans are elastically reconfigured into a group mean configuration, using surface-based warping to match 84 surface models (including gyral pattern elements) across all subjects. Reconfigured scans are then averaged voxel by voxel, after intensity normalization, to produce a group image template with the average geometry *and* average image intensity for the group.

points that fall on the same cortical locations across the set of brains, and it ensures that corresponding cortical features are averaged together.

J. Cortical Complexity

Modeling of anatomy using computational surfaces also allows measures of surface complexity to be defined and compared across groups. Traditional post-mortem and *in vivo* morphometric studies have examined cortical complexity as an index of the spatial frequency of the sulcal and gyral convolutions. Early approaches compared the length of an inner and an outer cortical contour in 2D coronal sections, to obtain a "gyrification index" (Zilles *et al.*, 1988; see also Bartley *et al.*, 1997; Kulynych *et al.*, 1997; Vogeley *et al.*, 2000). We recently extended the notion of fractal complexity to 3D cortical surfaces (Thompson *et al.*, 2005). In this approach, a surface model representing the cortex is remeshed at a range of spatial frequencies and its surface area measured at each spatial frequency. The rate of increase of surface area commensurate with spatial frequency is estimated by least-squares fitting of a linear model to the estimated surface area versus frequency, on a log–log plot; this plot is termed a multifractal plot in the fractal literature (see Kiselev *et al.*, 2003, for a discussion of this concept). If $A\{M(N)\}$ represents the surface area of the cortical surface mesh $M(N)$, the fractal dimension or complexity is computed as $\mathrm{Dim_F} = 2+\{d(A\{M(N)\})/d \ln N\}$. The gradient of the multifractal plot is obtained by regression of $\ln A\{M(N)\}$ against $\ln N$. For a flat surface, this slope is zero, and the dimension is 2; representation of the surface at a higher spatial frequency adds no detail. Values above 2 indicate increasing surface detail and greater gyral complexity. Intuitively, higher complexity means the area increases rapidly as finer scale details are included. We identified a profile of increased cortical complexity in Williams syndrome (Thompson *et al.*, 2005), which agrees with direct anatomical observations (Schmitt *et al.*, 2001). Also, in a study of 60 healthy normal subjects, females were found to exhibit greater gyrification than males in the frontal and parietal brain regions (Luders *et al.*, 2004), which may reflect sex-specific rates of cortical maturation.

K. Infancy

The earliest stages of development are often accompanied by dramatic changes in behavior, including advances in motor skills, visual processing, sensory–motor coordination, speech, language, and social learning. To provide a

---

Vector-field transformations of extremely high spatial dimension (bottom row) are required in order to resolve cortical features, in their mean configuration, after scans are averaged together. (See Color Insert.)

comprehensive assessment of infant cortical and volumetric anatomy in addition to establishing normative reference ranges, we examined regional cortical patterns as well as lobar and total intracranial volumes in 43 normal 6-month-old infants (22 females and 21 males). A significant effect of sex was found in the left-parietal and right-temporal lobe volumes: males showed increased volumes relative to females ($p < .001$). However, no significant differences in total intracranial volume were found. 3D cortical representations of infant anatomy showed differences primarily in the parietal and temporal regions. These results suggest the presence of early biological sex differences in infant maturation that may relate to observed behaviors, setting the stage for the sex differences seen later in development.

L. Species

There is an immense variety of brain maps and atlases that represent nonhuman primates and other species. Despite the advent of modern neuroimaging, our knowledge of the molecular, cellular, systems, and functional organization of the human brain remains rudimentary compared to the vast number of neuroscientific studies conducted in other species. The systematic electrophysiological mapping of the neocortex in a wide range of mammals has revealed how cortical specialization varies across species (Manzoni, 1997). Many specialized stereotaxic systems, atlases, and parcellation schemes have also been developed for brain mapping in nonhuman species (Felleman and Van Essen, 1991; Martin and Bowden, 1996; Paxinos and Watson, 1986; Shantha *et al.*, 1968; Szabo and Cowan, 1984). In parallel with the development of international electronic registries for human brain-mapping data, such systems are under active development for the archival and meta-analysis of brain-mapping data obtained in other species.

M. Multimodality Maps

Neuroimaging investigations in nonhuman primates have also included multimodality studies. In our laboratory (Cannestra *et al.*, 1997), we created a 3D multimodality computerized map of the *Nemestrina* monkey brain by combining premortem computed tomography (CT), PET, and MRI scans with a reconstructed postmortem volume of high-resolution full-color cryosection images captured at 50 micron increments. The resulting data sets were repositioned into the Horsley and Clark (1908) stereotaxic coordinate system, and labeled

3D surface models were reconstructed to represent nuclei, tracts, and other neuroanatomical features.

In another important example that combines data from multiple modalities, the Mouse Atlas Project (MAP) created a framework for organizing and analyzing large volumes of gene-expression data from genetically modified animals (MacKenzie-Graham *et al.*, 2004). The atlases produced as part of this project include MR, blockface imaging, histology, and gene-expression maps. A digital atlas used as the hub of an informatics network correlates imaging data, such as structural imaging and histology, with text-based data, such as nomenclature, connections, and references. Specially designed viewers were developed in order to visualize multiple data sets simultaneously and to coordinate between textual and image data. The atlas also allows the independent entry of other types of data, the facile retrieval of information, and the straightforward display of images.

In the future, statistical characterization of the relationships among cortical regions in human and nonhuman primate populations will be invaluable in investigations of how architectural patterns vary across individuals and species (see Van Essen *et al.*, 2004, for comparisons of human and Macaque cortices). Since neuronal connectivity patterns have been mapped extensively in several species, brain-mapping tools that support cross-species comparisons of brain maps are likely to help elucidate the complex functional geography of the human cortex.

The mathematical strategies employed in the construction of these atlases were needed to encode comprehensive information on structural variability in human populations. Particularly relevant is 3D statistical information on group-specific patterns of variation and how these patterns are altered in disease. This information can be exploited by expert diagnostic systems whose goal is to detect subtle or diffuse structural alterations in disease.

N. 4D Coordinate Systems

Atlasing of data from the developing or degenerating brain presents unique challenges (Toga *et al.*, 1996). However, warping algorithms can be applied to serial-scan data to track disease and growth processes in their full spatial and temporal complexity. Maps of anatomical change can be generated through the warping of scans acquired from the same subject over time (Thirion and Calmon, 1997; Thompson *et al.*, 2000). Serial scanning of human subjects (Fox *et al.*, 1996; Freeborough and Fox, 1998; Subsol *et al.*, 1997; Thompson *et al.*, 1998) or experimental animals (Jacobs and Fraser, 1994) in dynamic states of disease or development offers the potential to create 4D models of brain structure. Patterns of structural brain change over time can be visualized by nonlinear warping of a baseline MRI scan to match a subsequent scan from the same subject (Janke *et al.*, 2001; Rey *et al.*, 1999; Thompson *et al.*, 2000). Brain regions undergoing rapid growth or atrophy can be identified, and the resulting maps can be analyzed with tensor-based morphometry techniques (Chiang *et al.*, 2005; Chung *et al.*, 2001; Shen and Davatzikos, 2003; Studholme *et al.*, 2004; Thirion *et al.*, 2000). These models incorporate dynamic descriptors of how the brain changes during maturation or disease. They are therefore of interest in investigation and staging of brain development. In an atlas setting, these 4D maps can act as normative data to define aberrant growth rates and their modulation by therapy (Haney *et al.*, 2000).

In our initial human studies (Thompson *et al.*, 2000), we developed several algorithms to create 4D quantitative maps of growth patterns in the developing human brain. Time series of high-resolution pediatric MRI scans were analyzed. The resulting tensor maps of growth provided spatially detailed information on local growth patterns, quantifying rates of tissue maturation, atrophy, shearing, and dilation in the dynamically changing brain architecture. Pairs of scans were selected to determine patterns of structural change across the interscan interval. Deformation processes recovered by a high-dimensional warping algorithm were then analyzed, using vector-field operators, to produce a variety of tensor maps. These maps were designed to reflect the magnitude and principal directions of dilation or contraction, the rate of strain, and the local curl, divergence, and gradient of flow fields representing the growth processes recovered by the transformation. Rey *et al.* (2000) have done related work on tumor growth and multiple sclerosis lesion tracking.

The growth maps obtained in these studies exhibit several striking characteristics. First, foci of rapid growth at the callosal isthmus appeared consistently across puberty. These rates appeared to attenuate as subjects progressed into adolescence (Chung *et al.*, 2001; Sowell *et al.*, 2003; Thompson *et al.*, 2000). Rapid rates of tissue loss were also revealed at the head of the caudate nucleus in an earlier phase of development.

O. Dynamic Maps of Brain Change

Statistical brain maps from large populations (Fig. 13) are likely to help assess how different drug treatments affect the time course of aging and dementia. In developing dynamic atlases for clinical applications, researchers take a particular interest in modeling atrophic processes that speed up or slow down. As we noted before, diseases can accelerate, or their rate of progression can be slowed down by therapy. If individuals are scanned more than twice over large timespans (e.g., Fox *et al.*, 2001; Gogtay *et al.*, 2004; Janke *et al.*, 2001), their brain changes can be modeled more accurately. Nonlinear mixed models can be used to compare

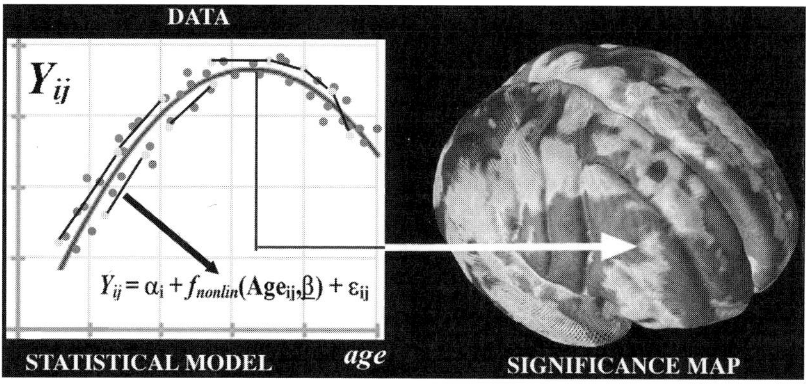

FIG. 13. Data, statistical models, and maps. This schematic (Thompson *et al.*, 2004) shows some of the steps used in mapping cortical change. First, measures (Y_{ij}) are defined that can be obtained longitudinally (green dots) or only once (red dots) in a group of subjects at different ages. Fitting of statistical models to these data (Statistical Model, left panel) produces estimates of parameters that can be plotted onto the cortex, using a color code. These parameters can include age at peak (see the arrow at the peak of the curve), significance values, estimated statistical parameters such as rates of change, or effects of drug treatment or risk genes. (See Color Insert.)

atrophic processes in different groups of subjects (Giedd *et al.*, 1999; Thompson *et al.*, 2003) and to analyze the registered degenerative profiles. For the *i*th individual's *j*th measure, we have

$$Y_{ij} = f(\mathrm{Age}_{ij}, \beta) + \varepsilon_{ij}$$

Here Y_{ij} signifies the outcome measure at a voxel or surface point, such as growth or tissue loss, f denotes a constant, linear, quadratic, cubic, or other function of the individual's age for that scan, and β denotes the regression/ANOVA coefficients to be estimated. In models whose fit is confirmed as significant, e.g., by permutation, loadings on nonlinear parameters may be visualized as attribute maps $\beta(\mathbf{x})$. This reveals the topography of accelerated or decelerated brain change (Thompson *et al.*, 2003). The result is a formal approach to the assessment of whether and where brain change is speeding up or slowing down, a key feature in medication studies.

In this statistical model, age (Age_{ij}) can be replaced by time from onset of disease (Gogtay *et al.*, 2004) or time from start of medication. This flexibility in the parameterization of the time axis allows one to temporally register dynamic patterns using criteria that are expected to bring into line temporal features of interest that appear systematically in a group (Janke *et al.*, 2001). For example, the independent variable could be a cognitive score, such as mini-mental status, which declines over time in Alzheimer's disease. Parameterization of dynamic effects using measures other than time (e.g., clinical status) is another way to align new patients' time series with a dynamic atlas, potentially even further increasing the power to reveal systematic effects.

In the near future, 4D atlases will be able to map growth and degeneration in their full spatial and temporal complexity. Despite the logistic and technical challenges, these mapping approaches hold tremendous promise for analysis of the dynamics of degenerative (Fox *et al.*, 1996; Leow *et al.*, 2005) or neoplastic (Haney *et al.*, 2000) diseases. They will ultimately play a role in detecting how different therapeutic approaches modulate the course of disease.

V. Conclusion

The uses of brain atlases are as varied as their construction. They provide the ability to measure, visualize, compare, and summarize brain images. They encompass descriptions of structure or function of the whole brain to maps of groups or populations. Individual systems of the brain can be mapped, as can changes over time (as in development or degeneration). An atlas enables comparison across individuals, modalities, or states. But in most cases, the value added by brain atlases is the unique and critical ability to integrate information from multiple sources. The utility of an atlas is dependent upon appropriate coordinate systems, registration methods, and deformation methods, along with useful visualization strategies. The probabilistic systems described will likely lead to advances in the encoding of patterns of anatomical variation in large image databases, detection of pathology in individuals and groups, and determination of effects in space and time on brain structure of age, gender, handedness, and other demographic or genetic factors.

Acknowledgments

This work was supported by research grants from the National Institutes of Health's Roadmap Initiative for Bioinformatics and Computational Biology (U54-RR21813), the National Center for Research Resources (P41 RR13642 and M01 RR00865), the National Institute of Mental Health (NIMH) and the National Institute of Neurological Disorders and Stroke (P20 MH/NS65166), a *Human Brain Project* grant to the International Consortium for Brain Mapping, funded jointly by NIMH and the National Institute on Drug Abuse (P20 MH/DA52176). Additional support was provided by the National Institute for Biomedical Imaging and Bioengineering, the National Center for Research Resources, and the National Institute on Aging (R21 EB01651, R21 RR019771, P50 AG016570), the National Library of Medicine (LM/MH05639), and the Biomedical Informatics Research Network (http://www.nbirn.net), which is funded by the National Center for Research

Resources at the National Institutes of Health. Portions of this manuscript, including several figures, are reprinted with permission from Toga *et al.* (2001).

References

Ashburner, J., and Friston, K. J. (2000). Voxel-based morphometry–the methods. *Neuroimage* **11,** 805–821.

Aso, M., Kurachi, M., Suzuki, M., Yuasa, S., Matsui, M., and Saitoh, O. (1995). Asymmetry of the ventricle and age at the onset of schizophrenia. *European Arch. Psych. Clin. Neurosci.* **245,** 142–144.

Avants, B., and Gee, J. C. (2004). Geodesic estimation for large deformation anatomical shape averaging and interpolation. *Neuroimage* **23**(Suppl. 1), S139–150.

Avoli, M., Hwa, G. C., Kostopoulos, G., Oliver, A., and Villemure, J. G. (1991). Electrophysiological analysis of human neocortex *in vitro*: Experimental techniques and methodological approaches. *Can. J. Neurol. Sci.* **18,** 636–639.

Bakircioglu, M., Joshi, S., and Miller, M. I. (1999). Landmark matching on brain surfaces via large deformation diffeomorphisms on the sphere. *Proc. SPIE Medical Imaging.* **3661,** 710–715.

Baron, J. C., Chetelat, G., Desgranges, B., Perchey, G., Landeau, B., de la Sayette, V., and Eustache, F. (2001). *In vivo* mapping of gray matter loss with voxel-based morphometry in mild Alzheimer's disease. *Neuroimage* **14,** 298–309.

Bartley, A. J., Jones, D. W., and Weinberger, D. R. (1997). Genetic variability of human brain size and cortical gyral patterns. *Brain* **120**(Pt 2), 257–269.

Berger, H. (1929). Uber das Elektrenkephalogramm des Menschen. *Arch. Psychiatr. Nervenkr.* **87,** 527–580.

Bohm, C., Greitz, T., Kingsley, D., Berggren, B. M., and Olsson, L. (1983). Adjustable computerized brain atlas for transmission and emission tomography. *Am. J. Neuroradiol.* **4,** 731–733.

Brodmann, K. (1909). Vergleichende Lokalisationslehre der Grosshirnrinde in Ihren Prinzipien Dargestellt auf Grund des Zellenbaues. *In* "Some Papers on the Cerebral Cortex" (translated as *On the Comparative Localization of the Cortex*). Thomas, Springfield, Illinois, 1960.

Bro-Nielsen, M., and Gramkow, C. (1996). Fast fluid registration of medical images. *In* "Visualization in Biomedical Computing" (K. H. Höhne and R. Kikinis, Eds.). Lecture Notes in Computer Science. Springer Verlag.

Brun, A., and Englund, E. (1981). Regional Pattern of Degeneration in Alzheimer's Disease: Neuronal Loss and Histopathologic Grading. *Histopathology* **5,** 549–564.

Bullmore, E. T., Suckling, J., Overmeyer, S., Rabe-Hesketh, S., Taylor, E., and Brammer, M. J. (1999). Global, voxel, and cluster tests, by theory and permutation, for a difference between two groups of structural MR images of the brain. *IEEE Trans Med Imag* **18,** 32–42.

Cachier, P., and Ayache, N. (2003). Isotropic energies, filters and splines for vectorial regularization. *J. Math. Imag. Vision* **20**(3), 251–265.

Cannestra, A. F., Santori, E. M., Holmes, C. J., and Toga, A. W. (1997). A three-dimensional multimodality brain map of the nemestrina monkey. *Brain Research Bulletin* **43**(2), 141–148.

Cannon, T. D., Thompson, P. M., van Erp, T. G., Toga, A. W., Poutanen, V. P., Huttunen, M., Lonnqvist, J., Standerskjold-Nordenstam, C. G., Narr, K. L., Khaledy, M., Zoumalan, C. I., Dail, R., and Kaprio, J. (2002). Cortex mapping reveals regionally specific patterns of genetic and disease-specific gray-matter deficits in twins discordant for schizophrenia. *Proc. Natl. Acad. Sci. USA* **99**(5), 3228–3233.

Cannon, T. D., Thompson, P. M., van Erp, T. G. M., Huttunen, M., Lonnqvist, J., Kaprio, J., and Toga, A. W. (2005). Mapping Heritability and Molecular Genetic Associations with Cortical Features using Probabilistic Brain Atlases: Methods and Initial Applications to Schizophrenia. *Neuroinformatics* (in press).

Chiang, M. C., Reiss, A. L., Eckert, M. A., Dutton, R. A., Lee, A. D., Alaghband, Y., Bellugi, U., Galaburda, A., Korenberg, J., Mills, D. L., Toga, A. W., and Thompson, P. M. (2005). Tensor-Based Morphometry of the Corpus Callosum in Williams Syndrome. *In* "11th Annual Meeting of the Organization for Human Brain Mapping (OHBM)." Toronto, Canada, June 12–16, 2005.

Christensen, G. E., Rabbitt, R. D., and Miller, M. I. (1996). Deformable Templates using Large Deformation Kinematics. *IEEE Trans. on Image Processing* **5**(10), 1435–1447.

Chung, M. K., Worsley, K. J., Paus, T., Cherif, C., Giedd, J. N., Rapoport, J. L., and Evans, A. C. (2001). A Unified Statistical Approach to Deformation-Based Morphometry. *NeuroImage* **14,** 596–606.

Chung, M. K., Worsley, K. J., Robbins, S., Paus, T., Taylor, J., Giedd, J. N., Rapoport, J. L., and Evans, A. C. (2001). Statistical Analysis of Cortical Surface Area Change, with an Application to Brain Growth. *Neuroimage* **13S,** 96.

Chung, M. K., Worsley, K. J., Taylor, J., and Ramsay, J. O. (2001). Diffusion Smoothing in Brain Imaging. *Proc. 53rd Session of the International Statistical Institute*, Seoul.

Collins, D. L., Neelin, P., Peters, T. M., and Evans, A. C. (1994). Automatic 3D Intersubject Registration of MR Volumetric Data into Standardized Talairach Space. *J. Comp. Assisted Tomography* **18**(2), 192–205.

Collins, D. L., Peters, T. M., and Evans, A. C. (1994). An Automated 3D Non-Linear Image Deformation Procedure for Determination of Gross Morphometric Variability in the Human Brain. *Proc. Visualization in Biomed. Comp.* (SPIE) **3,** 180–190.

Damasio, H. (1995). *Human Brain Anatomy in Computerized Images*, Oxford Univ. Press, Oxford and New York.

Davatzikos, C. (1996). Spatial Normalization of 3D Brain Images using Deformable Models. *J. Comp. Assisted Tomography* **20**(4), 656–665.

Davatzikos, C., Shen, D., Mohamed, A., and Kyriacou, S. K. (2001). A framework for predictive modeling of anatomical deformations. *IEEE Trans. Med. Imaging* **20**(8), 836–843.

Davidian, M., and Giltinan, D. (1995). "Nonlinear Models for Repeated Measurement Data." Chapman and Hall.

Dinov, I. D., Mega, M. S., Thompson, P. M., Lee, L., Woods, R. P., Holmes, C. J., Sumners, D. L., and Toga, A. W. (2000). Analyzing functional brain images in a probabilistic atlas: A validation of sub-volume thresholding. *J. Comp. Asst. Tomog.* **24**(1), 128–138.

Drury, H. A., Van Essen, D. C., Joshi, S. C., and Miller, M. I. (1996). Analysis and Comparison of Areal Partitioning Schemes Using Two-Dimensional Fluid Deformations, Poster Presentation. *2nd Int. Conf. on Functional Mapping of the Human Brain*, Boston, Massachusetts USA, June 17–21 1996, *NeuroImage* **3,** S130.

Duara, R., Kushch, A., Gross-Glenn, K., Barker, W. W., Jallad, B., Pascal, S., Loewenstein, D. A., Sheldon, J., Rabin, M., and Levin, B. (1991). Neuroanatomic Differences between Dyslexic and Normal Readers on Magnetic Resonance Imaging Scans. *Arch. Neurol.* **48,** 410–416.

Duncan, J. S., Papademetris, X., Yang, J., Jackowski, M., Zeng, X., and Staib, L. H. (2004). Geometric strategies for neuroanatomic analysis from MRI. *NeuroImage* **23**(Suppl. 1), S34–S45.

Duvernoy, H. M. (1991). The Human Brain Springer-Verlag, New York.

Evans, A. C., Collins, D. L., and Milner, B. (1992). An MRI-based Stereotactic Brain Atlas from 300 Young Normal Subjects. "Proceedings of the 22nd Symposium of the Society for Neuroscience." Anaheim, 408.

Evans, A. C., Dai, W., Collins, D. L., Neelin, P., and Marrett, S. (1991). Warping of a Computerized 3D Atlas to Match Brain Image Volumes for Quantitative Neuroanatomical and Functional Analysis. *SPIE Med. Imaging* **1445,** 236–247.

Felleman, D. J., and Van Essen, D. C. (1991). Distributed hierarchical processing in the primate cerebral cortex. *Cereb Cortex* **1**(1), 1–47.

Fischl, B., and Dale, A. M. (2001). Measuring the thickness of the human cerebral cortex from magnetic resonance images. *Proc. Natl. Acad. Sci. USA* **97**(20), 11050–11055.

Fischl, B., Liu, A., and Dale, A. M. (2001). Manifold surgery: Constructing geometrically accurate and topologically correct models of the human cerebral cortex. *IEEE Trans. Med. Imaging* **20**(1), 70–80.

Fischl, B., Salat, D. H., van der Kouwe, A. J., Makris, N., Segonne, F., Quinn, B. T., and Dale, A. M. (2004). Sequence-independent segmentation of magnetic resonance images. *NeuroImage* **23** (Suppl. 1), S69–S84.

Fischl, B., Sereno, M. I., Tootell, R. B. H., and Dale, A. M. (1999). High-Resolution Inter-Subject Averaging and a Coordinate System for the Cortical Surface. *Hum. Brain Mapp.* **8**(4), 272–284.

Fox, N. C., Crum, W. R., Scahill, R. I., Stevens, J. M., Janssen, J. C., and Rossor, M. N. (2001). Imaging of onset and progression of Alzheimer's disease with voxel-compression mapping of serial magnetic resonance images. *Lancet* **358,** 201–205.

Fox, N. C., Freeborough, P. A., and Rossor, M. N. (1996). Visualization and Quantification of Rates of Cerebral Atrophy in Alzheimer's Disease. *Lancet* **348**(9020), 94–97.

Fox, P. T., Perlmutter, J. S., and Raichle, M. (1985). A Stereotactic Method of Localization for Positron Emission Tomography. *J. Comp. Assist. Tomogr.* **9**(1), 141–153.

Freeborough, P. A., and Fox, N. C. (1997). The boundary shift integral: An accurate and robust measure of cerebral volume changes from registered repeat MRI. *IEEE Trans. Med. Imaging* **16,** 623–629.

Freeborough, P. A., and Fox, N. C. (1998). Modeling brain deformations in Alzheimer's disease by fluid registration of serial 3D MR images. *J. Comput. Assist. Tomogr.* **22,** 838–843.

Friston, K. J., Ashburner, J., Poline, J. B., Frith, C. D., Heather, J. D., and Frackowiak, R. S. (1995). Spatial registration and normalisation of images. *Human Brain Mapping* **2,** 16.

Friston, K. J., Holmes, A. P., Worsley, K. J., Poline, J. P., Frith, C. D., and Frackowiak, R. S. J. (1995). Statistical Parametric Maps in Functional Imaging: A General Linear Approach. *Human Brain Mapping* **2,** 189–210.

Gee, J. C., Le Briquer, L., Barillot, C., Haynor, D. R., and Bajcsy, R. (1995). Bayesian Approach to the Brain Image Matching Problem. *SPIE Medical Imaging: Image Processing* **2434,** 154–156.

Gerig, G., Styner, M., Shenton, M. E., and Lieberman, J. A. (2001). Shape versus Size: Improved Understanding of the Morphology of Brain Structures. *Proc. MICCAI 2001* Springer LNCS. **2208,** 24–32.

Giedd, J. N., Blumenthal, J., Jeffries, N. O., Castellanos, F. X., Liu, H., Zijdenbos, A., Paus, T., Evans, A. C., and Rapoport, J. L. (1999). Brain development during childhood and adolescence: A longitudinal MRI study. *Nature Neurosci.* **2**(10), 861–863.

Giedd, J. N., Jeffries, N. O., Blumenthal, J., Castellanos, F. X., Vaituzis, A. C., Fernandez, T., Hamburger, S. D., Liu, H., Nelson, J., Bedwell, J., Tran, L., Lenane, M., Nicolson, R., and Rapoport, J. L. (1999). Childhood-onset schizophrenia: Progressive brain changes during adolescence. *Biol. Psychiatry* **46**(7), 892–898.

Gogtay, N., Lusk, L., Hayashi, K. M., Giedd, J. N., Greenstein, D., Vaituzis, C., Herman, D. H., Nugent, T. F., Classen, L., Sporn, A., Toga, A. W., Rapoport, J. L., and Thompson, P. M. (2004). Dynamic mapping of human cortical development during childhood and adolescence. *Proc. Nat. Acad. Sci.* **101**(21), 8174–8179.

Gogtay, N., Ordonez, A., Herman, D., Hayashi, K. M., Vaituzis, C., Sporn, A., Giedd, J. N., Greenstein, D., Thompson, P. M., and Rapoport, J. L. (2004). Dynamic Mapping of Cortical Brain Development in Pediatric Bipolar Illness. *International Conference of the Organization for Human Brain Mapping* Budapest, Hungary, June 13–17, 2004.

Good, C. D., Johnsrude, I. S., Ashburner, J., Henson, R. N., Friston, K. J., and Frackowiak, R. S. J. (2001). A voxel-based morphometric study of ageing in 465 normal adult human brains. *Neuroimage* **14**(1 Pt 1), 21–36.

Gramkow, C. (1996). Registration of 2D and 3D Medical Images. *Denmark Tech. Univ.* M.Sc. Thesis.

Greitz, T., Bohm, C., Holte, S., and Eriksson, L. (1991). A Computerized Brain Atlas: Construction, Anatomical Content and Application. *J. Comp. Assist. Tomogr.* **15**(1), 26–38.

Grenander, U., and Miller, M. I. (1998). Computational Anatomy: An Emerging Discipline, Technical Report Dept. of Mathematics, Brown University.

Haller, J. W., Banerjee, A., Christensen, G. E., Gado, M., Joshi, S., Miller, M. I., Sheline, Y., Vannier, M. W., and Csernansky, J. G. (1997). Three-Dimensional Hippocampal MR Morphometry with High-Dimensional Transformation of a Neuroanatomic Atlas. *Radiology* **202**(2), 504–510.

Haney, S., Thompson, P. M., Cloughesy, T. F., Alger, J. R., Frew, A., and Toga, A. W. (2000). Cross-Validation of Tissue Classification and Surface Modeling Algorithms for Determining Growth Rates of Malignant Gliomas: Prognostic Value of Growth Rates and MR Spectroscopy. The *2000 International Conference on Mathematics and Engineering Techniques in Medicine and Biological Sciences*, Las Vegas, NV, June 2000.

Höhne, K. H., Bomans, M., Pommert, A., Riemer, M., Schiers, C., Tiede, U., and Wiebecke, G. (1990). 3D Visualization of Tomographic Volume Data using the Generalized Voxel Model. *Visual Comput.* **6,** 28–36.

Höhne, K. H., Bomans, M., Riemer, M., Schubert, R., Tiede, U., and Lierse, W. (1992). A 3D Anatomical Atlas Based on a Volume Model. *IEEE Comput. Graphics Appl.* **12,** 72–78.

Horsley, V., and Clark, R. H. (1908). The structure and function of cerebellum examined by a new method. *Brain* **31,** 45–124.

Hurdal, M. K., and Stephenson, K. (2004). Cortical cartography using the discrete conformal approach of circle packings. *NeuroImage* **23**(Suppl. 1), S119–S128.

Janke, A. L., Zubicaray, G. d., Rose, S. E., Griffin, M., Chalk, J. B., and Galloway, G. J. (2001). 4D deformation modeling of cortical disease progression in Alzheimer's dementia. *Magn. Reson. Med.* **46**(4), 661–666.

Joshi, A., Shattuck, D. W., Thompson, P. M., and Leahy, R. M. (2005). Thin-Plate Spline Registration in the Intrinsic Geometry of the Cortical Surface. "11th Annual Meeting of the Organization for Human Brain Mapping (OHBM)." Toronto, Canada, June 12–16, 2005.

Joshi, S., Grenander, U., and Miller, M. I. (1997). On the geometry and shape of brain sub-manifolds. *IEEE Transactions on Pattern Analysis and Machine Intelligence* **11,** 1317–1343.

Kikinis, R., Shenton, M. E., Iosifescu, D. V., McCarley, R. W., Saiviroonporn, P., Hokama, H. H., Robatino, A., Metcalf, D., Wible, C. G., Portas, C. M., Donnino, R., and Jolesz, F. (1996). A Digital Brain Atlas for Surgical Planning, Model-Driven Segmentation, and Teaching. *IEEE Trans. on Visualization and Comp. Graphics* **2**(3), 232–241.

Kindlmann, G. L., Weinstein, D. M., Lee, A. D., Toga, A. W., and Thompson, P. M. (2004). Visualization of Anatomic Covariance Tensor Fields. "Proc. IEEE Engineering in Medicine and Biology Society (EMBS)." San Francisco, CA, Sept 1–5, 2004.

Kiselev, V. G., Hahn, K. R., and Auer, D. P. (2003). Is the brain cortex a fractal? *NeuroImage* **20**(3), 1765–1774.

Kulynych, J. J., Luevano, L. F., Jones, D. W., and Weinberger, D. R. (1997). Cortical abnormality in schizophrenia: An *in vivo* application of the gyrification index. *Biol. Psychiatry* **41**(10), 995–999.

Larsen, J. P., Høien, T., and Ödegaard, H. (1992). Magnetic Resonance Imaging of the Corpus Callosum in Developmental Dyslexia. *Cognitive Neuropsychology* **9,** 123–134.

Le Bihan, D. (1996). Functional MRI of the Brain: Principles, Applications and Limitations. *Neuroradiol.* **23**(1), 1–5.

Le Goualher, G., Procyk, E., Collins, D. L., Venugopal, R., Barillot, C., and Evans, A. C. (1999). Automated extraction and variability analysis of sulcal neuroanatomy. *IEEE Trans. Med. Imaging* **18**(3), 206–217.

Lehmann, E. D., Hawkes, D., Hill, D., Bird, C., Robinson, G., Colchester, A., and Maisley, M. (1991). Computer-Aided Interpretation of SPECT Images of the Brain using an MRI-Derived Neuroanatomic Atlas. *Med. Informatics* **16,** 151–166.

Leow, A., Yu, C. L., Lee, S. J., Huang, S. C., Protas, H., Nicolson, R., Hayashi, K. M., Toga, A. W., and Thompson, P. M. (2005). Brain structural mapping using a novel hybrid implicit/explicit framework based on the level-set method. *NeuroImage* **24**(3), 910–927.

Lerch, J. P., and Evans, A. C. (2005). Cortical thickness analysis examined through power analysis and a popluation simulation. *NeuroImage* **24**(1), 163–173.

Levitt, J. G., O'Neill, J., Blanton, R. E., Smalley, S., Fadale, D., McCracken, J. T., Guthrie, D., Toga, A. W., and Alger, J. R. (2003). Proton magnetic resonance spectroscopic imaging of the brain in childhood autism. *Biol. Psychiatry* **54**(12), 1355–1366.

Lin, C. L., Dumont, A. S., Lieu, A. S., Yen, C. P., Hwang, S. L., Kwan, A. L., Kassell, N. F., and Howng, S. L. (2003). Characterization of perioperative seizures and epilepsy following aneurysmal subarachnoid hemorrhage. *J. Neurosurg.* **99**(6), 978–985.

Lohmann, G., Preul, C., and Hund-Georgiadis, M. (2003). Morphology-based cortical thickness estimation. *Inf. Process Med. Imaging* **18,** 89–100.

Lu, L. H., Leonard, C. M., Thompson, P. M., Kan, E., Toga, A. W., and Sowell, E. R. (2005). Mapping Structural Brain Changes Related to Improving Phonological Awareness in Normal Children. "11th Annual Meeting of the Organization for Human Brain Mapping (OHBM)." Toronto, Canada, June 12–16, 2005.

Luders, E., Narr, K. L., Thompson, P. M., Rex, D. E., Jancke, L., Steinmetz, H., and Toga, A. W. (2004). Gender differences in cortical complexity. *Nature Neuroscience* **7**(8), 799–800.

Luders, E., Narr, K. L., Thompson, P. M., Rex, D. E., Woods, R. P., De Luca, H., Jancke, L., and Toga, A. W. (2005a). Asymmetry effects on cortical thickness (in preparation).

Luders, E., Narr, K. L., Thompson, P. M., Rex, D. E., Woods, R. P., De Luca, H., Jancke, L., and Toga, A. W. (2005b). Gender Effects on Cortical Thickness. *Mapping* (in press).

Luders, E., Narr, K. L., Woods, R. P., Rex, D. E., Thompson, P. M., Jancke, L., Steinmetz, H., and Toga, A. W. (2005c). Mapping cortical gray matter in the young adult brain: Effects of gender. *NeuroImage* **26,** 493–501.

Luders, E., Thompson, P.M., Narr, K.L., Toga, A.W., Jancke, L., Gaser, C. (2005d). A curvature-based approach to estimate local gyrification on the cortical surface (submitted).

Mac Donald, D., Kabani, N., Avis, D., and Evans, A. C. (2000). Automated 3-D extraction of inner and outer surfaces of cerebral cortex from MRI. *NeuroImage* **12**(3), 340–356.

MacKenzie-Graham, A., Lee, E. F., Dinov, I. D., Bota, M., Shattuck, D. W., Ruffins, S., Yuan, H., Konstantinidis, F., Pitiot, A., Ding, Y., Hu, G., Jacobs, R. E., and Toga, A. W. (2004). A multimodal, multidimensional atlas of the C57BL/6J mouse brain. *J. Anatomy* **204,** 93–102.

Mangin, J. F., Poupon, F., Duchesnay, E., Riviere, D., Cachia, A., Collins, D. L., Evans, A. C., and Regis, J. (2004). Brain morphometry using 3D moment invariants. *Med. Image Anal.* **8**(3), 187–196.

Mansour, A., Fox, C. A., Burke, S., Akil, H., and Watson, S. J. (1995). Immunohistochemical Localization of the Cloned Mu Opioid Receptor in the Rat CNS. *J. Chem. Neuroanat.* **8**(4), 283–305.

Manzoni, T. (1997). The callosal connections of the hierachically organized somatosensory areas of primates. *J. Neurosurg. Sci.* **41**(1), 1–22.

Martin, R. F., and Bowden, D. M. (1996). A stereotaxic template atlas of the macaque brain for digital imaging and quantitative neuroanatomy. *NeuroImage* **4**(2), 119–150.

Matsui, T., and Hirano, A. (1978). "An Atlas of the Human Brain for Computerized Tomography." Igako-Shoin.

Mazziotta, J. C., Toga, A. W., Evans, A. C., Fox, P., and Lancaster, J. (1995). A Probabilistic Atlas of the Human Brain: Theory and Rationale for its Development. *NeuroImage* **2,** 89–101.

Mazziotta, J. C., Toga, A. W., Evans, A. C., Fox, P. T., Lancaster, J., Zilles, K., Woods, R. P., Paus, T., Simpson, G., Pike, B., Holmes, C. J., Collins, D. L., Thompson, P. M., MacDonald, D., Iacoboni, M., Schormann, T., Amunts, K., Palomero-Gallagher, N., Geyer, S., Parsons, L., Narr, K. L., Kabani, N., Le Goualher, G., Boomsma, D., Cannon, T., Kawashima, R., and Mazoyer, B. (2001). A Probabilistic Atlas and Reference System for the Human Brain: International Consortium for Brain Mapping (ICBM). *Philosophical Transactions of the Royal Society of London*, Series B (Biological Sciences) **356**(1412), 1293–1322.

Mega, M. S., Chen, S., Thompson, P. M., Woods, R. P., Karaca, T. J., Tiwari, A., Vinters, H., Small, G. W., and Toga, A. W. (1997). Mapping Pathology to Metabolism: Coregistration of Stained Whole Brain Sections to PET in Alzheimer's Disease. *NeuroImage* **5,** 147–153.

Mega, M. S., Thompson, P. M., Cummings, J. L., Back, C. L., Xu, L. Q., Zohoori, S., Goldkorn, A., Moussai, J., Fairbanks, L., Small, G. W., and Toga, A. W. (1998). Sulcal Variability in the Alzheimer's Brain: Correlations with Cognition. *Neurology* **50,** 145–151.

Mega, M. S., Thompson, P. M., Toga, A. W., and Cummings, J. L. (2000). Brain Mapping in Dementia. *In* "Brain Mapping: The Disorders" (A. W. Toga and J. C. Mazziotta, Eds.). Academic Press.

Meltzer, C. C., and Frost, J. J. (1994). Partial Volume Correction in Emission-Computed Tomography: Focus on Alzheimer Disease. *In* "Functional Neuroimaging: Technical Foundations" (R. W. Thatcher, M. Hallett, T. Zeffiro, E. R. John, and M. Huerta, Eds.), pp. 163–170. Academic Press.

Memoli, F., Sapiro, G., and Thompson, P. (2004). Implicit brain imaging. *NeuroImage.* **23**(Suppl. 1), S179–S188.

Miller, M. (2004). Computational anatomy: Shape, growth and atrophy comparison via diffeomorphisms. *NeuroImage* **23**(Suppl. 1), S19–S133.

Miller, M. I., Hosakere, M., Barker, A. R., Priebe, C. E., Lee, N., Ratnanather, J. T., Wang, L., Gado, M., Morns, J. C., and Csernanasky, J. G. (2003). Labeled cortical mantle distance maps of the cingulate quantify differences between dementia of the Alzheimer type and healthy aging. *Proc. Natl. Acad. Sci. USA* **100**(25), 15172–15177.

Miller, M. I., Younes, L., and Trouve, A. (2002). On the Metrics and Euler-Lagrange Equations of Computational Anatomy. *Annu. Rev. Biomed. Eng.* **4,** 375–405.

Minoshima, S., Koeppe, R. A., Frey, K. A., Ishihara, M., and Kuhl, D. E. (1994). Stereotactic PET Atlas of the Human Brain: Aid for Visual Interpretation of Functional Brain Images. *J. Nucl. Med.* **35,** 949–954.

Mummery, C. J., Patterson, K., Price, C. J., Ashburner, J., Frackowiak, R. S., and Hodges, J. R. (2000). A voxel-based morphometry study of semantic dementia: Relationship between temporal lobe atrophy and semantic memory. *Ann. Neurol.* **47**(1), 36–45.

Narr, K., Green, M. F., Capetillo-Cunliffe, L., Toga, A. W., and Zaidel, E. (2003). Lateralized lexical decision in Schizophrenia: Hemispheric specialization and interhemispheric lexicality priming. *J. Abnormal Psy.* **112**(4), 623–632.

Narr, K. L., Bilder, R. M., Kim, S., Thompson, P. M., Szeszko, P., Robinson, D., Luders, E., and Toga, A. W. (2004). Abnormal gyral complexity in first episode schizophrenia. *Biological Psychiatry* **55,** 859–867.

Narr, K. L., Cannon, T. D., Woods, R. P., Thompson, P. M., Kim, S., Asunction, D., van Erp, T. G., Poutanen, V. P., Huttunen, M., Lonnqvist, J., Standerksjold-Nordenstam, C. G., Kaprio, J., Mazziotta, J. C., and Toga, A. W. (2002). Genetic contributions to altered callosal morphology in schizophrenia. *J. Neurosci.* **22,** 3720–3729.

Narr, K. L., Thompson, P. M., Sharma, T., Moussai, J., Blanton, R., Anvar, B., Edris, A., Krupp, R., Rayman, J., Khaledy, M., and Toga, A. W. (2001). 3D mapping of temporo-limbic regions and the lateral ventricles in schizophrenia: Sex effects. *Biological Psychiatry* **50,** 84–97.

Narr, K. L., Thompson, P. M., Sharma, T., Moussai, J., Cannestra, A. F., and Toga, A. W. (2000). Mapping Corpus Callosum Morphology in Schizophrenia. *Cerebral Cortex* **10**(1), 40–49.

Narr, K. L., Thompson, P. M., Sharma, T., Moussai, J., Cannestra, A. F., and Toga, A. W. (2000). Mapping morphology of the corpus callosum in schizophrenia. *Cerebral Cortex* **10,** 40–49.

Narr, K. L., Thompson, P. M., Sharma, T., Moussai, J., Zoumalan, C. I., Rayman, J., and Toga, A. W. (2001). 3D Mapping of Gyral Shape and Cortical Surface Asymmetries in Schizophrenia: Gender Effects. *Am. J. Psychiatry* **158**(2), 244–255.

Narr, K. L., Thompson, P. M., Sharma, T., Moussai, J., Zoumalan, C. I., Wang, W., Rayman, J., and Toga, A. W. (1999). Cortical and Subcortical Asymmetries: Sex Effects in Schizophrenic and Normal Populations. *Proc. Society for Neuroscience*, Miami, FL, Oct. 1999.

Nielsen, M., Florack, L., and Deriche, R. (1994). Regularization and Scale Space. *INRIA Tech. Rep.* RR-2352.

Ono, M., Kubik, S., and Abernathey, C. D. (1990). Atlas of the Cerebral Sulci Stuttgart, Thieme.

Palovcik, R. A., Reid, S. A., Principe, J. C., and Albuquerque, A. (1992). 3D Computer Animation of Electrophysiological Responses. *J. Neurosci. Methods* **41,** 1–9.

Paxinos, G., and Watson, C. (1986). The Rat Brain in Stereotaxic Coordinates Academic Press, San Diego, CA.

Penfield, W., and Boldrey, E. (1937). Somatic Motor and Sensory Representation in the Cerebral Cortex of Man as Studied by Electrical Stimulation. *Brain* **60,** 389–443.

Pitiot, A., Delingette, H., Thompson, P. M., and Ayache, N. (2004). Expert knowledge-guided segmentation system for brain MRI. *NeuroImage* **23**(Suppl. 1), S85–S96.

Pommert, A., Schubert, R., Riemer, M., Schiemann, T., Tiede, U., and Höhne, K. H. (1994). Symbolic Modeling of Human Anatomy for Visualization and Simulation. *IEEE Vis. Biomed. Comp.* **2359,** 412–423.

Rasser, P. E., Johnston, P. J., Lagopoulos, J., Ward, P. B., Schall, U., Thienel, R., Bender, S., Toga, A. W., and Thompson, P. M. (2005). Analysis of fMRI BOLD activation during the Tower of London Tasking using Gyral Pattern and Intensity Models of Cerebral Cortex. *NeuroImage* **26**(3), 941–951.

Ratnanather, J. T., Botteron, K. N., Nishino, T., Massie, A. B., Lal, R. M., Patel, S. G., Peddi, S., Todd, R. D., and Miller, M. I. (2001). Validating cortical surface analysis of medical prefrontal cortex. *NeuroImage* **14**(5), 1058–1069.

Rex, D. E., Pouratian, N., Sicotte, N. L., and Toga, A. W. (2001). Locational effect of brain masses on fMRI of tongue movement. "7th Annual Meeting of the Organization for Human Brain Mapping." Brighton, UK, *NeuroImage Abs.* **13**(6), S232.

Rey, D., Subsol, G., Delingette, H., and Ayache, N. (2002). Automatic Detection and Segmentation of Evolving Processes in 3D Medical Images: Application to Multiple Sclerosis. *Medical Image Analysis* **6**(2), 163–179.

Rey, R., Sabourin, J. C., Venara, M., Long, W. Q., Jaubert, F., Zeller, W. P., Duvillard, P., Chemes, H., and Bidart, J. M. (2000). Anti-mullerian hormone is a specific marker of sertoli- and granulosa-cell origin in gonadal tumors. *Hum. Pathol.* **31**(10), 1202–1208.

Roland, P. E., and Zilles, K. (1994). Brain Atlases-A New Research Tool. *Trend. Neurosci.* **17**(11), 458–467.

Rombouts, S. A., Barkhof, F., Witter, M. P., and Scheltens, P. (2000). Unbiased whole-brain analysis of gray matter loss in Alzheimer's disease. *Neurosci. Lett.* **285**(3), 231–233.

Roybal, D. J., Dutton, R. A., Hayashi, K. M., de Zubicaray, G. I., O'Dowd, B. S., Chalk, J. B., Janke, A. L., McMahon, K. L., Cowin, G. J., Strudwick, M., Bryant, M. K., Doddrell, D. M., and Thompson, P. M. (2005). Mapping ApoE4 and Gender Effects on Hippocampal Atrophic Rates:

A Longitudinal MIR Study of Normal Ageing. "Annual Scientific Meeting of the American Geriatric Society (AGS)." Orlando, FL, May 11–15, 2005.

Schaltenbrand, G., and Bailey, P. (1959). "Introduction to Stereotaxis with an Atlas of the Human Brain." Stuttgart, Thieme, New York.

Schaltenbrand, G., and Wahren, W. (1977). "Atlas for Stereotaxy of the Human Brain," 2nd edn. Stuttgart, Thieme.

Schiemann, T., Nuthmann, J., Tiede, U., and Höhne, K. H. (1996). Segmentation of the Visible Human for High-Quality Volume-Based Visualization. *Vis. Biomed. Comp.* **4,** 13–22.

Schmitt, J. E., Eliez, S., Warsofsky, I. S., Bellugi, U., and Reiss, A. L. (2001). Enlarged cerebellar vermis is Williams syndrome. *J. Psyhciatr. Res.* **35**(4), 225–229.

Sereno, M. I., Dale, A. M., Liu, A., and Tootell, R. B. H. (1996). A Surface-Based Coordinate System for a Canonical Cortex. *Proc. 2nd Int. Conf. Hum. Brain Mapping,* Neuroimage **3**(3), S252. Boston, MA.

Shantha, T. R., Manocha, S. L., and Bourne, G. H. (1968). Enzyme histochemistry of the cerebral cortex of squirrel, monkey and rat. *Acta Histochem* **30**(2), 218–233.

Shenton, M. E., Kikinis, R., Jolesz, F. A., Pollack, S. D., LeMay, M., Wible, C. G., Hokama, H., Martin, J., Metcalf, D., Coleman, M., and McCarley, R. (1992). Abnormalities of the left temporal lobe and thought disorder in schizophrenia. *New England J. Med.* **327**(9), 604–612.

Sowell, E. R., Mattson, S. N., Thompson, P. M., Jernigan, T. L., Riley, E. P., and Toga, A. W. (2001). Mapping callosal morphology and cognitive correlates: Effects of heavy prenatal alcohol exposure. *Neurology* **57,** 235–244.

Sowell, E. R., Thompson, P. M., Holmes, C. J., Jernigan, T. L., and Toga, A. W. (1999). Progression of Structural Changes in the Human Brain during the First Three Decades of Life: *In Vivo* Evidence for Post-Adolescent Frontal and Striatal Maturation. *Nature Neuroscience* **2**(10), 859–861.

Sowell, E. R., Thompson, P. M., Leonard, C. M., Welcome, S. E., Kan, E., and Toga, A. W. (2004). Longitudinal mapping of cortical thickness and brain growth in normal children. *J. Neurosci.* **24**(38), 8223–8231.

Sowell, E. R., Thompson, P. M., Mattson, S. N., Tessner, K. D., Jernigan, T. L., Riley, E. P., and Toga, A. W. (2002). Regional Brain Shape Abnormalities Persist into Adolescence after Heavy Prenatal Alcohol Exposure. *Cerebral Cortex* **12**(8), 856–865.

Sowell, E. R., Thompson, P. M., Peterson, B. S., Welcome, S. E., Henkenius, A. L., and Toga, A. W. (2003). Mapping cortical change across the human life span. *Nat. Neurosci.* **6**(3), 309–315.

Sowell, E. R., Thompson, P. M., Tessner, K. D., and Toga, A. W. (2001). Accelerated Brain Growth and Cortical Gray Matter Thinning are Inversely Related during Post-Adolescent Frontal Lobe Maturation,. *J. Neurosci.* **21**(22), 8819–8829.

Sowell, E. R., Thompson, P. M., and Toga, A. W. (2004). Mapping changes in the human cortex throughout the span of life. *Neuroscientist* **10**(4), 372–392.

Sowell, E. R., Thompson, P. M., Welcome, S. E., Henkenius, A. L., Toga, A. W., and Peterson, B. S. (2003). Cortical abnormalities in children and adolescents with attention deficit hyperactivity disorder. *The Lancet* **362**(9397), 1699–1707.

Sowell, E. R., Thompson, P. M., Yoshii, J., Kan, E., Toga, A. W., and Peterson, B. S. (2004). Gray matter thickness abnormalities mapped in children with Tourette Syndrome. "34th International meeting of the Society for Neuroscience." San Diego, CA, Oct. 23–27, 2004.

Spitzer, V., Ackerman, M. J., Scherzinger, A. L., and Whitlock, D. (1996). The Visible Human Male: A Technical Report. *J. Amer. Med. Informatics Assoc.* **3**(2), 118–130.

Stewart, J. E., Broaddus, W. C., and Johnson, J. H. (1996). Rebuilding the Visible Man. *Vis. Biomed. Comp.* **4,** 81–86.

Studholme, C., Cardenas, V., Blumenfeld, R., Schuff, N., Rosen, H. J., Miller, B., and Weiner, M. (2004). Deformation tensor morphometry of semantic dementia with quantitative validation. *NeuroImage* **21**(4), 1387–1398.

Subsol, G., Roberts, N., Doran, M., Thirion, J. P., and Whitehouse, G. H. (1997). Automatic Analysis of Cerebral Atrophy. *Magn. Reson. Imaging* **15**(8), 917–927.

Szeszko, P. R., Bates, J., Robinson, D., Kane, J., and Bilder, R. M. (2004). Investigation of unirhinal olfactory identification in antipsychotic-free patients experiencing a first-episode schizophrenia. *Schizophr Res* **67**(2-3), 219–225.

Talairach, J., and Szikla, G. (1967). "Atlas d'Anatomie Stereotaxique du Telencephale: Etudes Anatomo-Radiologiques." Masson and Cie, Paris.

Talairach, J., and Tournoux, P. (1988). "Co-planar Stereotaxic Atlas of the Human Brain." Thieme, New York.

Thirion, J.-P. (1995). Fast Non-Rigid Matching of Medical Images. "INRIA Internal Report." Projet Epidaure, INRIA, France.

Thirion, J.-P., and Calmon, G. (1997). Deformation Analysis to Detect and Quantify Active Lesions in 3D Medical Image Sequences. *INRIA Technical Report*, 3101.

Thompson, P., Cannon, T. D., Narr, K. L., van Erp, T., Poutanen, V. P., Hutteunen, M., Lonnqvist, J., Standertskjold-Nordenstam, C. G., Kaprio, J., Khaledy, M., Dail, R., Zoumalen, C., and Toga, A. W. (2001). Genetic influences on brain structure. *Nature Neuroscience* **4**(12), 1253–1258.

Thompson, P., Hayashi, K. M., de Zubicaray, G., Janke, A. L., Rose, S. E., Semple, J., Hong, M. S., Herman, D., Gravano, D., Doddrell, D. M., and Toga, A. W. (2004). Mapping Hippocampal and Ventricular Change in Alzheimer's Disease. *NeuroImage* **22**(4), 1754–1766.

Thompson, P., Hayashi, K. M., Simon, S. L., Geaga, J. A., Hong, M. S., Sui, Y., Lee, J. Y., Toga, A. W., Ling, W., and London, E. D. (2004). Structural abnormalities in the brains of human subjects who use methamphetamine. *J. Neurosci.* **24**(26), 6028–6036.

Thompson, P. M., Cannon, T. D., and Toga, A. W. (2002). Mapping Genetic Influences on Human Brain Structure. *Ann. Med.* **34**(7–8), 523–536.

Thompson, P. M., Dutton, R. A., Hayashi, K. M., Lu, A., Lee, S. E., Lee, J. Y., Toga, A. W., Lopez, O. L., Aizenstein, H. J., and Becker, J. T. (2005). 3D Cortical Thickness Reductions Mapped in HIV/AIDS Correlate with Cognitive Impairment and Immune System Deterioration. "11th Annual Meeting of the Organization for Human Brain Mapping (OHBM)." Toronto, Canada, June 12–16, 2005.

Thompson, P. M., Giedd, J. N., Woods, R. P., MacDonald, D., Evans, A. C., and Toga, A. W. (2000). Growth Patterns in the Developing Brain Detected By Using Continuum-Mechanical Tensor Maps. *Nature* **404**(6774), 190–193.

Thompson, P. M., Hayashi, K. M., de Zubicaray, G., Janke, A. L., Rose, S. E., Semple, J., Herman, D., Hong, M. S., Dittmer, S. S., Doddrell, D. M., and Toga, A. W. (2003). Dynamics of gray matter loss in Alzheimer's disease. *J. Neurosci.* **23,** 994–1005.

Thompson, P. M., Lee, A. D., Dutton, R. A., Geaga, J. A., Hayashi, K. M., Eckert, M. A., Bellugi, U., Galaburda, A. M., Korenberg, J. R., Mills, O. J., Toga, A. W., and Reiss, A. L. (2005). Abnormal cortical complexity and thickness profiles mapped in Williams syndrome. *J. Neurosci.* **25**(16), 4146–4158.

Thompson, P. M., Mac Donald, D., Mega, M. S., Holmes, C. J., Evans, A. C., and Toga, A. W. (1997). Detection and Mapping of Abnormal Brain Structure with a Probabilistic Atlas of Cortical Surfaces. *J. Comp. Asst. Tomog.* **21**(4), 567–581.

Thompson, P. M., Mega, M. S., and Toga, A. W. (2000). Disease-Specific Brain Atlases. *In* "Brain Mapping: The Disorders" (A. W. Toga and J. C. Mazziotta, Eds.). Academic Press (June 2000).

Thompson, P.M., Mega, M.S., Vidal, C., Rapoport, J.L. and Toga, A.W. (2001). Detecting Disease-Specific Patterns of Brain Structure using Cortical Pattern Matching and a Population-Based Probabilistic Brain Atlas, IEEE Conference on Information Processing in Medical Imaging (IPMI), UC Davis, 2001. *In* "Lecture Notes in Computer Science (LNCS)" (M. Insana and R. Leahy, Eds.), Vol. 2082, pp. 488–501. Springer-Verlag.

Thompson, P. M., Mega, M. S., Woods, R. P., Blanton, R. E., Moussai, J., Zoumalan, C. I., Aron, J., Cummings, J. L., and Toga, A. W. (2001). Early Cortical Change in Alzheimer's Disease Detected with a Disease-Specific Population-Based Brain Atlas. *Cerebral Cortex* **11**(1), 1–16.

Thompson, P. M., Moussai, J., Khan, A. A., Zohoori, S., Goldkorn, A., Mega, M. S., Small, G. W., Cummings, J. L., and Toga, A. W. (1998). Cortical Variability and Asymmetry in Normal Aging and Alzheimer's Disease. *Cerebral Cortex* **8**(6), 492–509.

Thompson, P. M., Narr, K. L., Blanton, R. E., and Toga, A. W. (2000). Mapping Structural Alterations of the Corpus Callosum during Brain Development and Degeneration. *In* "The Corpus Callosum" (M Iacoboni and E Zaidel, Eds.). Kluwer Academic Press.

Thompson, P. M., Rapoport, J. L., Cannon, T. D., and Toga, A. W. (2003). Imaging the Brain as Schizophrenia Develops: Dynamic and Genetic Brain Maps. *Primary Psychiatry* **9**(11), 40–47.

Thompson, P. M., Schwartz, C., Lin, R. T., Khan, A. A., and Toga, A. W. (1996). 3D Statistical Analysis of Sulcal Variability in the Human Brain. *J. Neurosci.* **16**(13), 4261–4274.

Thompson, P. M., Schwartz, C., and Toga, A. W. (1996). High-Resolution Random Mesh Algorithms for Creating a Probabilistic 3D Surface Atlas of the Human Brain. *NeuroImage* **3,** 19–34.

Thompson, P. M., and Toga, A. W. (1996). A Surface-Based Technique for Warping 3-Dimensional Images of the Brain. *IEEE Trans. Med. Imag.* **15**(4), 1–16.

Thompson, P. M., and Toga, A. W. (1997). Detection, Visualization and Animation of Abnormal Anatomic Structure with a Deformable Probabilistic Brain Atlas based on Random Vector Field Transformations. *Medical Image Analysis* **1**(4), 271–294.

Thompson, P. M., and Toga, A. W. (1998). Anatomically-Driven Strategies for High-Dimensional Brain Image Warping and Pathology Detection. *In* "Brain Warping" (A. W. Toga, Ed.), pp. 311–336. Academic Press.

Thompson, P. M., and Toga, A. W. (2002). A framework for computational anatomy. *Comp. Vis. Scie.* **5,** 1–12.

Thompson, P. M., Vidal, C., Giedd, J. N., Gochman, P., Blumenthal, J., Nicolson, R., Toga, A. W., and Rapoport, J. L. (2001). Mapping Adolescent Brain Change Reveals Dynamic Wave of Accelerated Gray Matter Loss in Very Early-Onset Schizophrenia. *Proc. Nat. Acad Sci. USA* **98** (20), 11650–11655. September 25, 2001.

Thompson, P. M., Woods, R. P., Mega, M. S., and Toga, A. W. (2000). Mathematical/ Computational Challenges in Creating Population-Based Brain Atlases. *Human Brain Mapping* **9**(2), 81–92.

Tiede, U., Bomans, M., Höhne, K. H., Pommert, A., Riemer, M., Schiemann, T., Schubert, R., and Lierse, W. (1993). A Computerized 3D Atlas of the Human Skull and Brain. *Am. J. Neuroradiol.* **14,** 551–559.

Toga, A. W. (1998). "Brain Warping." Academic Press.

Toga, A. W., Ambach, K., Quinn, B., Hutchin, M., and Burton, J. S. (1994). Postmortem Anatomy from Cryosectioned Whole Human Brain. *J. Neurosci. Methods* **54**(2), 239–252.

Toga, A. W., and Mazziotta, J. C. (1996). Introduction to Cartography of the Brain. *In* "Brain Mapping: The Methods" (A. W. Toga and J. C. Mazziotta, Eds.), pp. 3–25. Academic Press, San Diego.

Toga, A. W., and Thompson, P. M. (1997). Measuring, Mapping, and Modeling Brain Structure and Function. *SPIE Medical Imaging Symposium*, Feb. 1997, Newport Beach, CA, USA SPIE Lecture Notes Volume 3033.

Toga, A. W., and Thompson, P. M. (1998). Multimodal Brain Atlases, Chapter. *In* "Medical Image Databases" (S. T. C. Wong, Ed.), pp. 53–88. Kluwer Academic Press.

Toga, A. W., and Thompson, P. M. (2003). Mapping brain asymmetry. *Nature Review Neuroscience* **4,** 37–38.

Toga, A. W., and Thompson, P. M. (2005). Genetics of Brain Structure and Intelligence. *Annual Review of Neuroscience* **28,** 1–23.

Toga, A. W., Thompson, P. M., and Payne, B. A. (1996). Modeling Morphometric Changes of the Brain during Development. *In* "Developmental Neuroimaging: Mapping the Development of Brain and Behavior" (R. W. Thatcher, G. Reid Lyon, J. Rumsey, and N. Krasnegor, Eds.), pp. 15–27. Academic Press.

Toga, A. W., Thompson, P. M., Mega, M. S., Narr, K. L., and Blanton, R. E. (2001). Probabilistic approaches for atlasing normal and disease-specific brain variability. *Anatomy and Embryology* **204,** 267–282.

Van Buren, J. M., and Borke, R. C. (1972). Variations and Connections of the Human Thalamus Vols. 1 and 2. Springer, New York.

Van Buren, J. M., and Maccubin, D. (1962). An Outline Atlas of Human Basal Ganglia and Estimation of Anatomic Variants. *J. Neurosurg.* **19,** 811–839.

Van Essen, D. C. (2004). Surface-based approaches to spatial localization and registration in primate cerebral cortex. *NeuroImage* **23**(Suppl 1), S97–S107.

Van Essen, D. C., Drury, H. A., Joshi, S. C., and Miller, M. I. (1997). Comparisons between Human and Macaque using Shape-Based Deformation Algorithms Applied to Cortical Flat Maps. "3rd Int. Conference on Functional Mapping of the Human Brain." Copenhagen, May 19–23, 1997, NeuroImage **5**(4), S41.

Van Essen, D. C., and Maunsell, J. H. R. (1983). Hierarchical Organization an Functional Streams in the Visual Cortex. *Trends Neurol. Sci.* **6,** 370–375.

Van Leemput, K., Maes, F., Vandermeulen, D., and Suetens, P. (1999). Automated model-based tissue classification of MR images of the brain. *IEEE Trans. Med. Imaging* **10,** 897–908.

Verbeke, G., and Molenberghs, G. (2000). Linear Mixed Models for Longitudinal Data Springer-Verlag, New York.

Vogeley, K., Schneider-Axmann, T., Pfeiffer, U., Tepest, R., Bayer, T. A., Bogerts, B., Honer, W. G., and Falkai, P. (2000). Disturbed gyrification of the prefrontal region in male schizophrenic patients: A morphometric postmortem study. *Am. J. Psychiatry* **157**(1), 34–39.

Wang, Y., Gu, X., Hayashi, K. M., Chan, T. F., Thompson, P. M., and Yau, S. T. (2005). Brain Surface Parametrization with Holomorphic Differential Forms. "11th Annual Meeting of the Organization for Human Brain Mapping (OHBM)." Toronto, Canada, June 12–16, 2005.

Wang, Y. L., Gu, X., Chan, T., Thompson, P. M., and Yau, S. T. (2004). Genus zero surface conformal mapping and its application to brain surface mapping. *IEEE Transactions on Medical Imaging* **23**(8), 949–958.

Warfield, S., Robatino, A., Dengler, J., Jolesz, F., and Kikinis, R. (1998). Nonlinear Registration and Template Driven Segmentation. *In* "Brain Warping" (A. W. Toga, Ed.), pp. 67–84. Academic Press Ch.4, 67–84.

Wilke, M., Holland, S. K., Myseros, J. S., Schmithorst, V. J., and Ball, W. S., Jr. (2003). Functional magnetic resonance imaging in pediatrics. *Neuropediatrics* **34**(5), 225–233.

Wong, G., Haney, S., Thompson, P. M., Cloughesy, T. F., Alger, J. R., Frew, A., and Toga, A. W. (2000). Prognostic Value of Growth Rates and Spectroscopic Data in Patients with Malignant Gliomas. *Proc. Society for Neuroscience*, 189.5.

Woods, R. P. (1996). Modeling for Intergroup Comparisons of Imaging Data. *NeuroImage* **4**(3), 84–94.

Worsley, K. J., Andermann, M., Koulis, T., MacDonald, D., and Evans, A. C. (1999). Detecting changes in non-isotropic images. *Human Brain Mapping* **8,** 98–101.

Wright, I. C., McGuire, P. K., Poline, J. B., Travere, J. M., Murray, R. M., Frith, C. D., Frackowiak, R. S. J., and Friston, K. J. (1995). A voxel-based method for the statistical analysis of gray and white matter density applied to schizophrenia. *NeuroImage* **2,** 244–252.

Xu, M., Thompson, P.M., and Toga, A.W. (2005). Adaptive reproducing kernel particle method for extraction of the cortical surface (submitted).

Zeineh, M. M., Engel, S. A., Thompson, P. M., and Bookheimer, S. Y. (2003). Dynamic Changes Within the Human Hippocampus During Memory Consolidation. *Science* Jan. 2003.

Zhou, Y., Thompson, P. M., and Toga, A. W. (1999). Automatic Extraction and Parametric Representations of Cortical Sulci. *Computer Graphics and Applications* **19**(3), 49–55.

Zilles, K., Armstrong, E., Schleicher, A., and Kretschmann, H. J. (1988). The human pattern of gyrification in the cerebral cortex. *Anat Embryol* **179**(2), 173–179.

Zilles, K., Palomero-Gallagher, N., Grefkes, C., Scheperjans, F., Boy, C., Amunts, K., and Schleicher, A. (2002). Architectonics of the human cerebral cortex and transmitter receptor fingerprints: Reconciling functional neuroanatomy and neurochemistry. *Eur. Neuropsychopharmacol* **12**(6), 587–599.

Further Reading

Ballmaier, M., Kumar, A., Thompson, P. M., Narr, K. L., Lavretsky, H., Estanol, L., DeLuca, H., and Toga, A. W. (2004). Localizing gray matter deficits in late onset depression using computational cortical pattern matching methods. *Amer. J. Psyc.* **161,** 11.

Gee, J. C., and Bajscy, R. K. (1998). Elastic Matching: Continuum-Mechanical and Probabilistic Analysis. *In* "Brain Warping" (A. W. Toga, Ed.). Academic Press, San Diego.

Haney, S., Thompson, P. M., Cloughesy, T. F., Alger, J. R., Frew, A., and Toga, A. W. (2000). Prognostic Value of Growth Rates and Spectroscopic Data in Patients with Malignant Gliomas. *Proc. Society for Neuroscience* 2000.

Haney, S., Thompson, P. M., Cloughesy, T. F., Alger, J. R., and Toga, A. W. (2000). Tracking Tumor Growth Rates in Patients with Malignant Gliomas: A Test of Two Algorithms. *Am. J. Neuroradiol.* **22,** 73–82.

Narr, K. L., Bilder, R. M., Toga, A. W., Woods, R. P., Rex, D. E., Szeszko, P. R., Robinson, D., Sevy, S., Gunduz-Bruce, H., Wang, Y. P., De Luca, H., and Thompson, P. M. (2005). Mapping cortical thickness and gray matter concentration in first episode schizophrenia. *Cerebral Cortex* (in press).

Osher, S. J., and Sethian, J. A. (1998). Fronts propagation with curvature dependent speed: Algorithms based on Hamilton-Jacobi formulations. *J. Comput. Phy.* **79,** 12–49.

Régis, J. (1994). Anatomie sulcale profonde et cartographie fonctionnelle du cortex cérébral. Ph. D. Thesis. University of Marseille [in French].

Shattuck, D. W., and Leahy, R. M. (2001). Automated graph-based analysis and correction of cortical volume topology. *IEEE Trans. Med. Imaging* **20**(11), 1167–1177.

Thompson, P. M., and Toga, A. W. (2000). Elastic Image Registration and Pathology Detection. *In* "Handbook of Medical Image Processing" (I. Bankman, R. Rangayyan, A. C. Evans, R. P. Woods, E. Fishman, and H. K. Huang, Eds.). Academic Press.

Vaillant, M., and Davatzikos, C. (1999). Hierarchical Matching of Cortical Features for Deformable Brain Image Registration. "Proceedings Information Processing in Medical Imaging." Budapest, June 1999.

NEUROIMAGING DATABASES AS A RESOURCE FOR SCIENTIFIC DISCOVERY

John Darrell Van Horn, John Wolfe, Autumn Agnoli, Jeffrey Woodward, Michael Schmitt, James Dobson, Sarene Schumacher, and Bennet Vance

The fMRI Data Center, Dartmouth College
Hanover, New Hampshire 03755

The field of neuroscience has an increasing need for access to primary research data in order to more thoroughly explore fundamental neural function beyond those examined in the original published article. For instance, functional magnetic resonance imaging (fMRI) studies of the human brain during the performance of cognitive tasks involve the collection of several gigabytes of image volume time course data as well as detailed meta-data concerning subject, experimental, and scanner protocols. Much of this data is unseen by anyone other than the original study authors but could be used by others to gain new insights into basic cognitive processes. We describe how several efforts have sought to archive the primary data from brain imaging studies and make them available to researchers in the community. We detail several aspects of neuroscientific data sharing that can help promote new inquiry. Essential in this process is the design of hierarchical frameworks for encapsulating fMRI study data for the purposes of extensible study organization that helps to encourage data sharing between collaborators or centralized data archives. In this chapter, we feature our own effort, the fMRI Data Center, as an example of large-scale archiving of fMRI study data from the peer-reviewed literature and how this is being used to explore data beyond the scope of the original article. Through such efforts, brain imaging has begun following the lead of the biological sciences by leveraging its accumulated data into new knowledge about fundamental brain processes.

INTERNATIONAL REVIEW OF NEUROBIOLOGY, VOL. 66
DOI: 10.1016/S0074-7742(05)66002-3

I. Introduction

The ambitious attempts being made across the biological sciences to promote the sharing of data and to facilitate meta-analysis (Becker, 2001; Mavroudis and Jacobs, 2000; Mirnics, 2001; Nowinski *et al.*, 2002; Reidpath and Allotey, 2001; Richard and Williams, 2002) have now transformed the process of science in the digital age. Large-scale scientific databases and collaborating networks of researchers are being developed to enhance scientific interaction and promote the sharing of primary research information (Collins and Mansoura, 2001; Shepherd *et al.*, 1998). For example, scientific databases containing experimental data and results, such as the Protein Data Bank (Berman *et al.*, 2000), permit re-examination or comparison of data on 3D protein structures in order to test new hypotheses and enable data mining from across research centers to reveal trends that may give rise to new avenues of research. Much excitement has occurred in the neurosciences about large-scale databasing (Koslow, 2000) and the promise that they hold for understanding cellular properties (Martone *et al.*, 2002), neuronal models (Marenco *et al.*, 1999), and underlying biochemical pathways (Karp *et al.*, 2002).

In the 1990s, use of fMRI during performance of neurocognitive tasks rapidly emerged, surpassing the use of positron emission tomography (PET), as the fundamental tool for mapping brain function (D'Esposito, 2000; Detre and Floyd, 2001; Savoy, 2001). The data collected in these studies are exceedingly rich and hold potential for understanding the complex neurobiological mechanisms underlying brain systems such as memory (Cabeza *et al.*, 2002), language (Mechelli *et al.*, 2003a), and motor function (Kawato *et al.*, 2003) among many cognitive domains. Most contemporary fMRI studies routinely include multiple functional scan runs (i.e., on the order of five to ten minutes per run), the collection of high-resolution anatomical images of the brain, involve multiple levels of experimental manipulation, and increasingly involve several experimental groups. Stimulus input and behavioral output files also contribute to the body of study meta-data, 4-dimension image volumes, and images of brain structure. However, despite their being summarized and interpreted in the peer-reviewed literature, the raw image data and accompanying meta-data from these studies are often not accessible to other researchers in the neuroscientific community. This makes it difficult for others to rigorously inspect results, verify published claims, or to conduct novel analyses of these rich datasets. In the field of functional neuroimaging, despite being available for over a decade, databases of systems-level functional brain imaging studies of cognition have begun to receive increasing attention (Van Horn and Gazzaniga, 2002).

The field of neuroimaging has for some time recognized this need for archiving functional neuroimaging results (Fox and Lancaster, 1994; Fox *et al.*, 1994b). However, databasing and data sharing remain relatively unfamiliar ideas to many in the field of functional neuroimaging. Coupled with the fact that the size of fMRI studies is growing rapidly, few tools exist that assist investigators in managing the data they collect and that facilitate efficient data sharing. The needs of the community are particularly unique and extant software models are not broadly applicable for dealing with many of the issues a large archive like this requires as well as helping users manage data at their own sites. Particular problems include: Users of databases that store experimental results are often faced with entering data retrospectively; certain models for sophisticated high-throughput data analysis (Roland *et al.*, 2001) put certain constraints on the type and form of data that can be stored; the management of study data is left up to the investigator until completion (e.g., publication) of the study; and finally, the tools available from existing databases are largely bibliographic and not entirely useful until all information about a study has been collected. In other words, software tools that help an investigator gather, manage, and share fMRI data that preserves that data's utility to the investigator beyond its being shared do not widely exist.

Simply managing these large amounts of data, however, is an increasing challenge for many in the field. Future investigations of *in vivo* brain function using fMRI can be expected to continue gaining in sophistication, as the questions being asked about the brain, and the methods themselves, become more elaborate. With possibly hundreds or thousands of data files, each recorded under conditions that vary over time, the amount of data may soon overwhelm the abilities of investigators to keep track of this information from even a single fMRI investigation. Being able to manage and organize these large and varied collections of data is a prelude to subsequent data sharing and databasing.

In this chapter, we review the form, size, and means of processing of functional neuroimaging data but highlight the potential of these data to be used beyond the initial interpretation by the original investigators; how these data may be used to develop and test new ideas about human brain function as well as for developing useful computational algorithms for extracting new results; we discuss the emerging role of high-performance and Grid computing; the use of clever visualization techniques, and the availability of online databases of neuroscientific data. Finally, we discuss some of the challenges that must be overcome for these and other resources to fully leverage the wealth of neuroscientific information into new discovery.

II. Examining Cognitive Function with fMRI

The signal of interest using positron emission tomography (PET) is based on the fact that changes in the cellular activity of the brain of normal, awake humans and laboratory animals are accompanied almost invariably by changes in local blood flow (Raichle, 1975, 2001b; Sokoloff, 1981). Early PET studies of the brain's response to cognitive tasks provided a level of precision in the measurement of blood flow that opened up the modern era of functional human brain mapping (Raichle, 2003). Functional MRI, on the other hand, distinguishes itself from PET by capitalizing on endogenous magnetic properties of deoxygenated to oxygenated hemoglobin in order to track regional cerebral blood flow (Hoppel *et al.*, 1993; Rosen *et al.*, 1993). Using fMRI to visualize brain function *in vivo*, neuroscientists have demonstrated that the mental operations carried out by the human brain can be empirically and repeatedly measured (Bandettini and Wong, 1997; D'Esposito, 2000) and, since the early 1990s, fMRI has taken the place of PET as the most widely used method for brain mapping and studying the neural basis of human cognition. Though now enjoying widespread practice throughout the world, an incomplete understanding of the physiological basis of the fMRI signal has remained to confidently interpret the data with respect to neuronal activity. The biological origins for these signals is an area of much interest for the application of tools for cognitive neuroscience research and modeling (Raichle, 2001a; Woo and Hathout, 2001). Understanding the origins of the BOLD signal is useful for informing models of the hemodynamic response function (Buxton and Frank, 1997; Buxton *et al.*, 1998) or to guide characterization of the neurophysiological processes that occur in advance of BOLD signal change as a result of many higher-order cognitive models (Friston, 2002; Friston and Price, 2001; Price and Friston, 2002).

New insights into higher cognitive functions, such as episodic and working memory (Cabeza *et al.*, 2002; Carpenter *et al.*, 2000), linguistic processes (Binder *et al.*, 1997; Buchel *et al.*, 1998; Crosson *et al.*, 1999), and object visual processing (Beauchamp *et al.*, 2002) have been described. Face perception is one particular cognitive operation to be extensively examined using fMRI (Haxby *et al.*, 2000) and appears to be governed principally in the ventral portion of the temporal lobe—the "fusiform face area" (Kanwisher *et al.*, 1997). Brain areas bordering this region may be sensitive to the spatial properties of pictures of other objects, such as chairs and houses (Ishai *et al.*, 2000) with spatially distributed but overlapping portions (Haxby *et al.*, 2001). Additional research has indicated that this region may, in fact, be specialized for visual recognition expertise which includes processing for faces (Gauthier and Nelson, 2001; Gauthier *et al.*, 1999). Work has also investigated the social context of face perception, in particular with respect to the perception of threat (Adolphs, 2003; Haxby *et al.*, 2002; Richeson *et al.*, 2003), the familiarity of faces (Leveroni *et al.*, 2000), and the processing of faces in diseases such as autism (Adolphs *et al.*, 2001). Visuospatial attention has also been explored using fMRI (Binkofski *et al.*, 2002; Culham *et al.*, 2001; Hamalainen *et al.*, 2002; Kanwisher and Wojciulik, 2000). fMRI studies have pointed toward a network of cortical visuospatial and oculomotor control areas, specifically the lateral occipital cortex, precentral sulcus, and intraparietal sulcus, as being active in covert shifts of spatial attention (Beauchamp *et al.*, 2001). In parietal and frontal cortical areas, BOLD activation increased with attentional load, suggesting that these areas are directly involved in attentional processes, though this was not evident in the fusiform gyrus (Culham *et al.*, 2001), indicating possibly separate but complimentary systems underlying attention to stimuli such as human faces.

A number of factors have been implicated in the origins of the BOLD response including energetics, oxygen consumption, as well as parameters such as blood volume and flow (Buxton *et al.*, 1998). The question of whether the BOLD response is the result of neuronal output or if it is due to the internal communication among localized populations of cells has also been recently addressed. Logothetis and coworkers (2001) conducted the first simultaneous intra-cortical recordings of neural signals and hemodynamic responses. Varying the temporal characteristics of the stimulus, they observed a moderate to strong association between the neural activity measured with microelectrodes and the pooled BOLD signal from around a small area near the microelectrode tips. However, the BOLD signal showed significantly higher variability than the neural activity, indicating that human fMRI coupled with traditional statistical methods underestimates the reliability of the neuronal activity. To further characterize the relative contribution of several types of neuronal signals to the hemodynamic response, they compared local field potentials (LFPs), single- and multi-unit activity (MUA) with high spatiotemporal fMRI responses recorded simultaneously in primate visual cortex. Selecting recording sites having transient responses, only the LFP signal showed significant correlation with the hemodynamic response and were superior to MUA at predicting the fMRI response. Thus, BOLD signal is a putative measure of the input and processing of neuronal information within brain foci, not the output signal transmitted to other brain areas.

Epoch or "block" experimental designs have been the work horse of fMRI experimentation and are those in which stimuli are presented for some period of seconds (several TRs or brain volume sampling intervals) and alternated randomly or pseudo-randomly over the course of the data acquisition period. They are the easiest to conduct and tend to provide robust activation in most tasks but may limit the number of stimulus types that can be presented. Conversely, event-related experimental designs are characterized by having a baseline time course that is punctuated with stimulus events. Event-related methods, conversely, have permitted a broad array of task designs to be explored with brain imaging techniques (Buckner, 1998; Buckner *et al.*, 1996; Rosen *et al.*, 1998). Individual trial events can be presented rapidly, in randomly or intermixed order, and the hemodynamic responses associated with each trial event type reliably estimated (Dale and Buckner, 1997). The basis of event-related studies is that the hemodynamic response tracks neuronal activity on the temporal scale of seconds and, in many situations, summates over trials in a manner well predicted by a linear model that is sufficient even for very briefly spaced stimuli (e.g., $\sim$2 seconds). With this increased interest in event-related paradigms in fMRI, there has been considerable effort in identifying the optimal stimulus timing, especially when the inter-stimulus interval is varied during the imaging acquisition run (Birn *et al.*, 2002; Dreher *et al.*, 2002).

Experimental designs for event-related functional magnetic resonance imaging can be characterized by both their detection power, a measure of the ability to detect activation, and their estimation efficiency, a measure of the ability to estimate the shape of the hemodynamic response. Computer simulation studies have indicated that estimation of the hemodynamic response function is optimized when stimuli are frequently alternated between task and control states, having shorter interstimulus intervals and stimulus durations, while the overall detection ability of activated areas is optimized when using blocked designs (Birn *et al.*, 2002; Mohamed *et al.*, 2000). This suggests that event-related designs may provide more accurate estimates of the HRF than epoch-related designs, with the maximal response to events occurring sooner and returning to baseline later than in a stimulus epoch (Mechelli *et al.*, 2003b). The choice of data processing operations, however, can affect statistical inference in all designs and means for optimizing data processing pipelines is an area of active research (LaConte *et al.*, 2003; Lukic *et al.*, 2002; Strother *et al.*, 2002).

Functional neuroimaging using MRI promises to continue growing as the principal method for examining *in vivo* brain function. Though individual studies using fMRI promise to reveal much about such basic brain processes, there also exists great potential for contrasting, comparing, and combining these studies to explore fundamental properties of cognitive function as well as the properties of the BOLD signal itself. The large amount of information collected in an individual study, however, and how this could be mined by others to produce novel research is an under-appreciated aspect of this work that is worth addressing further. There is often more information contained in a neuroimaging study that can be adequately described in a single neuroimaging article. The expertise required to extract this information, however, is often not necessarily possessed by the original study authors. Finally, new imaging facilities are very costly and to install MRI scanners in psychology departments across the country may not be cost effective in contrast to providing an open archive of such data where researchers can readily obtain the original fMRI time series and subject them to new analyses.

III. Large-Scale Archiving of fMRI Study Data

The advent and development of fMRI has resulted in a quantum leap in the ability to visualize the brain's capabilities. However, this has also vastly increased the amount of information that brain researchers must manipulate, manage, and store. fMRI study data sets are large, often exceeding several gigabytes (GB) in size. As advances are made in MRI scanner technology to permit the more rapid acquisition of data, functional imaging experiments will consist of more data per unit time over the same scan duration. As cognitive neuroscientists ask ever more sophisticated questions about fundamental brain processes, they will undoubtedly collect data on a greater number of subjects and more fMRI time courses per subject. Indeed, archives equivalent in size to that of several petabytes are not out of the question and will likely be the norm within the next decade. A number of individual fMRI data sets already rival the full size of many extant large genetic (Ackerman, 1999; Ackerman and Banvard, 2000) and protein (Chen and Xu, 2003; Legato *et al.*, 2003; Noguchi and Akiyama, 2003) science data archives (see Table II, below, for comparison). For instance, the complete study data from Buckner *et al.* (2000) (fMRIDC Accession#: 2-2000-1118W) represents a study in excess of 20GB. It can be expected that as technological advances are made in MRI scanner technology which improve the spatiotemporal resolution of the data obtained, the amount of brain image data collected in published articles will routinely rival the size of the human genomic database. A challenge therefore exists in devising efficient means for comparing and contrasting these data on a large-scale but within a reasonable time frame.

As fMRI use in cognitive, clinical, and social neuroscience grows and becomes more widespread, individual researchers must be prepared for the large disk storage requirements that are needed to contain the data and their analyses. A greater number of subjects, for instance, improves the inferential power of the statistical tests performed and helps researchers to be confident in the effects they observe (Van Horn *et al.*, 1998). However, increases in sample size readily require increased costs associated with each fMRI study, in terms of scanner time, subject reimbursement, among other expenses. Publicly accessible archives of these data (for example see Table I) can help spread the costs of this research over the community, whereby the researcher may perform re-analyses on existing data at a greatly reduced cost compared to collecting the data themselves.

With these issues in mind, the fMRI Data Center (fMRIDC; *http://www.fmridc.org*) was established as a public archive for fMRI study data and the associated experimental meta-data. The fMRIDC began receiving data from researchers in 2000 and began making datasets publicly available in 2001. At present, the archive contains over 100 complete data sets which researchers may request online and have shipped to them free of charge. Authors of fMRI studies have been asked to provide the details of their experiments across several levels: description of the subjects taking part in the experiment (e.g., their age, handedness, clinical diagnosis, etc.); the description of the MRI scanner (e.g., manufacturer, model, software revision, field strength, etc.) as well as the scanning session protocols used during the study (e.g., number of slices acquired, echo time (TE), relaxation time (TR), etc.); and, finally, the details of the experimental design (e.g., stimulus time course information, number of experimental

TABLE I
BRAIN DATABASE RESOURCES AVAILABLE ONLINE

Database name	Principle modality	Data sets provided	Public access?	Species	Country and funding source[a]	Web site URL
Allen Brain Atlas	Anatomical sections	Photomicrographs of gene expression	Limited	Mouse	US; Private	http://www.brainatlas.org
BIRN	MRI	Images (MRI and cell photomicrograph)	Limited; greater access to participating BIRN centers	Human, Mouse	US; NCRR	http://nbirn.net/
BRAID	MRI, fMRI	Image volume data	Limited	Human	US; N/A	http://www.rad.upenn.edu/sbla/braid/publications/all.shtml
Brain Gene Expression Map (BGEM)	Anatomical sections	Photomicrographs of gene expression	Open	Mouse	NIH/NINDS, ALSAC	http://www.stjudebgem.org/web/mainPage/mainPage.php
BrainMapDBJ	PET/fMRI	Results local maxima	Limited; greater access to participating ICBM centers	Human	US; NLM	http://www.brainmapDBJ.org
BrainWeb	MRI	Simulated MRI image volume data	Open; part of the LONI/ICBM consortium	Human	Canada; NIMH/HBP Non-US	http://www.bic.mni.mcgill.ca/brainweb/
BREDE	fMRI	Results local maxima, VRML, XML	Open	Human	Denmark; NIMH/HBP Non-US	http://hendrix.imm.dtu.dk/software/brede/

CoCoMac	Single and multi-unit recordings	Neural connectivity data	Open	Non-Human Primate	Germany; Non-US	http://www.mon-kunden.de/cocomac/
EarLab	Single/multi-unit recording	Cell recording time series	Open	Non-Human	US; NIMH/HBP	http://earlab.bu.edu/
fMRIDC	fMRI, MRI	Raw, processed, results, anatomical brain images and study meta-data	Open	Human	US; NSF, Keck, NIMH/HBP	http://www.fmridc.org
International Brain Volume Database (IBDV)	MRI	High resolution structural Image volumes	Open	Human	US; NIMH/HBP	http://www.cma.mgh.harvard.edu/ibvd/
LONI/ICBM	PET, MRI, fMRI, EEG, MEG	Image data	Limited; greater access to ICBM centers	Human	International; NIMH/HBP NCRR, Private funding	http://www.ioni.ucla.edu
Mouse Brain Library (MBL)	Anatomical sections	Photomicrographs	Open	Mouse	US; NIMH/HBP	http://www.mbl.org
Neurodatabase.org	Single/multi-unit recording	Cortical neuron electrical recordings	Open	Multiple	US; NIMH/HBP NINDS	http://neurodatabase.org
Neurogenerator	PET, fMRI	Imaging data submitted by users is organized into a database that is returned to the user	Limited	Human	Sweden; The European Commission	http://www.neurogenerator.org
SenseLab	Single/multi-unit recording	Cell recordings from multiple sources	Open	Non-Human Primate	US; NIMH/HBP	http://www.senselab.yale.edu
Surface Management System (SuMS)	MRI	Digitally-based cortical surface models	Open	Human, Non-Human Primate	US; NIMH/HBP	http://brainmap.wustl.edu/sumshome/

[a]Where evident from the database Web site

runs, etc.). Table I presents a summary of the items requested from authors which describe their experimental data. The principle intent of obtaining this degree of information about each study is that it should be complete such that another researcher could take the information and the accompanying brain image data and reconstruct the results reported in the literature by the original authors (see Van Horn *et al.*, 2001 for review).

IV. The Emergence of "Discovery Science"

The collection of biological data into large databases has led to a change in thinking about the potentially restrictive nature of strictly hypothesis-based research. Increasingly, researchers are beginning to move toward a science of discovery—examining vast and disparate collections of data and hunting for unseen patterns that might provide clues to underlying biological mechanisms. The mountains of data being collected in many fields provide input for pattern-seeking and other relevant algorithms that can provide additional insights into complex, multidimensional data (Jones and Swindells, 2002; Ma *et al.*, 2002; Schutte *et al.*, 2002). These patterns can suggest mechanisms, and the mechanisms can, in turn, suggest testable biological experiments to foster new hypothesis-driven research. Confirmed mechanisms add to the knowledge base of the biological sciences and provide the basis for further discoveries including those that will improve quality of life and provide the means for attacking disease.

Such mining of the integrated resources developed and disseminated by the NCBI, Genbank, and the Human Genome Project has led to several scientific advances. The discovery of the genes for hereditary nonpolyposis colorectal cancer (HNPCC) is one such example. HNPCC is thought to account for one-sixth of all colon cancer cases (van Stolk, 2002). Although most forms of cancer appear to be nongenetic, there are certain forms where a person has a hereditary risk attributable to a single altered gene (Calvert and Frucht, 2002). Using the tools developed through the Human Genome Project, notably Genbank, an international research team tracked the gene to a specific region of chromosome 2 (Lindblom *et al.*, 1993). Researchers then identified a second gene on chromosome 3 that was also associated with this form of cancer (Peltomaki, 1994). Together, mutations within these two genes are responsible for the majority of cases of HNPCC. Researchers have used this new knowledge to develop blood tests to screen select individuals for these gene mutations (Ramesar *et al.*, 2000; Thomas, 1994). Detecting the presence of the mutated genes for HNPCC within a family allows clinicians to target relatives most likely to benefit from treatment. By identifying an unaffected family member at risk for HNPCC, physicians may then more closely monitor them for signs of disease development. Family members determined to be noncarriers no longer have to suffer through extensive medical examinations. Most importantly, patients demonstrating early signs of cancer and determined to carry a gene mutation may undergo prompt medical treatment. Due to the role played by informatics, when diagnosed and treated early, HNPCC is nearly 100 percent curable (Boardman, 2002).

Several such large-scale, infrastructural, and discovery-focused database research efforts that are already seeing considerable scientific payoff are underway in the biological and astrophysical sciences. The successes of these molecular biological, biomedical, and astrophysics infrastructures are well known. They have provided the means for experts in computer science, mathematics, and statistics to make significant contributions to these fields from which most of their expertise would have been excluded without the infrastructure. These successes are not necessarily unique but building upon them and extending them to a wider set of scientific research arenas is an ever-present theme (Altman, 2003; Brookes, 2001; Persson, 2000).

Before this process can begin for any particular field of science, however, an infrastructure must be laid down that will support these new approaches. The path blazed by the molecular biologists is, once again, illustrative. Likening these efforts to civil engineering projects, Eric Lander, Director of the Whitehead-MIT Center for Genome Research, has noted that programs to develop computational infrastructure represent "very important roads." (Incyte Genomics interview (2001) (http://www.incyte.com/)).

V. Data Sharing in Neuroscience

The driving force behind many biological and physical science informatics, data mining, and research initiatives has revolved around the sharing of primary research data (Becker, 2001; Ilioudis and Pangalos, 2001; Reidpath and Allotey, 2001). The National Institutes of Health (NIH) in the United States have recognized the benefits that the sharing of primary research data has for advancing science and has recently implemented policy requiring data sharing for grants in excess of US$500K/yr in direct costs (*Final NIH Data Sharing Policy Notice*: http://grants.nih.gov/grants/guide/notice-files/NOT-OD-03-032.html).[1] Likewise, the National Science Foundation (NSF) in the US has encouraged data sharing for several years, in the social and economic sciences in particular, (*http://www.nsf.gov/sbe/ses/common/archive.htm*). The Medical Research Council (MRC;

[1]Further information on the NIH data sharing policy may be found on the NIH Data Sharing Web Page: *http://grants.nih.gov/grants/policy/data_sharing/*

http://www.mrc.ac.uk) in Great Britain also strongly encourages open scientific data sharing.

The NIH's position on coordinated scientific data sharing, particularly as it relates to neuroscience, has recently been underscored in an essay by several NIH institute directors: "Efforts driven by collaboration, coordination, and computation should yield the data, tools, and resources that neuroscientists will need in the coming decades." (Insel *et al.*, 2003). The sharing of primary research data is needed to provide a record of the scientific body of work, permit comparison of various approaches to studying brain function, and enable large-scale analyses across data sets. There are several models for the sharing of research data that form a spectrum of complexity and detail. These include models for data archive access; the simple model of an anonymous FTP site, where a data set is placed and openly available but with no guarantee that the data have not been subjected to quality control, been published in a peer-reviewed publication, or that the dataset will be maintained; peer-to-peer models, wherein individual investigators set up and personally maintain private data sharing relationships with colleagues and co-investigators of both published and unpublished data; the conforming site model, in which consortiums of several research centers agree to exchange data through conventions established and governed by one of the consortium member centers, but with no guarantee that nonconsortium members may have access to the data archive or will have access to consortium-derived software tools needed for interacting with the data; and the centralized repository model, in which complete datasets are contributed, curated, and maintained in a central site by dedicated personnel and made openly available to the entire research community.

Data sharing models also focus on the type and amount of data that should be shared. Recent commentary has suggested that the value of shared neuroimaging data is greatest only after processing has been applied and interpretation provided by study authors (Fox and Lancaster, 2002). However, the information content of the image voxel time course data remains the same or is reduced by every step of processing (Van Horn and Gazzaniga, 2004). Therefore, it is unclear as to the amount of added value when archiving only statistical local maxima tables obtained after the data are heavily processed.

The model for sharing peer-reviewed study data in which the potential benefits of a data set are likely to be greatest is when the data are curated in a centralized location by a dedicated staff, complete study data have been indexed, and are freely available to the entire community (Van Horn and Gazzaniga, 2004). Accompanied by detailed, ontologically-structured, study meta-data, and a comprehensive description of data processing methods, experiment image data may be examined by other researchers at various points in the processing chain (raw, processed, or results) depending upon the needs and interests of colleagues or independent researchers. Through centralized curation and open distribution, efforts to subject functional data to re-analysis or perform mega-analyses across data sets may be maximally successful and thereby promote unique scientific discovery and advance education. This model of data sharing helps promote the cycle of science by adding an extra component to the publication process that may enhance new research and education, foster new avenues for research, and contribute back into the collective body of knowledge.

VI. The Role of Computation in Neuroscience

The NIH Roadmap (*http://nihroadmap.nih.gov/*) stresses the importance of computational biology, bioinformatics, and the establishment of digital science libraries. A recent NSF Blue Ribbon Advisory Report emphasizes that computers, computer science, and technology are at the heart of the future of a range of research fields that have "profound broader implications for education, commerce, and social good." (*http://www.communitytechnology.org/nsf_ci_report/*, (Blue Ribbon Advisory Panel On Cyberinfrastructure, 2003)). Though perhaps overdue in fully recognizing the potential of computers and the internet, the field of neuroscience is now growing in its dependence on high-end computational infrastructure. Neuroscience, in particular cognitive neuroscience, has emerged over the past decade as a cross-cutting aggregate of these key areas with an emphasis on the human brain information mining, modeling, and visualization (Adolphs, 2003; Casey, 2002; Corchs and Deco, 2002; Toga, 2002a). To broaden participation in understanding brain function derived from technologies such as brain imaging, greater reliance upon computational infrastructure to facilitate research collaboration is required. The sharing of large data sets via the internet; being able to collect, archive, and index these data; and subject these data to high-throughput analysis is not an option, but a mandatory next step in the advancement of understanding of normal brain function. Via this route, large, culturally-, and gender-valid norms must be established, for example, forming a benchmark against which to provide diagnosis in brain illness and disease.

Several novel concepts have been borne out of this interest in large-scale, scientific collaborative infrastructure. Most notably is the concept of Grid computing (Butler, 2003), the basis of which has existed for several years in the form of distributed computing, and the emerging need for high-speed Internet connectivity. The middleware for the development of global interconnected computer systems has made great progress in the last few years (Avery, 2002). The Grid software enables users, tools, and computer hardware to interact and share resources over high speed connections in an Internet- and Internet standards-compliant fashion (Foster, 2003). Users of Grid-enabled systems will be able to write applications to these published interfaces and will expect to be able to run on large-scale heterogeneous systems (Fig. 1). The Web, and in particular

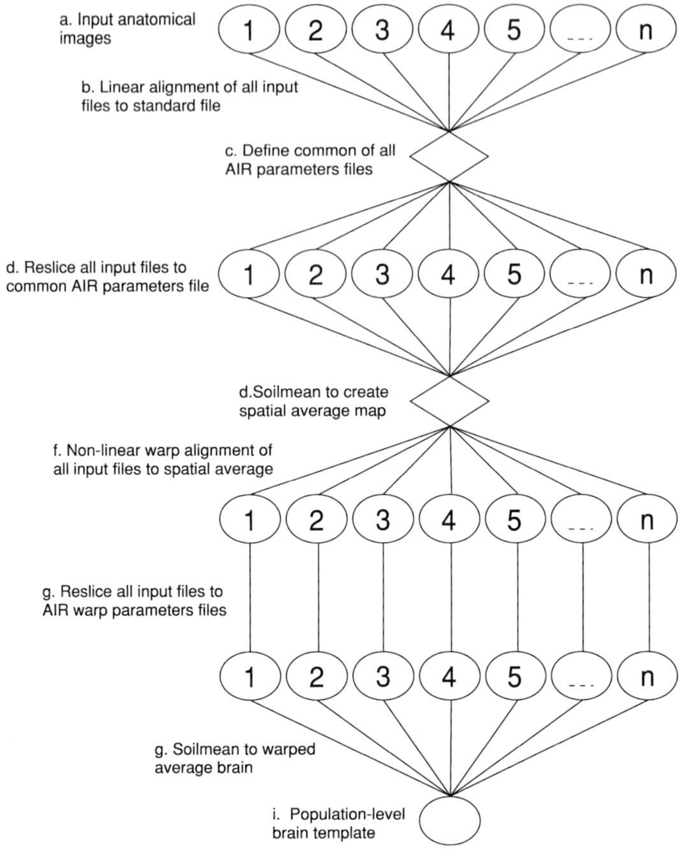

FIG. 1. Neuroimaging data processing pipelines are ideally suited to Grid-based distributed computing.

Web – services, can provide a model for this large-scale computational system. Various fields have been earlier adopters of Grid computing, notably the High Energy Physics community. Several worldwide and multi-institutional Grid projects are underway to enable simulations on large distributed datasets. The TeraGrid (*http://www.teragrid.org*) is a NSF-funded project to promote distributed scientific computing using Grid infrastructure to connect the nation's largest supercomputer centers. Using the Grid as a backbone, The Globus Toolkit (http://www.globus.org) has become the *de facto* standard for Global Grid communications. The Globus Alliance leads the development of the toolkit. The Global Grid Forum (GGF) is currently leading the standards efforts which Globus implements. Shared computational infrastructure, tools, and tool development, as well as collaborative research on archival data leading to new, testable hypotheses is becoming easier each day. These technological advances in large-scale computing on shared infrastructure have important implications for neuroscientific research dealing with massive amounts of data as is the case in functional neuroimaging.

VII. Brain Data Repositories as a Shared Resource for Neuroscience

When the fMRI Data Center effort was initiated, one of the central intents was to build an Internet-accessible platform through which researchers based at other institutions might access the growing collection of fMRI study data to evaluate methodologies for data processing (LaConte *et al.*, 2003; Lukic *et al.*, 2002); guide the design open-source software tools for data management (Van Horn *et al.*, 2002); construct means to summarize these large data sets to facilitate rapid search, visualization, and discovery; all with a view toward driving new hypothesis-based fMRI research. The fMRIDC team, in particular, has worked to construct a shared community access cluster Grid system for the analysis of functional datasets from the fMRIDC archive. This system has seen increased utilization as an analysis platform to mitigate the current CD delivery of datasets. Accounts on the system are available to any member of the fMRI community who wishes to perform a large analysis of data from the fMRIDC (see *http://www.fmridc.org/grid*).

The software infrastructure provided by Grid services are needed to perform meta-analyses on the multi-gigabyte fMRI datasets housed in the fMRIDC archive. The fMRIDC has implemented a number of components from the Globus Toolkit 3.0, including the Grid-FTP service. Current plans exist for the full toolkit to be installed to enable multi-institutional scheduling of resources. The Globus software will abstract the fMRIDC's own Grid scheduling system and the systems of our collaborators as well as those of other systems. With advanced computational resources dedicated to neuroimaging and that are deployed using Globus, new methods can be developed through testing on what is fast becoming the world's largest data warehouse of functional neuroimaging data.

VIII. fMRI Data Archiving, Mining, and Visualization

Like much of neuroscience, literature-driven, hypothesis-based approaches have been the underlying approach to most cognitive neuroscientific investigations over the past 100 years. However, streamlined, computationally efficient approaches to examining large amounts of data are emerging as advantageous where volumes of information from diverse sources impede a straightforward test of experimental hypotheses (Baumgartner *et al.*, 2000; Friman *et al.*, 2002; Jarmasz and Somorjai, 2002). Approaches used in the biological sciences include nonparametric clustering methods (Cordes *et al.*, 2002; Goutte *et al.*, 2001; Salli *et al.*, 2001), pattern searching (Cummings *et al.*, 2002; Jones and Swindells, 2002), as well as novel visualization techniques (Baumgartner and Somorjai, 2001; Teo *et al.*, 1997). In brain imaging, similar methods might be used on a database as a precursor to more thorough parametric hypothesis testing using more sophisticated modeling methods (Cox, 1996; Cox and Hyde, 1997; Friston *et al.*, 2002; Lohmann *et al.*, 2001) on subsets of the overall study space. Exploratory data analysis approaches (Tukey, 1977; Velleman, 1981) often assume little about the underlying data, which allows the data to more freely inform the investigator about itself using more elementary statistical approaches than might be true if the data were assessed using highly parameterized modeling procedures. These methods have been successfully applied to fMRI data with promising results (Baumgartner and Somorjai, 2001; Baumgartner *et al.*, 2000; Friman *et al.*, 2002; McKeown and Sejnowski, 1998). They can often be used to highlight relationships among the overall collection of data in ways not possible when viewing only a single data set. These relationships may not emerge until viewed in the light of a large number of other studies to which they may be compared. Thus, neuroscience, in particular studies using fMRI, has a great potential to become a discovery-based science, where exploratory analyses can lead to new ideas worth pursuing with hypotheses-generated experimentation (Van Horn and Gazzaniga, 2002). But to put such ideas into practice, however, requires an interdisciplinary approach, bringing together experts in cognitive, computer, and mathematical science to work jointly in solving the challenges inherent in large-data science.

Initiatives to archive neuroscience data form a unique collaboration between cognitive neuroscientists, mathematicians, and computer scientists to explore the challenges inherent in (1) data warehousing—*en mass* data storage; (2) data mining—how to apply efficient mathematical and computer algorithms to sift through large amounts of data to extract unique and interesting features from fMRI study data; and (3) data distribution—effective means to permit others to interact with the data archive. For example, the efficient retrieval of useful information from these large datasets poses many interesting problems for which computer scientists play an important role in providing answers. Methods for applying sophisticated search queries across multiple levels of neuroimaging study data (Table II) that could be investigated include: (1) searches for key text phrases across the published research article itself (i.e., the PDF version of the published study) and subsequent document clustering by assessing the usage of similar words at similar rates; (2) queries across the study "meta-data" composed of scanner protocol, experimental paradigm, subject demographic, and other information provided by the study authors; and (3) the 4D fMRI image time course data itself by, for instance, performing image timecourse-based clustering

TABLE II
BASIC fMRI STUDY INFORMATION COLLECTED FOR THE fMRIDC ARCHIVE

MR Scanner Protocol Information

- Scanner Protocol ID
- Scanner Head Coil Type
- Pulse Sequence Type
- Flip Angle (degrees)
- TE (in milliseconds)
- TR (in milliseconds)
- Number of time-points
- Number of acquisitions
- Number of dummy scans
- Number of slices
- Slice thickness (in millimeters)
- Slice skip (in millimeters)
- Interleaved or sequential slice acquisition
- Field of View (FOV)
- Receiver bandwidth (MHz)
- Original image acquisition matrix size
- Reconstructed image acquisition matrix size
- Full or partial K-space
- Image Acquisition Orientation
- Ramp sampling
- Echo train length
- Echo shift in asymmetric spin-echo
- Type of reference scan for reconstruction

Subject Information

- Subject ID
- Experimental group code
- Gender
- Age
- Health Status
- Assessments (e.g., handedness, etc.)
- Medication status
- Other (e.g., diagnostic, etc.)

Scan Session Information

- Scan Session ID
- Scanner Manufacturer
- Scanner Model
- Scanner software revision
- Magnet field strength
- Scanner Gradient Slew rate
- Date of scan session
- Duration of scan session
- Other

(*Continued*)

TABLE II (*Continued*)

Experimental Protocol

- Experimental Protocol ID
- Number of groups
- Number of subjects per group
- Number of functional runs
- Epoch-related conditions
- Event-related conditions
- Experimental methods
- Stimulus regressor files
- Other (e.g., additional condition descriptions, associated data files, etc.)

accompanied by *a posteriori* probabilistic classifier algorithm to measure classification reliability. The latter of these approaches has necessitated the application of summarizing signal processing and information theoretic methods for rapidly analyzing large fMRI data sets. The integration of these levels using leading-edge, computer-based IR algorithms permit "global" study clustering in order to "learn" what is needed to identify interesting patterns within and between these levels of data. The application of algorithms that permit fMRI data self-description, allowing the data to tell an investigator about itself rather than through the fitting of statistical models, as is common in functional neuroimaging, is a promising application of "machine learning" (Mitchell, 1997; Mitchell, 1999).

Another area of active interest in neuroimaging data representation lies in identifying unique approaches to the visualization of this massive amount of information. Traditional approaches to visualizing brain imaging study results have relied on overlay patterns of brain activation from functional scans on top of high-resolution structural images (Fig. 2). Popular methods of display include representing patterns of functional activity on flattened models of the cortical surface (Van Essen *et al.*, 2001a; Van Essen *et al.*, 2001b). Still other, novel methods for displaying and interacting with more abstract representations of large collections of information are needed that may reveal previously unseen relationships in the data. For instance, approaches centering on nontraditional and abstract methods of data exploration, such as taking the data out of "brain space" and placing it in some alternative parameter space, and examining patterns in the data that might have been invisible in the original anatomically-based space (Baumgartner and Somorjai, 2001; Cordes *et al.*, 2002). These may also include iconic representations of data endowed with synthetic physical properties that distort the relative coordinates of data in an abstracted space. Such models have been successfully employed in visualizing biochemical and metabolic pathways (Becker and Rojas, 2001; Karp and Paley, 1994; Ogata *et al.*, 2000). Coherent subspaces within this abstracted space can be identified and examined as those where something of neuropsychological interest might be

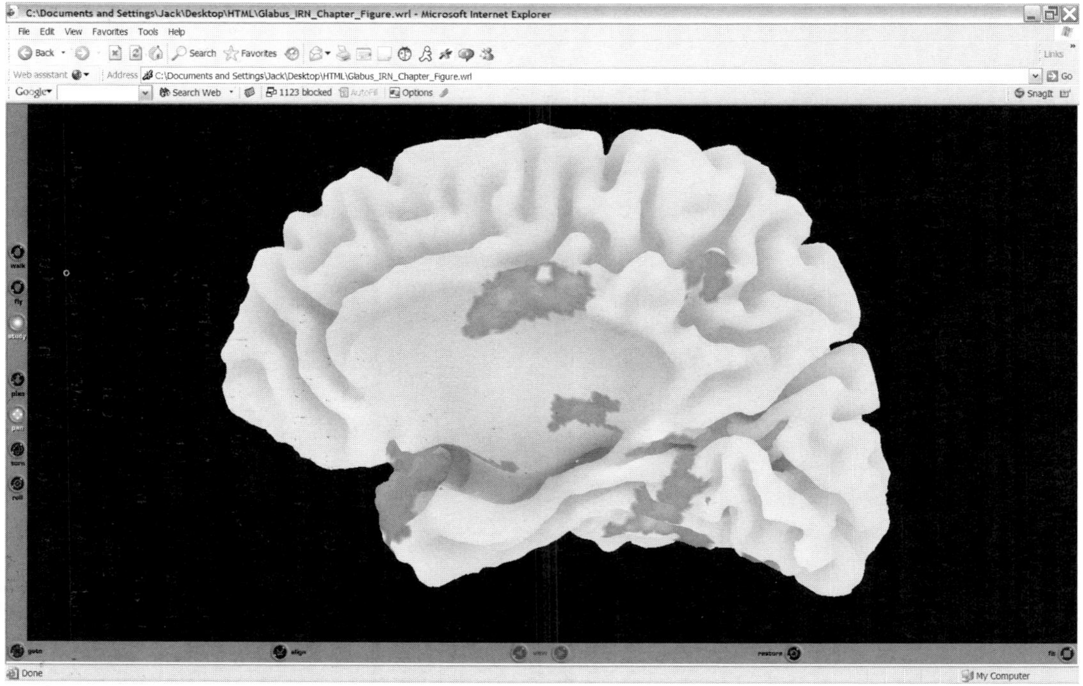

FIG. 2. The novel visualization of fMRI results can enhance or emphasize areas not fully appreciated in the original published article. This figure shows a VRML rendering, exported from the Caret software package, of the memory encoding statistical main effects from the study of Druzgal and D'Esposito (2001) viewed as interactive object using the Cortona Client for MS Windows XP, Version 4.2.R93, from Parallel Graphics, Inc. (*http://www.parallelgraphics.com*). This rendering highlights the robust activation of the fusiform gyrus, the superior portion of the anterior cingulate, inferior parietal lobule, as well as the temporal pole in the right cerebral hemisphere. Viewed as a virtual reality object would enable a student or researcher to interactively translate, rotate, or zoom in on regions of particular interest and to view the data as a whole, rather than as a collection of 2D overlay graphics. (See Color Insert.)

occurring. As fMRI archives continue to grow through active data sharing, working with the shear amount of data, as well as its direct visualization, becomes increasingly difficult. These computational and visualization methods will allow large amounts of data to be processed and visualized.

IX. Neuroinformatics—The Nexus of Brain, Computational, and Computer Sciences

Given recent success stories from the domains of genomics (Escribano and Coca-Prados, 2002; Feolo *et al.*, 2000; Rafalski *et al.*, 1998) and proteomics (Berman *et al.*, 2000; Ezzell, 2002; Persson, 2000) for organizing large amounts of data, cognitive neuroscientists are likewise becoming intimately familiar with large-scale data analysis, applying high performance computing systems, and using sophisticated computer science to extract information from large archives of neurophysiological data. The evolution of the cognitive neuroscience field is fast approaching the time when it forms a confluence of brain science, high-performance computing systems, and leading edge computer science (Beltrame and Koslow, 1999; Wong and Koslow, 2001). As such, a more thorough understanding of the brain and its cognitive processes will necessitate increased computational infrastructure, novel software technology to accelerate data analysis and to mine vastly larger amounts of data, and the sharing of primary research data. Moreover, these data must be understood on a level that permits the representation of the dynamic examination of brain data and brain systems required for cognitive processes such as memory function, visual abilities, and motor skill. This effort must reach beyond the level of the examination of individual loci of brain activity to that of identifying patterns of activity across individuals that speak to the dynamics and complexity of the neural processes that are not typically reported in the scientific literature though may be worthy of additional scrutiny and study.

In response, the field of neuroscience is rapidly moving beyond its roots as a theoretical and experimental science toward becoming a highly computational science ever more dependent upon lead edge technologies in computer science, engineering, and mathematics. This is the origin of *neuroinformatics*, a unifying discipline at the nexus of information technology, computer science, and the neurosciences. It also involves the incorporation of high performance computing, visualization, and data mining techniques with the fundamentals of experimental design, image processing, and spatial and temporal statistics for neurophysiological data, in particular, for functional neuroimaging (Douglas *et al.*, 1996; Smaglik, 2000; Young and Scannell, 2000). By using computers to organize, link, analyze, and examine large, complex sets of neuroscientific data, raw data may be converted into meaningful knowledge that can be used for further experimentation into cognitive function and the treatment of patients with neurological and neuropsychiatric disease (Beltrame and Koslow, 1999).

The Human Brain Project (HBP) funds many of the current database and neuroinformatics efforts. The HBP is a broad-based initiative which supports research and development of advanced technologies and infrastructure support through cooperative efforts among neuroscientists and information scientists (computer scientists, physicists, mathematicians, and engineers) (Brinkley and Rosse, 2002; Shepherd *et al.*, 1998). The principle aim of the NIMH-based HBP is to guide the production of new digital capabilities that provide Internet-driven information management systems in the form of interoperable databases and associated neuroscience data management tools (Shepherd *et al.*, 1998). Such software tools include graphical interfaces, querying and mining approaches, information retrieval, statistical analysis, visualization and manipulation, integrating tools for data analysis, biological modeling and simulation, and tools for electronic collaboration. The effort strongly supports open data sharing, believing that the primary data from neuroscience investigations has continued value to the field long after its initial publication (Huerta and Koslow, 1996; Koslow, 2000; Koslow, 2002). The HBP seeks to make neuroinformatics efforts funded under its auspices interoperable with other databases, tools, and centers, similar to some genomic and protein databases, and, thereby, create the capability to explore brain functional and structural interactions in even greater detail. The HBP also encourages researchers to leverage the emerging Internet capabilities for opening novel channels of communication and collaboration between geographically distinct sites.

A number of neuroscience databases exist that provide a variety of information and data pertaining to neural function. For neuroimaging, in addition to the fMRIDC, two other notable efforts exist: (1) The BrainMapDBJ database from the University of South Texas Health Sciences Center (USTHSC), and its current incarnation BrainMapDBJ (*http://www.brainmapdbj.org*),—pioneering efforts to provide access to the human brain-mapping literature and its results-based data in a manner to promote quantitative meta-analysis of related studies (Fox *et al.*, 1994a). It is comprised of a multi-level indexing scheme describing the study of experimental protocol as well as the derived Talairach-normalized local maxima from brain-mapping studies. However, raw, processed, or results image data are not provided to users of the database. Submissions to BrainMapDBJ are voluntary but peer-reviewed by the database editorial board, independently of the journal peer-review process, to review content suitability and correct coding. Access to the database is open to the public but its contents are limited only to the provided meta-data and the reported study local maxima. (2) The UCLA Laboratory of Neuroimaging (LONI) structural image database (Toga, 2002b) and the International Consortium of Brain Mapping (ICBM) Probabilistic Brain Atlas (Mazziotta *et al.*, 2001) have been constructed to provide a rigorous means for data archiving and protection of collaborator-collected image data (*http://www.loni.ucla.edu/; http://www.loni.ucla.edu/ICBM/index.html*). Database query mechanisms ensure that no image data or identifying patient information is accessible to the public or to any others without the appropriate authorization and the expressed permission to release data from the ICBM collaborator that acquired and provided the data. The LONI database provides an integrated access and security mechanism such that perusal through archives is organized by authorized scientific groups within each ICBM laboratory. LONI maintains a large-scale computer infrastructure to maintain this archive and for use in pipelined processing of data (Rex *et al.*, 2003). Each of these efforts is a rich resource for finding information about brain function from both published and unpublished neuroimaging studies at multiple levels of detail. However, access to complete study information may be limited either due to proprietary restrictions from study investigators or through a limited scope of the data that is available.

The fMRIDC effort has helped to catalyze the field in considering the benefits of neuroscientific databasing and its potential for further advancing progress in understanding cognitive function through the open sharing of these large data sets (Van Horn, 2002; Van Horn and Gazzaniga, 2002; Van Horn *et al.*, 2001). The fMRIDC has committed to a policy of open science and provides its archive contents and software to the scientific community free of charge. The neuroscientific outcomes derived from the services offered by the fMRIDC via this novel computational resource are now beginning to bear fruit in the peer-reviewed scientific literature (Table III) and add markedly to the knowledge base of brain research.

X. Current Challenges for Neuroscience Databases

As with many neuroscience data archives, the challenge they are now presented with is how to best utilize the information contained in their database toward novel scientific outcomes that could not have existed without that large collection of data. Finding useful, rigorous, and timely answers to these and other questions will serve to demonstrate the promise of large-scale databasing and neuroinformatics methods and their utility in the study of brain function. In coming years, it can be expected that database-driven research will, indeed, help to supplement hypothesis-based experimentation, spur the formation of novel lines of research, and help to educate the next generation of neuroscientists.

It should be recognized, however, that a single online resource will not be capable of organizing and indexing all possible types of brain data. By linking information from one online resource with that contained in another, the wealth and richness of information provided by them both is increased. These linkages need

TABLE III

RECENT REANALYSES OF fMRI STUDY DATA

New authors	New journal	Purpose of new analysis	Original dataset reference (fMRIDC accession number)
Carlson *et al.*	J. Cog. Neuro., (2003)	Used canonical discriminant analysis to examine object categories	Ishai *et al.* (2000) JOCN, 12 Suppl 2, 35–51
Greicius and Menon	J. Cog. Neuro., (2004)	Used ICA to assess default-mode activity in auditory processing	Laurienti *et al.* (2002) JOCN, 14(3), 420–429
Greicius *et al.*	Proc. Nat. Acad. Sci., (2004)	Used ICA to assess alterations in default-mode activity in normal, older, and demented subjects	Buckner *et al.* (2000) JOCN, 12 Suppl 2, 24–34
Liou *et al.*	J. Cog. Neuro., (2003)	Characterize the statistical reproducibility of fMRI block design results	Ishai *et al.* (2000) JOCN, 12 Suppl 2, 35–51
Lloyd	J. Cog. Neuro., (2002)	Data assessed for patterns relevant to human consciousness	Ishai *et al.* (2000) JOCN, 12 Suppl 2, 35–51 Hazeltine, Poldrack, and Gabrieli (2000) JOCN, 12 Suppl 2, 118–129 Postle *et al.* (2000) JOCN, 12 Suppl 2, 2–24 Mechelli *et al.* (2000) JOCN, 12 Suppl 2, 145–156
Mechelli *et al.*	J. Cog. Neuro., (2003c)	To assess functional connectivity using dynamic causal modeling	Ishai *et al.* (2000) JOCN, 12 Suppl 2, 35–51
Penny *et al.*	Neuroimage, (2004)	Data used to compare dynamic causal models	Ishai *et al.* (2000) JOCN, 12 Suppl 2, 35–51

a

b

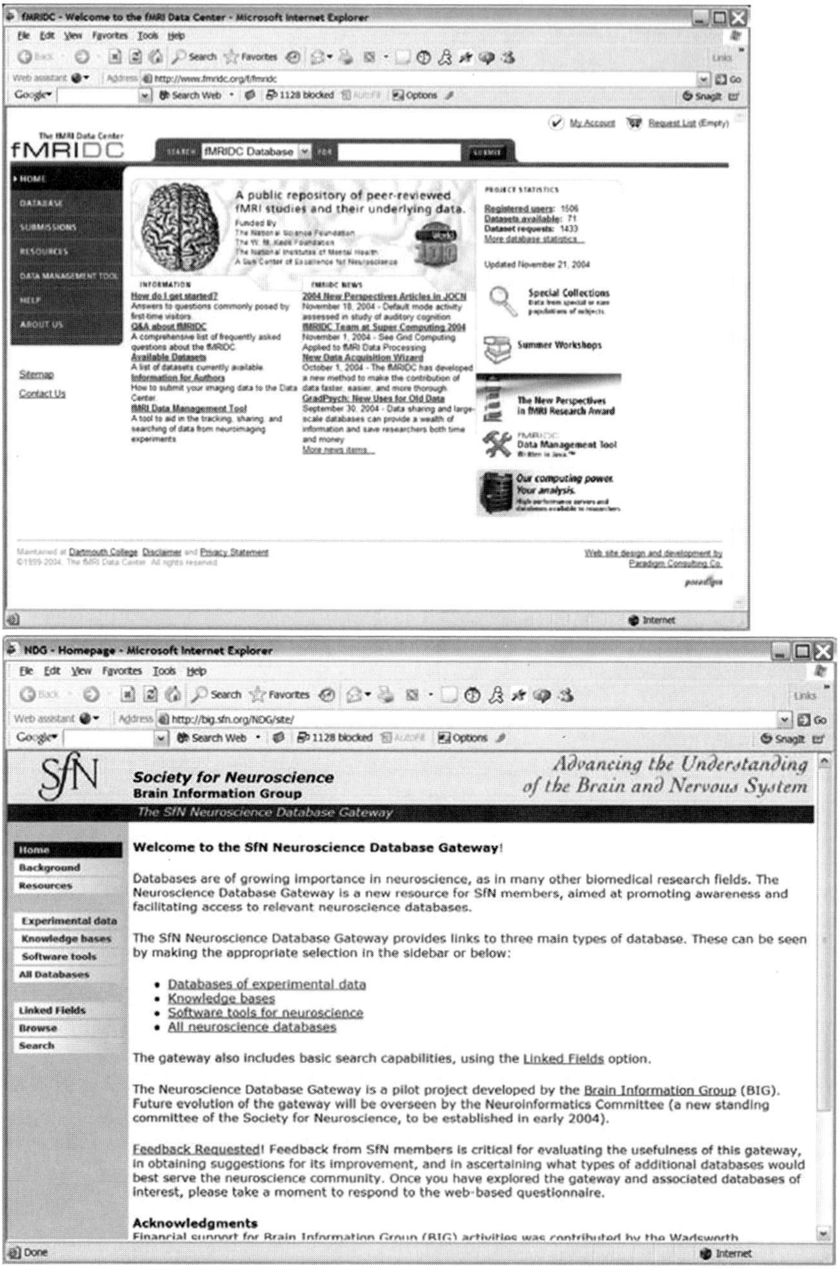

FIG. 3. (*Continued*)

C

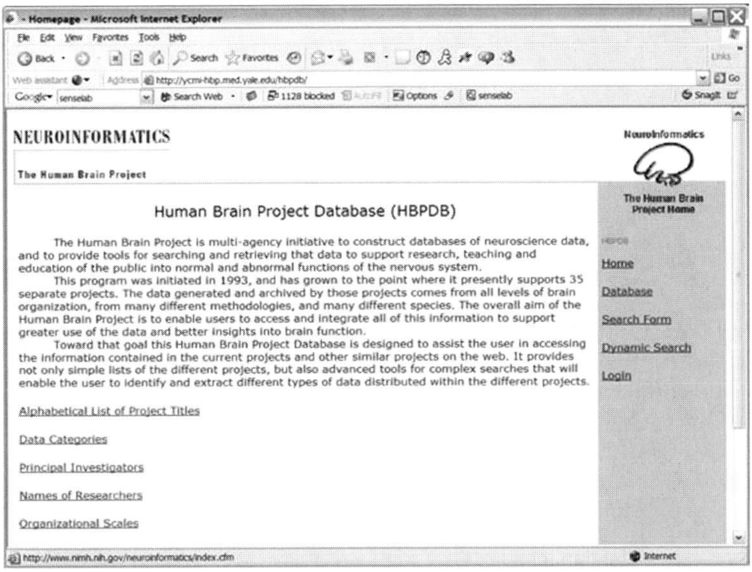

FIG. 3. (a) The fMRI Data Center web site permits researchers to browse and request complete fMRI data sets from peer reviewed, published journal articles; (b) The Society for Neuroscience provides an ever increasing portal to online neuroscience data archives (http://web.sfn.org/content/Programs/NeuroscienceDatabaseGateway/index.html), from molecular- to systems-levels, in which workers may obtain data for novel analysis and visualization; and (c) The NIMH Human Brain Project actively promotes the sharing of primary research data and lists the database efforts of investigators funded under its program (http://ycmi-hbp.med.yale.edu/hbpdb/). (See Color Insert.)

not be part of a strictly federalized scheme of database participation but should span multiple independent archiving and data sharing efforts. In time, far-reaching linkages between individual resources, encouraged by the governing societies and organizations in neuroscience, will form a dynamic web of brain-related information spanning multiple temporal and spatial scales. "This notion is now being recognized by numerous funding awarding bodies. From their point of view, the sharing of primary data is now an integral part of science funding (for example, see Fig. 3)."

How the large amounts of data obtained in neuroscience experiments is best organized is another area of active interest. Ontologies have the advantage over other database frameworks that they have been developed to handle and search over qualitative information—often generally referred to as "knowledge"—as easily as the more traditional formats deal with quantitative information (Hendler, 2003). Knowledge bases, which are databases organized according to an ontology (Oliver *et al.*, 2002), rather than a strictly relational database schema, expose a middle ground between very loose and very rigid data architectures. However, they must possess the structure required for data re-use and sharing, while maintaining the flexibility required to accommodate variations from lab to lab, researcher to researcher, and as the field concerned evolves. This approach makes sense in a context of pre-existing data management tools that need merely to be interconnected.

XI. Conclusion

Neuroscience databases are a rapidly growing resource for scientific discovery whose role in everyday neuroscience can be expected to increase in coming years. These rich archives of physiological data, brain images, genomic information, and behavioral assessments can be mined by students wishing to leverage existing knowledge into new hypotheses or used by established investigators to explore unforeseen relationships not discussed in the original published research article. Linking these resources, thereby permitting an ever denser, more enriched collection of scientific knowledge, will serve to promote and enhance brain sciences by leveraging our previous understanding toward the collection of new and exciting knowledge about brain function.

References

Ackerman, M. J. (1999). The Visible Human Project: A resource for education. *Acad. Med.* **74,** 667–670.

Ackerman, M. J., and Banvard, R. A. (2000). Imaging outcomes from the national library of medicine's visible human project. *Comput. Med. Imaging Graph.* **24,** 125–126.

Adolphs, R. (2003). Cognitive neuroscience of human social behaviour. *Nat. Rev. Neurosci.* **4,** 165–178.

Adolphs, R., Sears, L., and Piven, J. (2001). Abnormal processing of social information from faces in autism. *J. Cogn. Neurosci.* **13,** 232–240.

Altman, R. B. (2003). The expanding scope of bioinformatics: Sequence analysis and beyond. *Heredity* **90,** 345.

Avery, P. (2002). Data Grids: A new computational infrastructure for data-intensive science. *Philos. Transact Ser. A Math. Phys. Eng. Sci.* **360,** 1191–1209.

Bandettini, P. A., and Wong, E. C. (1997). Magnetic resonance imaging of human brain function. Principles, practicalities, and possibilities. *Neurosurg. Clin. N Am.* **8,** 345–371.

Baumgartner, R., Ryner, L., Richter, W., Summers, R., Jarmasz, M., and Somorjai, R. (2000). Comparison of two exploratory data analysis methods for fMRI: Fuzzy clustering vs. principal component analysis. *Magn. Reson. Imaging* **18,** 89–94.

Baumgartner, R., and Somorjai, R. (2001). Graphical display of fMRI data: Visualizing multidimensional space. *Magn. Reson. Imaging* **19,** 283–286.

Beauchamp, M. S., Lee, K. E., Haxby, J. V., and Martin, A. (2002). Parallel visual motion processing streams for manipulable objects and human movements. *Neuron* **34,** 149–159.

Beauchamp, M. S., Petit, L., Ellmore, T. M., Ingeholm, J., and Haxby, J. V. (2001). A parametric fMRI study of overt and covert shifts of visuospatial attention. *Neuroimage* **14,** 310–321.

Becker, K. G. (2001). The sharing of cDNA microarray data. *Nat. Rev. Neurosci.* **2,** 438–440.

Becker, M. Y., and Rojas, I. (2001). A graph layout algorithm for drawing metabolic pathways. *Bioinformatics* **17,** 461–467.

Beltrame, F., and Koslow, S. H. (1999). Neuroinformatics as a megascience issue. *IEEE Trans. Inf. Technol. Biomed.* **3,** 239–240.

Berman, H. M., Westbrook, J., Feng, Z., Gilliand, G., Bhat, T. N., Weissig, H., Shindyalov, I. N., and Bourne, P. E. (2000). The protein data bank. *Nucleic Acids Res.* **28,** 235–242.

Binder, J. R., Frost, J. A., Hammeke, T. A., Cox, R. W., Rao, S. M., and Prieto, T. (1997). Human brain language areas identified by functional magnetic resonance imaging. *J. Neurosci.* **17,** 353–362.

Binkofski, F., Fink, G. R., Geyer, S., Buccino, G., Gruber, O., Shah, N. J., Taylor, J. G., Seitz, R. J., Zilles, K., and Freund, H. J. (2002). Neural activity in human primary motor cortex areas 4a and 4p is modulated differentially by attention to action. *J. Neurophysiol.* **88,** 514–519.

Birn, R. M., Cox, R. W., and Bandettini, P. A. (2002). Detection versus estimation in event-related fMRI: Choosing the optimal stimulus timing. *Neuroimage* **15,** 252–264.

Blue Ribbon Advisory Panel On Cyberinfrastructure (2003). Revolutionizing Science and Engineering Through Cyberinfrastructure. The National Science Foundation, Washington DC.

Boardman, L. A. (2002). Heritable colorectal cancer syndromes: Recognition and preventive management. *Gastroenterol Clin. North Am.* **31,** 1107–1131.

Brinkley, J. F., and Rosse, C. (2002). Imaging and the Human Brain Project: A review. *Methods Inf. Med.* **41,** 245–260.

Brookes, A. J. (2001). Rethinking genetic strategies to study complex diseases. *Trends Mol. Med.* **7,** 512–516.

Buchel, C., Price, C., and Friston, K. (1998). A multimodal language region in the ventral visual pathway. *Nature* **394,** 274–277.

Buckner, R. L. (1998). Event-related fMRI and the hemodynamic response. *Hum. Brain. Mapp.* **6,** 373–377.

Buckner, R. L., Bandettini, P. A., O'Craven, K. M., Savoy, R. L., Petersen, S. E., Raichle, M. E., and Rosen, B. R. (1996). Detection of cortical activation during averaged single trials of a cognitive task using functional magnetic resonance imaging. *Proc. Natl. Acad. Sci. USA* **93,** 14878–14883.

Buckner, R. L., Snyder, A. Z., Sanders, A. L., Raichle, M. E., and Morris, J. C. (2000). Functional brain imaging of young, nondemented, and demented older adults. *J. Cogn. Neurosci.* **12,** 24–34.

Butler, D. (2003). The Grid: Tomorrow's computing today. *Nature* **422,** 799–800.

Buxton, R. B., and Frank, L. R. (1997). A model for the coupling between cerebral blood flow and oxygen metabolism during neural stimulation. *J. Cereb. Blood Flow Metab.* **17,** 64–72.

Buxton, R. B., Wong, E. C., and Frank, L. R. (1998). Dynamics of blood flow and oxygenation changes during brain activation: The balloon model. *Magn. Reson. Med.* **39,** 855–864.

Cabeza, R., Dolcos, F., Graham, R., and Nyberg, L. (2002). Similarities and differences in the neural correlates of episodic memory retrieval and working memory. *Neuroimage* **16,** 317–330.

Calvert, P. M., and Frucht, H. (2002). The genetics of colorectal cancer. *Ann Intern Med* **137,** 603–612.

Carlson, T. A., Schrater, P., and He, S. (2003). Patterns of activity in the categorical representations of objects. *J. Cogn. Neurosci.* **15,** 704–717.

Carpenter, P. A., Just, M. A., and Reichle, E. D. (2000). Working memory and executive function: Evidence from neuroimaging. *Curr. Opin. Neurobiol.* **10,** 195–199.

Casey, B. J. (2002). Neuroscience. Windows into the human brain. *Science* **296,** 1408–1409.

Chen, Y., and Xu, D. (2003). Computational analyses of high-throughput protein–protein interaction data. *Curr. Protein Pept. Sci.* **4,** 159–181.

Collins, F. S., and Mansoura, M. K. (2001). The human genome project. *Cancer* **91,** 221–225.

Corchs, S., and Deco, G. (2002). Large-scale neural model for visual attention: Integration of experimental single-cell and fMRI data. *Cereb. Cortex.* **12,** 339–348.

Cordes, D., Haughton, V., Carew, J. D., Arfanakis, K., and Maravilla, K. (2002). Hierarchical clustering to measure connectivity in fMRI resting-state data. *Magn. Reson. Imaging.* **20,** 305–317.

Cox, R. W. (1996). AFNI: Software for analysis and visualization of functional magnetic resonance neuroimages. *Comput. Biomed. Res.* **29,** 162–173.

Cox, R. W., and Hyde, J. S. (1997). Software tools for analysis and visualization of fMRI data. *NMR Biomed.* **10,** 171–178.

Crosson, B., Rao, S. M., Woodley, S. J., Rosen, A. C., Bobholz, J. A., Mayer, A., Cunningham, J. M., Hammeke, T. A., Fuller, S. A., Binder, J. R., Cox, R. W., and Stein, E. A. (1999). Mapping of semantic, phonological, and orthographic verbal working memory in normal adults with functional magnetic resonance imaging. *Neuropsychology* **13,** 171–187.

Culham, J. C., Cavanagh, P., and Kanwisher, N. G. (2001). Attention response functions: Characterizing brain areas using fMRI activation during parametric variations of attentional load. *Neuron* **32,** 737–745.

Cummings, L., Riley, L., Black, L., Souvorov, A., Resenchuk, S., Dondoshansky, I., and Tatusova, T. (2002). Genomic BLAST: Custom-defined virtual databases for complete and unfinished genomes. *FEMS Microbiol. Lett.* **216,** 133–138.

Dale, A. M., and Buckner, R. L. (1997). Selective averaging of rapidly presented individual trials using fMRI. *Human Brain Mapping* **5,** 329–340.

D'Esposito, M. (2000). Functional neuroimaging of cognition. *Semin Neurol* **20,** 487–498.

Detre, J. A., and Floyd, T. F. (2001). Functional MRI and its applications to the clinical neurosciences. *Neuroscientist* **7,** 64–79.

Douglas, R., Mahowald, M., and Martin, K. (1996). Neuroinformatics as explanatory neuroscience. *Neuroimage* **4,** S25–S28.

Dreher, J. C., Koechlin, E., Ali, S. O., and Grafman, J. (2002). The roles of timing and task order during task switching. *Neuroimage* **17,** 95–109.

Escribano, J., and Coca-Prados, M. (2002). Bioinformatics and reanalysis of subtracted expressed sequence tags from the human ciliary body: Identification of novel biological functions. *Mol. Vis.* **8,** 315–332.

Ezzell, C. (2002). Proteins Rule. *Scientific American.* **286,** 40–47.

Feolo, M., Helmberg, W., Sherry, S., and Maglott, D. R. (2000). NCBI genetic resources supporting immunogenetic research. *Rev. Immunogenet.* **2,** 461–467.

Foster, I. (2003). The grid: Computing without bounds. *Sci. Am.* **288,** 78–85.

Fox, P., and Lancaster, J. (2002). Mapping context and content: The BrainMap model. *Nature Reviews Neuroscience* **3,** 319–321.

Fox, P. T., and Lancaster, J. L. (1994). Neuroscience on the net. *Science* **266,** 994–996.

Fox, P. T., Mikiten, S., Davis, G., and Lancaster, J. (1994a). BrainMap: A database of human function brain mapping. *In* "Functional Neuroimaging Technical Foundations" (R. W. Thatcher, M. Hallett, T. Zeffiro, E. R. John, and M. Heurta, Eds.), pp. 95–105. Academic Press, San Diego.

Fox, P. T., Mikiten, S., Davis, G., and Lancaster, J. L. (1994b). Brain-Map: A database of human functional brain mapping, pp. 95–105. Academic Press, San Diego.

Friman, O., Borga, M., Lundberg, P., and Knutsson, H. (2002). Exploratory fMRI analysis by autocorrelation maximization. *Neuroimage* **16,** 454–464.

Friston, K. (2002). Beyond phrenology: What can neuroimaging tell us about distributed circuitry? *Annu. Rev. Neurosci.* **25,** 221–250.

Friston, K. J., Glaser, D. E., Henson, R. N., Kiebel, S., Phillips, C., and Ashburner, J. (2002). Classical and Bayesian inference in neuroimaging: Applications. *Neuroimage* **16,** 484–512.

Friston, K. J., and Price, C. J. (2001). Generative models, brain function and neuroimaging. *Scand. J. Psychol.* **42,** 167–177.

Gauthier, I., and Nelson, C. A. (2001). The development of face expertise. *Curr. Opin. Neurobiol.* **11,** 219–224.

Gauthier, I., Tarr, M. J., Anderson, A. W., Skudlarski, P., and Gore, J. C. (1999). Activation of the middle fusiform 'face area' increases with expertise in recognizing novel objects. *Nature Neuroscience.* **2,** 568–573.

Goutte, C., Hansen, L. K., Liptrot, M. G., and Rostrup, E. (2001). Feature-space clustering for fMRI meta-analysis. *Hum. Brain Mapp.* **13,** 165–183.

Greicius, M. D., and Menon, V. (2004). Default-mode activity during a passive sensory task: Uncoupled from deactivation but impacting activation. *J. Cogn. Neurosci.* **16,** 1484–1492.

Greicius, M. D., Srivastava, G., Reiss, A. L., and Menon, V. (2004). Default-mode network activity distinguishes Alzheimer's disease from healthy aging: Evidence from functional MRI. *Proc. Natl. Acad. Sci. USA* **101,** 4637–4642.

Hamalainen, H., Hiltunen, J., and Titievskaja, I. (2002). Activation of somatosensory cortical areas varies with attentional state: An fMRI study. *Behav. Brain Res.* **135,** 159.

Haxby, J. V., Gobbini, M. I., Furey, M. L., Ishai, A., Schouten, J. L., and Pietrini, P. (2001). Distributed and overlapping representations of faces and objects in ventral temporal cortex. *Science* **293,** 2425–2430.

Haxby, J. V., Hoffman, E. A., and Gobbini, M. I. (2000). The distributed human neural system for face perception. *Trends Cogn. Sci.* **4,** 223–233.

Haxby, J. V., Hoffman, E. A., and Gobbini, M. I. (2002). Human neural systems for face recognition and social communication. *Biol. Psychiatry* **51,** 59–67.

Hazeltine, E., Poldrack, R., and Gabrieli, J. D. (2000). Neural activation during response competition. *J. Cogn. Neurosci.* **12,** 118–129.

Hendler, J. (2003). COMMUNICATION: Enhanced: Science and the Semantic Web. *Science* **299,** 520–521.

Hoppel, B. E., Weisskoff, R. M., Thulborn, K. R., Moore, J. B., Kwong, K. K., and Rosen, B. R. (1993). Measurement of regional blood oxygenation and cerebral hemodynamics. *Magn. Reson. Med.* **30,** 715–723.

Huerta, M. F., and Koslow, S. H. (1996). Neuroinformatics: Opportunities across disciplinary and national borders. *Neuroimage* **4,** S4–S6.

Ilioudis, C., and Pangalos, G. (2001). A framework for an institutional high level security policy for the processing of medical data and their transmission through the Internet. *J. Med. Internet Res.* **3,** E14.

Insel, T. R., Volkow, N. D., Li, T-K., Battey, J. F., and Landis, S. C. (2003). Neuroscience Networks: Data-sharing in an Information Age. *Public Library of Science: Biology* **1,** 9–11.

Ishai, A., Ungerleider, L. G., Martin, A., and Haxby, J. V. (2000). The representation of objects in the human occipital and temporal cortex. *J. Cogn. Neurosci.* **12,** 35–51.

Jarmasz, M., and Somorjai, R. L. (2002). Exploring regions of interest with cluster analysis (EROICA) using a spectral peak statistic for selecting and testing the significance of fMRI activation time-series. *Artif. Intell. Med.* **25,** 45–67.

Jones, D. T., and Swindells, M. B. (2002). Getting the most from PSI-BLAST. *Trends Biochem. Sci.* **27,** 161–164.

Kanwisher, N., McDermott, J., and Chun, M. M. (1997). The fusiform face area: A module in human extrastriate cortex specialized for face perception. *J. Neurosci.* **17,** 4302–4311.

Kanwisher, N., and Wojciulik, E. (2000). Visual attention: Insights from brain imaging. *Nat. Rev. Neurosci.* **1,** 91–100.

Karp, P. D., Paley, S., and Romero, P. (2002). The Pathway Tools software. *Bioinformatics* **18,** S225–S232.

Karp, P. D., and Paley, S. M. (1994). Representations of metabolic knowledge: Pathways. *Proc. Int. Conf. Intell. Syst. Mol. Biol.* **2,** 203–211.

Kawato, M., Kuroda, T., Imamizu, H., Nakano, E., Miyauchi, S., and Yoshioka, T. (2003). Internal forward models in the cerebellum: fMRI study on grip force and load force coupling. *Prog. Brain. Res.* **142,** 171–188.

Koslow, S. H. (2000). Should the neuroscience community make a paradigm shift to sharing primary data? *Nat. Neurosci.* **2,** 863–864.

Koslow, S. H. (2002). Opinion: Sharing primary data: A threat or asset to discovery? *Nat. Rev. Neurosci.* **3,** 311–313.

LaConte, S., Anderson, J., Muley, S., Ashe, J., Frutiger, S., Rehm, K., Hansen, L. K., Yacoub, E., Hu, X., Rottenberg, D., and Strother, S. (2003). The Evaluation of Preprocessing Choices in Single-Subject BOLD fMRI Using NPAIRS Performance Metrics. *Neuroimage* **18,** 10–27.

Laurienti, P. J., Burdette, J. H., Wallace, M. T., Yen, Y. F., Field, A. S., and Stein, B. E. (2002). Deactivation of sensory-specific cortex by cross-modal stimuli. *J. Cogn. Neurosci.* **14,** 420–429.

Legato, J., Knepper, M. A., Star, R. A., and Mejia, R. (2003). Database for renal collecting duct regulatory and transporter proteins. *Physiol. Genomics.* **13,** 179–181.

Leveroni, C. L., Seidenberg, M., Mayer, A. R., Mead, L. A., Binder, J. R., and Rao, S. M. (2000). Neural systems underlying the recognition of familiar and newly learned faces. *J. Neurosci.* **20,** 878–886.

Lindblom, A., Tannergard, P., Werelius, B., and Nordenskjold, M. (1993). Genetic mapping of a second locus predisposing to hereditary non-polyposis colon cancer. *Nat. Genet.* **5,** 279–282.

Liou, M., Su, H-R., Lee, J-D., and Cheng, P. E. (2003). Bridging functional MR images and scientific inference: Reproducibility maps. *J. Cog. Neurosci.* **15,** 934–945.

Lloyd, D. (2002). Functional MRI and the study of human consciousness. *J. Cogn. Neurosci.* **14,** 818–831.

Logothetis, N. K., Pauls, J., Augath, M., Trinath, T., and Oeltermann, A. (2001). Neurophysiological investigation of the basis of the fMRI signal. *Nature* **412,** 150–157.

Lohmann, G., Muller, K., Bosch, V., Mentzel, H., Hessler, S., Chen, L., Zysset, S., and von Cramon, D. Y. (2001). LIPSIA—a new software system for the evaluation of functional magnetic resonance images of the human brain. *Comput. Med. Imaging Graph* **25,** 449–457.

Lukic, A. S., Wernick, M. N., and Strother, S. C. (2002). An evaluation of methods for detecting brain activations from functional neuroimages. *Artif. Intell. Med.* **25,** 69–88.

Ma, B., Tromp, J., and Li, M. (2002). PatternHunter: Faster and more sensitive homology search. *Bioinformatics* **18,** 440–445.

Marenco, L., Nadkarni, P., Skoufos, E., Shepherd, G., and Miller, P. (1999). Neuronal database integration: The Senselab EAV data model. *Proc. AMIA Symp.* **10,** 102–106.

Martone, M., Gupta, A., Wong, M., Qian, X., Sosinsky, G., Ludascher, B., and Ellisman, M. (2002). A cell-centered database for electron tomographic data. *J. Struct. Biol.* **138,** 145.

Mavroudis, C., and Jacobs, J. P. (2000). Congenital heart surgery nomenclature and database project: Overview and minimum dataset. *Ann. Thorac. Surg.* **69,** S2–S17.

Mazziotta, J., Toga, A. W., Evans, A., Fox, P., Lancaster, J., Ziles, K., Woods, R., Paus, T., Simpson, G., Pike, B., Holmes, C., Collins, L., Thompson, P., MacDonald, D., Iacoboni, M., Schormann, T., Amunts, K., Palomero-Gallagher, N., Geyer, S., Parsons, L., Narr, K., Kabani, N., Le Goualher, G., Boomsma, D., Cannon, T., Kawashima, R., and Mazoyer, B. (2001). A probabilistic atlas and reference system for the human brain: International Consortium for Brain Mapping (ICBM). *Philos. Trans. R Soc. Lond. B Biol. Sci.* **356,** 1293–1322.

McKeown, M. J., and Sejnowski, T. J. (1998). Independent component analysis of fMRI data: Examining the assumptions. *Hum. Brain Mapp.* **6,** 368–372.

Mechelli, A., Friston, K. J., and Price, C. J. (2000). The effects of presentation rate during word and pseudoword reading: A comparison of PET and fMRI. *J. Cogn. Neurosci.* **12,** 145–156.

Mechelli, A., Gorno-Tempini, M. L., and Price, C. J. (2003a). Neuroimaging studies of word and pseudoword reading: Consistencies, inconsistencies, and limitations. *J. Cogn. Neurosci.* **15,** 260–271.

Mechelli, A., Henson, R. N., Price, C. J., and Friston, K. J. (2003b). Comparing event-related and epoch analysis in blocked design fMRI. *Neuroimage* **18,** 806–810.

Mechelli, A., Price, C., Noppeney, U., and Friston, K. (2003c). A dynamic causal modelling study on category effects: Bottom-up or top-down mediation? *J. Cogn. Neurosci.* **15,** 925–934.

Mirnics, K. (2001). Microarrays in brain research: The good, the bad and the ugly. *Nat. Rev. Neurosci.* **2,** 444–447.

Mitchell, T. (1997). "Machine Learning." McGraw-Hill, New York.

Mitchell, T. (1999). "Machine Learning and Data Mining." Communications of the ACM 42.

Mohamed, F. B., Tracy, J. I., Faro, S. H., Emperado, J., Koenigsberg, R., Pinus, A., and Tsai, F. Y. (2000). Investigation of alternating and continuous experimental task designs during single finger opposition fMRI: A comparative study. *J. Comput. Assist. Tomogr.* **24,** 935–941.

Noguchi, T., and Akiyama, Y. (2003). PDB-REPRDB: A database of representative protein chains from the Protein Data Bank (PDB) in 2003. *Nucleic Acids Res.* **31,** 492–493.

Nowinski, W. L., Belov, D., and Benabid, A. L. (2002). A community-centric internet portal for stereotactic and functional neurosurgery with a probabilistic functional atlas. *Stereotact. Funct. Neurosurg.* **79,** 1–12.

Ogata, H., Fujibuchi, W., Goto, S., and Kanehisa, M. (2000). A heuristic graph comparison algorithm and its application to detect functionally related enzyme clusters. *Nucleic Acids Res.* **28,** 4021–4028.

Oliver, D. E., Rubin, D. L., Stuart, J. M., Hewett, M., Klein, T. E., and Altman, R. B. (2002). Ontology development for a pharmacogenetics knowledge base. *Pac. Symp. Biocomput.* **6,** 5–76.

Peltomaki, P. T. (1994). Genetic basis of hereditary nonpolyposis colorectal carcinoma (HNPCC). *Ann. Med.* **26,** 215–219.

Penny, W. D., Stephan, K. E., Mechelli, A., and Friston, K. J. (2004). Comparing dynamic causal models. *Neuroimage* **22,** 1157–1172.

Persson, B. (2000). Bioinformatics in protein analysis. *Exs* **88,** 215–231.

Postle, B. R., Berger, J. S., Taich, A. M., and D'Esposito, M. (2000). Activity in human frontal cortex associated with spatial working memory and saccadic behavior. *J. Cogn. Neurosci.* **12,** 2–14.

Price, C. J., and Friston, K. J. (2002). Degeneracy and cognitive anatomy. *Trends Cogn. Sci.* **6,** 416–421.

Rafalski, J. A., Hanafey, M., Miao, G. H., Ching, A., Lee, J. M., Dolan, M., and Tingey, S. (1998). New experimental and computational approaches to the analysis of gene expression. *Acta Biochim. Pol.* **45,** 929–934.

Raichle, M. E. (1975). Cerebral blood flow and metabolism. *Ciba. Found Symp.* 85–96.

Raichle, M. E. (2001a). Cognitive neuroscience. Bold insights. *Nature* **412,** 128–130.

Raichle, M. E. (2001b). Functional Neuroimaging: A historical and physiological perspective. *In* "Handbook of Functional Neuroimaging of Cognition" (R. Cabeza and A. Kingstone, Eds.), pp. 3–26. MIT Press, Cambridge, MA.

Raichle, M. E. (2003). Functional brain imaging and human brain function. *J. Neurosci.* **23,** 3959–3962.

Ramesar, R. S., Madden, M. V., Felix, R., Harocopos, C. J., Westbrook, C. A., Jones, G., Cruse, J. P., and Goldberg, P. A. (2000). Molecular genetics improves the management of hereditary nonpolyposis colorectal cancer. *S Afr. Med. J.* **90,** 709–714.

Reidpath, D. D., and Allotey, P. A. (2001). Data sharing in medical research: An empirical investigation. *Bioethics* **15,** 125–134.

Rex, D. E., Ma, J. Q., and Toga, A. W. (2003). The LONI pipeline processing environment. *Neuroimage* **19,** 1033–1048.

Richard, A. M., and Williams, C. R. (2002). Distributed structure-searchable toxicity (DSSTox) public database network: A proposal. *Mutat. Res.* **499,** 27–52.

Richeson, J. A., Baird, A. A., Gordon, H. L., Heatherton, T. F., Wyland, C. L., Trawalter, S., and Shelton, J. N. (2003). An fMRI investigation of the impact of interracial contact on executive function. *Nat. Neurosci.* **6,** 1323–1328.

Roland, P., Svensson, G., Lindeberg, T., Risch, T., Baumann, P., Dehmel, A., Frederiksson, J., Halldorson, H., Forsberg, L., Young, J., and Zilles, K. (2001). A database generator for human brain imaging. *Trends Neurosci.* **24,** 562–564.

Rosen, B. R., Aronen, H. J., Kwong, K. K., Belliveau, J. W., Hamberg, L. M., and Fordham, J. A. (1993). Advances in clinical neuroimaging: Functional MR imaging techniques. *Radiographics* **13,** 889–896.

Rosen, B. R., Buckner, R. L., and Dale, A. (1998). Event-related functional MRI: Past, present, and future. *Proc. Nat. Acad. Sci. USA* **95,** 773–780.

Salli, E., Aronen, H. J., Savolainen, S., Korvenoja, A., and Visa, A. (2001). Contextual clustering for analysis of functional MRI data. *IEEE Trans. Med. Imaging* **20,** 403–414.

Savoy, R. L. (2001). History and future directions of human brain mapping and functional neuroimaging. *Acta Psychologica* **107,** 9–42.

Schutte, B. C., Mitros, J. P., Bartlett, J. A., Walters, J. D., Jia, H. P., Welsh, M. J., Casavant, T. L., and McCray, P. B., Jr. (2002). Discovery of five conserved beta -defensin gene clusters using a computational search strategy. *Proc. Natl. Acad. Sci. USA* **99,** 2129–2133.

Shepherd, G. M., Mirsky, J. S., Healy, M. D., Singer, M. S., Skoufos, E., Hines, M. S., Nadkarni, P. M., and Miller, P. L. (1998). The Human Brain Project: Neuroinformatics tools for integrating, searching and modeling multidisciplinary neuroscience data. *Trends Neurosci.* **21,** 460–468.

Smaglik, P. (2000). Internet gateway planned for neuroinformatics data. *Nature* **405,** 603.

Sokoloff, L. (1981). Relationships among local functional activity, energy metabolism, and blood flow in the central nervous system. *Fed. Proc.* **40,** 2311–2316.

Strother, S. C., Anderson, J., Hansen, L. K., Kjems, U., Kustra, R., Sidtis, J., Frutiger, S., Muley, S., LaConte, S., and Rottenberg, D. (2002). The quantitative evaluation of functional neuroimaging experiments: The NPAIRS data analysis framework. *Neuroimage* **15,** 747–771.

Teo, P. C., Sapiro, G., and Wandell, B. A. (1997). Creating connected representations of cortical gray matter for functional MRI visualization. *IEEE Trans. Med. Imaging* **16,** 852–863.

Thomas, G. (1994). Advances in the genetics and molecular biology of colorectal tumors. *Curr. Opin. Oncol.* **6,** 406–412.

Toga, A. W. (2002a). Imaging databases and neuroscience. *Neuroscientist* **8,** 423–436.

Toga, A. W. (2002b). The laboratory of neuro imaging: What it is, why it is, and how it came to be. *IEEE Trans. Med. Imaging* **21,** 1333–1343.

Tukey, J. W. (1977). Exploratory Data Analysis. Addison, Wesley.

Van Essen, D. C., Drury, H. A., Dickson, J., Harwell, J., Hanlon, D., and Anderson, C. H. (2001a). An integrated software suite for surface-based analyses of cerebral cortex. *J. Am. Med. Inform. Assoc.* **8,** 443–459.

Van Essen, D. C., Lewis, J. W., Drury, H. A., Hadjikhani, N., Tootell, R. B., Bakircioglu, M., and Miller, M. I. (2001b). Mapping visual cortex in monkeys and humans using surface-based atlases. *Vision Res.* **41,** 1359–1378.

Van Horn, J. D. (2002). Maturing as a science: The new perspectives in fMRI research award. *J. Cogn. Neurosci.* **14,** 817.

Van Horn, J. D., Ellmore, T. M., Esposito, G., and Berman, K. F. (1998). Mapping voxel-based statistical power on parametric images. *Neuroimage* **7,** 97–107.

Van Horn, J. D., and Gazzaniga, M. S. (2002). Databasing fMRI studies—Toward a 'Discovery Science' of brain function. *Nat. Rev. Neurosci.* **3,** 314–318.

Van Horn, J. D., and Gazzaniga, M. S. (2004). Maximizing information content in shared neuroimaging studies of cognitive function. *In* "Databasing the Brain: From Data to Knowledge" (A. Subramanian, Ed.). John Wiley and Sons, New York.

Van Horn, J. D., Grethe, J. S., Kostelec, P., Woodward, J. B., Aslam, J. A., Rus, D., Rockmore, D., and Gazzaniga, M. S. (2001). The functional magnetic resonance imaging data center (fMRIDC): The challenges and rewards of large-scale databasing of neuroimaging studies. *Philos. Trans. R. Soc. Lond B. Biol. Sci.* **356,** 1323–1339.

Van Horn, J. D., Woodward, J. B., Simonds, G., Vance, B., Grethe, J. S., Montague, M., Aslam, J. A., Rus, D., Rockmore, D., and Gazzaniga, M. S. (2002). The fMRI data center: Software tools for neuroimaging data management, inspection, and sharing. *In* "A Practical Guide to Neuroscience Databases and Associated Tools" (R. Kotter, Ed.), pp. 221–235. Kluwer, Amsterdam.

van Stolk, R. U. (2002). Familial and inherited colorectal cancer: Endoscopic screening and surveillance. *Gastrointest Endosc. Clin. N. Am.* **12,** 111–133.

Velleman, P., and Hoaglin, D. (1981). The ABC's of EDA: Applications, basics, and computing of exploratory data analysis: Duxbury.

Wong, S. T., and Koslow, S. H. (2001). Human brain program research progress in bioinformatics/neuroinformatics. *J. Am. Med. Inform. Assoc.* **8,** 103–104.

Woo, J. H., and Hathout, G. M. (2001). Systems analysis of functional magnetic resonance imaging data using a physiologic model of venous oxygenation. *J. Cereb. Blood. Flow. Metab.* **21,** 517–528.

Young, M. P., and Scannell, J. W. (2000). Brain structure-function relationships: Advances from neuroinformatics. *Philos. Trans. R. Soc. Lond. B. Biol. Sci.* **355,** 3–6.

Further Reading

Druzgal, T. J., and D'Esposito, M. (2001). Activity in fusiform face area modulated as a function of working memory load. *Brain. Res. Cogn. Brain Res.* **10,** 355–364.

MODELING BRAIN RESPONSES

Karl J. Friston, William Penny, and Olivier David

The Wellcome Department of Cognitive Neurology, University College London
London WC1N 3BG, United Kingdom

Inferences about brain function that are made based on neuroimaging data rest on models of how the data were caused. These models can be quite diverse, ranging from conceptual models of functional anatomy to nonlinear mathematical models of neuronal responses and hemodynamics. The aim of this review is to introduce the key models used in imaging neuroscience and explain how they relate to each other. We start with anatomical models of functional brain architectures, which motivate some of the theoretical foundations of neuroimaging. We then turn to basic statistical models (e.g., the general linear model) used to make classical and Bayesian inferences about *where* neuronal responses are expressed. When biophysical constraints are incorporated, these basic models can be finessed and, in a dynamic setting, rendered causal. This allows us to infer *how* interactions among brain regions are mediated.

We will cover models of brain responses, starting with general linear models of functional magnetic resonance imaging (fMRI) data, used for classical inferences about regionally specific responses. This model is successively refined until we arrive at neuronal mass models of electroencephalograph (EEG) responses.

INTERNATIONAL REVIEW OF
NEUROBIOLOGY, VOL. 66
DOI: 10.1016/S0074-7742(05)66003-5

The latter models allow mechanistic inferences about how evoked responses are caused, at the level of neuronal subpopulations and the coupling among them.

I. Introduction

Neuroscience depends on conceptual, anatomical, statistical, and causal models that link ideas about how the brain works with observed neuronal responses. The aim of this review is to highlight the relationships among the sorts of models that are employed in imaging. We will show how simple statistical models, used to identify *where* evoked brain responses are expressed (c.f. neophrenology) can be elaborated to provide models of *how* neuronal responses are caused (e.g., dynamic causal modeling). These successive elaborations rely, increasingly, on biological mechanisms. We will review a series of models, covering everything from conceptual models that motivate experimental design to detailed biophysical models of coupled neuronal ensembles that enable questions to be asked, at both a physiological and a computational level.

Anatomical models of functional brain architectures motivate the groundwork of neuroimaging. In Section II we start by reviewing the distinction between functional *specialization* and functional *integration* and how these principles together serve as the basis for most models of neuroimaging data. In Section III, we turn to simple statistical models (e.g., the general linear model) used to make classical and Bayesian inferences about functional specialization, in terms of *where* neuronal responses are expressed. Characterization of a region-specific effect rests on estimation and inference. Inferences in neuroimaging can be about differences seen when two groups of subjects are compared or when changes are observed over a sequence of observations from one subject. They may pertain to structural differences (e.g., in voxel-based morphometry [Ashburner and Friston, 2000]) or to neurophysiological indices of brain functions (e.g., fMRI or EEGs). The principles of data analysis are very similar for all these applications. We will focus on the analysis of fMRI time series, because this covers most of the issues encountered in other modalities. By incorporation of biological constraints, simple observation models can be made more realistic and, in a dynamic framework, causal. This section concludes by considering some of the recent advances in biophysical modeling of hemodynamic responses. All the models considered in this section pertain to regional responses. In the final section, we focus on models of distributed responses, where the interactions among cortical areas or neuronal subpopulations are modeled explicitly. This section covers the distinction between *functional connectivity* and *effective connectivity* and reviews dynamic causal modeling of functional integration using fMRI and EEG. We conclude with an example from event-related potential (ERP) research and show how the P300 component (a positive wave at about 300 ms) of an ERP can be explained by changes in coupling among neuronal sources that may underlie perceptual learning.

II. Anatomical Models

A. Functional Specialization and Integration

The brain appears to adhere to two key principles of functional organization, functional specialization and functional integration, which hold that the integration within and among specialized areas is mediated by effective connectivity. The distinction relates to that between *localisationism* and *[dis]connectionism*, two concepts that dominated thinking about cortical function in the nineteenth century. Since the early anatomical theories of Gall, the identification of particular brain regions with specific functions has become a central theme in neuroscience. However, functional localization *per se* was not easy to demonstrate. For example, a meeting that took place in August of 1881 addressed the difficulties of attributing function to a cortical area, given the dependence of cerebral activity on underlying connections (Phillips *et al.*, 1984). This meeting was entitled "Localization of function in the cortex cerebri." Despite accepting the results of electrical stimulation in dog and monkey cortices, Goltz (1881) considered that the excitation method was inconclusive, in that the movements elicited could have originated in related pathways, or the current could have spread to distant centers of the brain. In short, the excitation method could not be used to infer functional localization, because localizationism discounted interactions, or what we now call functional integration among different brain areas. It was proposed that lesion studies could supplement excitation experiments. Ironically, observations on patients with brain lesions some years later (Absher and Benson, 1993) were the basis for the concept of *disconnection syndromes* and for the refutation of localizationism as a complete or sufficient explanation of cortical organization. Functional localization implies that a function can be localized in a cortical area, whereas specialization suggests that a cortical area is specialized for some aspects of perceptual or motor processing and that this specialization is anatomically segregated within the cortex. The cortical infrastructure supporting a single function may, therefore, involve many specialized areas whose union is mediated by the functional integration among them. In this view, functional specialization is only meaningful in the context of functional integration, and vice versa.

B. Functional Specialization and Segregation

The functional role of any component (e.g., cortical area, cortical subarea, or neuronal population) of the brain is defined largely by its connections. Certain patterns of cortical projections are so common that they could amount to rules of cortical connectivity. According to Zeki (1990), "These rules revolve around one, apparently, overriding strategy that the cerebral cortex uses—that of functional segregation." Functional segregation demands that cells with common functional properties be grouped together. This architectural constraint necessitates both convergence and divergence of cortical connections. Extrinsic connections among cortical regions are not continuous but occur in patches or clusters. This patchiness has, in some instances, a clear relationship to functional segregation. For example, visual area V2 has a distinctive cytochrome oxidase architecture consisting of thick stripes, thin stripes, and interstripes. When recordings are made in V2, directionally selective (but not wavelength or color selective) cells are found exclusively in the thick stripes. Retrograde (i.e., backward) labeling of cells in the V5 area is limited to these thick stripes. All of the available physiological evidence suggests that V5 is a functionally homogeneous area that is specialized for visual motion. The evidence of this nature also supports the notion that patchy connectivity is the anatomical infrastructure that mediates functional segregation and specialization. If it is the case that neurons in a given cortical area share a common responsiveness (by virtue of their extrinsic connectivity) to some sensorimotor or cognitive attribute, then this functional segregation is also an anatomical one.

In summary, functional specialization suggests that challenging a subject with an appropriate sensorimotor attribute or cognitive process should lead to activity changes in, and only in, the specialized areas. This is the anatomical and physiological model upon which the search for regionally specific effects is based. We will deal first with models of regionally specific responses and return to models of functional integration later.

III. Statistical Models of Regional Responses

A. Statistical Parametric Mapping

Functional mapping studies are usually analyzed with some form of statistical parametric mapping (Friston *et al.*, 1995a). Statistical parametric mapping entails the construction of spatially extended statistical processes to test hypotheses about regionally specific effects (Friston *et al.*, 1991). Statistical parametric maps (SPMs) are image processes with voxel values that are, under the null hypothesis, distributed according to a known probability density function, usually Student's T- or F-distribution. These are known colloquially as a T-map or an F-map. The success of statistical parametric mapping is due largely to the simplicity of the idea. Namely, one analyzes each and every voxel using any standard (univariate) statistical test. These usually test for activation or for a significant regression on some explanatory variable. The resulting statistical parameters are assembled into an image, the SPM. SPMs are interpreted as statistical processes that are continuous in space (or sometimes time) by referring to the probabilistic behavior of random fields (Adler, 1981; Friston *et al.*, 1991; Worsley *et al.*, 1992, 1996). Random fields model both the univariate probabilistic characteristics of an SPM and any nonstationary spatial covariance structure under the null hypothesis. "Unlikely" excursions of the SPM are interpreted as regionally specific effects attributable to the sensorimotor or cognitive process that has been manipulated experimentally.

Over the years, the term "statistical parametric mapping" has come to refer to the conjoint use of the *general linear model* (GLM) and r*andom field theory* (RFT) to analyze and make classical inferences about spatially extended data through SPMs. The GLM is used to *estimate* some parameters that could explain the spatially continuous data in exactly the same way as would conventional analyses of discrete data. RFT is used to resolve the multiple-comparisons problem that occurs when *inferences* are made over a volume of the brain. RFT provides a method for adjusting *p* values for the search volume of an SPM to control false-positive rates. RFT plays the same role for *continuous* data (i.e., images or time series) as the Bonferonni correction plays for a family of discontinuous or *discrete* statistical tests.

Later we will consider the Bayesian alternative to classical inference with SPMs. This rests on conditional inferences about an effect, given the data, as opposed to classical inferences about the data, given a zero effect. Bayesian inferences about effects that are continuous in space use posterior probability maps (PPMs). Although less established than SPMs, PPMs are potentially very useful, not least because they do not have to contend with the multiple-comparisons problem induced by classical inference (Berry and Hochberg, 1999). In contrast with SPM, this means that inferences about a given regional response do not depend on inferences about responses elsewhere. Before looking at the models underlying Bayesian inference, we first consider estimation and classical inference in the context of the GLM.

B. The GLM

Statistical analysis of imaging data essentially involves (1) modelling the data to partition observed neurophysiological responses into components of interest, confounds, and error, and (2) making inferences about interesting effects, using the variances of the partitions. A brief review of the literature may give the impression that there are numerous ways to analyze PET and fMRI time series, with a diversity of statistical and conceptual approaches. This is not the case. With very few exceptions, every analysis is a variant of the GLM. These include (1) simple T-tests on scans assigned to one condition or another; (2) correlation coefficients between observed responses and boxcar stimulus functions in fMRI (Bandettini *et al.*, 1993); (3) inferences made using multiple linear regression; (4) evoked responses estimated using linear time-invariant models; and (5) selective averaging to estimate event-related responses. Mathematically, they are all identical and can be implemented with the same equations and algorithms. The only thing that distinguishes one from another is the *design matrix* encoding the experimental design.

The GLM is an equation

$$y = X\beta + \varepsilon \qquad (1)$$

that expresses the observed response y in terms of a linear combination of explanatory variables in the matrix X plus a well-behaved error term. The GLM is variously known as "analysis of [co]variance" or "multiple regression" and subsumes simpler variants, like the T-test for a difference in means, to more elaborate linear convolution models such as finite impulse-response (FIR) models. The matrix X that contains the explanatory variables (e.g., designed effects or confounds) is called the design matrix. Each column of the design matrix corresponds to some effect that one has built into the experiment or that may confound the results. These are referred to as explanatory variables, covariates, or regressors. Sometimes the design matrix contains covariates or indicator variables that take values of 0 or 1, to indicate the presence of a particular level of an experimental factor (c.f. analysis of variance [ANOVA]). The example in Fig. 1 relates to an fMRI study of visual stimulation under four conditions. The effects on the response variable are modeled in terms of functions of the presence of these conditions (i.e., box or stick functions smoothed with components of a hemodynamic response function). Note that this is more complicated than a simple ANOVA, because the design matrix is modeling a time series, as opposed to discrete observations. The relative contribution of each of these columns to the response is controlled by the parameters. These are estimated using standard least squares. Inferences about the parameter estimates are made using T or F statistics, depending upon whether one is looking at one or more than one linear combination of them.

In simple analyses, the design matrix contains indicator variables or parametric variables encoding the experimental manipulations. These are formally identical to classical ANOVA or multiple linear regression models. However, when the observations correspond to time series, convolution models are often used: An important instance of the GLM, from the perspective of fMRI, is the

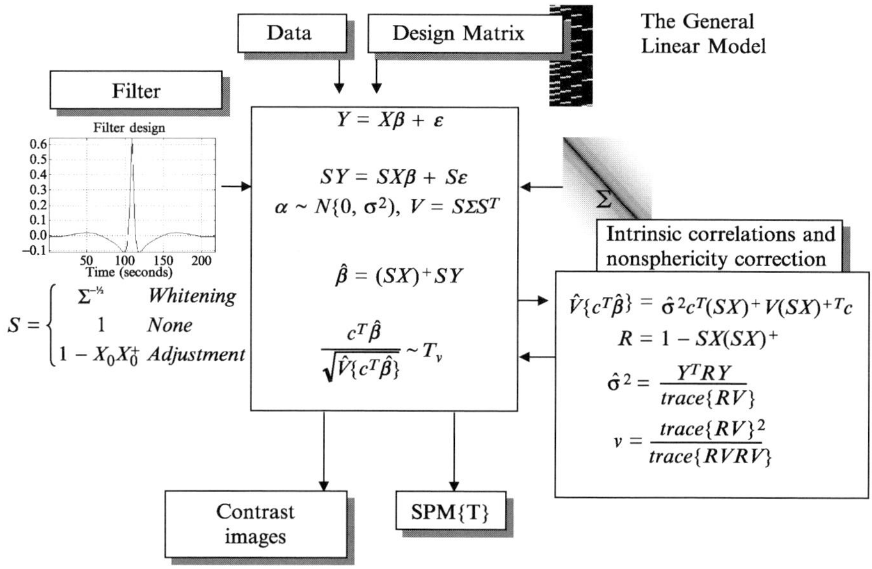

FIG. 1. The general linear model (GLM) is an equation expressing the response variable Y in terms of a linear combination of explanatory variables in a design matrix X and an error term with assumed or known autocorrelation Σ. In fMRI, the data can be filtered with a convolution or residual forming matrix (or a combination) S, leading to a generalized linear model that includes (intrinsic) serial correlations and applied (extrinsic) filtering. Different choices of S correspond to different estimation schemes, as indicated on the upper left. The parameter estimates obtain in a least-squares sense using the pseudoinverse (denoted by $+$) of the filtered design matrix. An effect of interest is specified by a vector of contrast weights c that give a weighted sum or compound of parameter estimates, referred to as a *contrast*. The T statistic is simply this contrast divided by the standard error (i.e.,the square root of its estimated variance). The ensuing T statistic is distributed with v degrees of freedom. The equations for estimating the variance of the contrast and the degrees of freedom are provided in the righthand panel and accommodate the nonsphericity implied by Σ.

linear time invariant (LTI) convolution model. Mathematically, this is no different from any other GLM. However, it explicitly treats the data sequence as an ordered time series and allows for the use of a signal-processing perspective that can be useful.

Each column of X has an associated but unknown parameter. Some of these parameters will be of interest (e.g., the effect of a sensorimotor or cognitive condition, or the regression coefficient of hemodynamic responses on reaction time). The remaining parameters will be of no interest and pertain to nuisance or confounding effects (e.g., the effect of being a particular subject or the regression slope of regional activity on global activity). The statistical test is directed to interesting effects by specifying the null hypothesis with a *contrast*. A contrast is simply a linear mixture of parameter estimates. The T statistic allows one to test the null hypothesis that some contrast (e.g., a subtraction) of the estimates is zero. The T statistic is obtained by division of the contrast (specified by contrast weights) of the parameter estimates by its standard error. Sometimes, several contrasts are tested jointly, for example, during the use of polynomial (Büchel *et al.*, 1996) or basis function expansions of some experimental factor. In these instances, the SPM{F} (i.e., an SPM of the F-statistic) is used and is specified with a matrix of contrast weights that can be thought of as a collection of "T contrasts" that one wants to test *en masse*.

After the statistic is computed, RFT is used to assign adjusted *p*-values to topological features of the SPM, such as the height of peaks or the spatial extent of regions above a threshold. This *p*-value is a function of the search volume and smoothness of the residuals (see Fig. 1). The intuition behind RFT is that it allows one to control the false-positive rate of peaks or "blobs" corresponding to regional effects. A Bonferonni correction would control the false-positive rate of voxels, which is inexact and unnecessarily severe. The *p*-value is the probability of getting a peak in the SPM, or higher, by chance over the search volume. If sufficiently small (usually less than 0.05), the regional effect can be declared significant.

The equations summarized in Fig. 1 can be used to implement a vast range of statistical analyses. The issue is therefore not so much the mathematics but the formulation of a design matrix X appropriate to the study design and inferences that are sought. Before considering GLMs as biophysical or causal models of brain responses, we will focus on the design matrix as a device to specify experimental design. Here, the explanatory variables encode treatment effects that we assume are expressed in a linear and instantaneous fashion in the data, without reference to any particular mechanism.

C. Experimental Design

This section considers the different sorts of designs employed in neuroimaging studies. Experimental designs can be classified as *single-factor* or *multifactorial*, and within that classification the levels of each factor can be *categorical* or *parametric*.

1. *Categorical Designs, Cognitive Subtraction, and Conjunctions*

The tenet of cognitive subtraction is that the difference between two tasks can be formulated as a separable cognitive or sensorimotor component. Regionally specific differences in the responses evoked by the two tasks identify the corresponding functionally specialized area. Early applications of subtraction range from the functional anatomy of word processing (Petersen *et al.*, 1989) to functional specialization in extrastriate cortex (Lueck *et al.*, 1989). The latter studies involved presenting visual stimuli with and without some sensory attribute (e.g., color, motion, etc.). The areas highlighted by subtraction were identified with homologous areas in monkeys that showed selective electrophysiological responses to equivalent visual stimuli.

Cognitive conjunctions (Price and Friston, 1997) can be thought of as an extension of the subtraction technique, in the sense that they combine a series of subtractions. In subtraction, one tests a single hypothesis pertaining to the activation in one task relative to another. In conjunction analyses, several contrasts are tested, asking whether all the activations, in a series of task pairs, are expressed conjointly. Consider the problem of identifying regionally specific activations due to a particular cognitive component (e.g., object recognition). If one can identify a series of task pairs whose differences have only that component in common, then the region that activates, in all the corresponding subtractions, can be associated with the common component. In short, conjunction analysis allows one to disclose context-invariant regional responses.

2. *Parametric Designs*

The premise behind parametric designs is that regional physiology will vary systematically with the degree of cognitive or sensorimotor processing, or deficits thereof. Examples of this approach include the PET experiments of Grafton *et al.* (1992), which demonstrated significant correlations between hemodynamic responses and the performance of a visually guided motor tracking task. In sensory studies, Price *et al.* (1992) demonstrated a remarkable linear relationship between perfusion in periauditory regions and frequency of aural word presentation. This correlation was not observed in Wernicke's area, where perfusion appeared to correlate not with the discriminative attributes of the stimulus but with the presence or absence of semantic content. These relationships or *neurometric functions* can be linear or nonlinear. Using polynomial regression, in the context of the GLM, one can identify nonlinear relationships between stimulus parameters (e.g., stimulus duration or presentation rate) and evoked responses. To do this, one usually uses a SPM{F} (Büchel *et al.*, 1996).

The example provided in Fig. 2 illustrates both the categorical and the parametric aspect of design and analysis. These data were obtained from an fMRI study of visual motion processing using radially moving dots. The stimuli were presented over a range of speeds using *isoluminant* and *isochromatic* stimuli. To identify areas involved in visual motion, a stationary-dots condition was subtracted from a moving-dots condition (see the contrast weights on the upper right of the figure). To ensure significant motion-sensitive responses, using both color and luminance cues, a conjunction of the equivalent subtractions was assessed under both viewing contexts. Areas V5 and V3a are seen in the ensuing SPM{T}. The T values in this SPM are simply the minimum of the T values for each

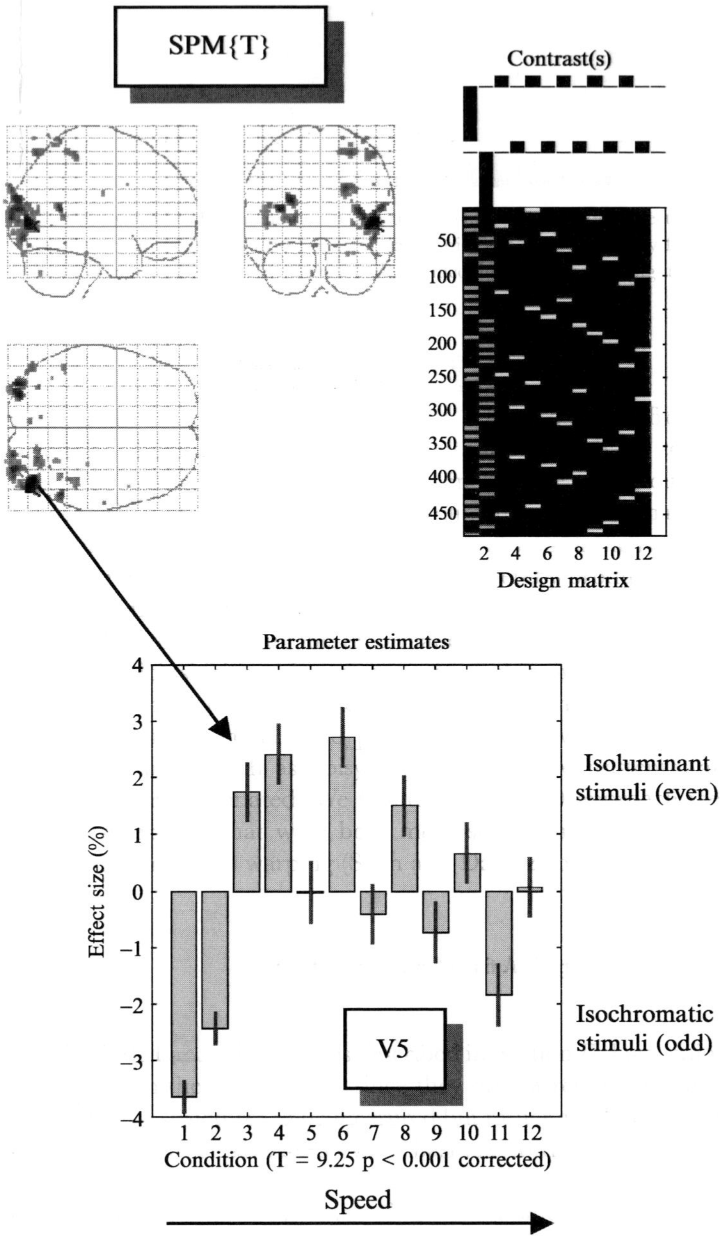

FIG. 2. Top right: An image representation of the design matrix. Contrasts: These are the vectors of contrast weights defining the linear compounds of parameters tested. The contrast weights are displayed over the column of the design matrix encoding the effects tested. The design matrix here

subtraction. By thresholding this SPM, we ensure that all voxels survive a threshold in each subtraction separately. This *conjunction* SPM has an equivalent interpretation; it represents the intersection of the excursion sets, defined by the threshold of each *component* SPM. This intersection is the essence of a conjunction.

The responses in the left V5 are shown in the lower panel of Fig. 2 and speak to a compelling inverted-U relationship between speed and evoked response that peaks at around six degrees per second. It is this sort of relationship that parametric designs try to characterize. Interestingly, the form of these speed-dependent responses was similar in both stimulus types, although luminance cues are seen to elicit a greater response. From the point of view of a factorial design, there is a *main effect* of cue (isoluminant vs. isochromatic) and there is a main (nonlinear) effect of speed, but there is no speed by cue *interaction.*

3. *Multifactorial Designs*

Factorial designs are more prevalent than single-factor designs because they enable inferences about interactions. At its simplest, an interaction represents a change in a change. Interactions are associated with factorial designs where two or more factors are combined in the same experiment. The effect of one factor on the effect of the other is assessed by the interaction term. Factorial designs have a wide range of applications. An early application in neuroimaging examined physiological adaptation and plasticity during motor performance, by assessing time by condition interactions (Friston *et al.*, 1992). Factorial designs have an important role in the context of cognitive subtraction and additive factor logic by virtue of being able to test for interactions or context-sensitive activations (i.e., to demonstrate the fallacy of "pure insertion" [see Friston *et al.*, 1996a]). These interaction effects can sometimes be interpreted as (1) the integration of two or more (cognitive) processes or (2) the modulation of one (perceptual) process by another.

In summary, the design matrix encodes the causes of observed data and, in particular, the treatment effects caused by changes in the level of various experimental factors. These factors can have categorical or parametric levels, and most experiments nowadays use multiple factors to test for both main effects and interactions. Before turning to mechanistically more-informed formulations of

---

includes condition-specific effects (boxcars convolved with a hemodynamic response function). Odd columns correspond to stimuli shown under isochromatic conditions, and even columns model responses to isoluminant stimuli. The first two columns are for stationary stimuli and the remaining columns are for stimuli of increasing speed. The final column is a constant term. Top left: Statistical parametric map (SPM){T}. This is a maximum-intensity projection of the SPM{T}, conforming to the standard anatomical space of Talairach and Tournoux (1988). The T values here are the minimum T values from both contrasts, thresholded at $p = 0.001$ uncorrected. The most significant conjunction is seen in left V5. Lower panel: Plot of the condition-specific parameter estimates for this voxel. The T value was 9.25 ($p < 0.001$, adjusted according to random field theory).

the GLM, we will consider briefly the two sorts of inferences that can be made about the parameter estimates.

D. Classical and Bayesian Inference

To date, inference in neuroimaging has been restricted largely to classical inferences based on statistical parametric maps. The statistics that comprise these SPMs are essentially functions of the data. The probability distribution of the chosen statistic, under the null hypothesis (i.e., the null distribution) is used to compute a *p*-value. This *p*-value is the probability of obtaining the statistic, or the data, given that the null hypothesis is true. If the *p*-value is sufficiently small, the null hypothesis can be rejected and an inference made. The alternative approach is to use Bayesian or conditional inference based upon the posterior distribution of the activation given the data (Holmes and Ford, 1993). This necessitates the specification of priors (i.e., the probability distribution of the activation). Bayesian inference requires a posterior distribution and therefore rests upon a posterior-density analysis. A useful way to summarize this posterior density is to compute the probability that the activation exceeds some threshold. This represents a Bayesian inference about the effect, in relation to the specified threshold. By computing posterior probability for each voxel, we can construct PPMs that are useful complements to classical SPMs.

The motivation for using conditional or Bayesian inference is that it has high face validity. This is because the inference is about an effect, or activation, greater than some specified size, that has some meaning in relation to underlying neurophysiology. This contrasts with classical inference, in which the inference is about the effect being significantly different from zero. The problem for classical inference is that trivial departures from the null hypothesis can be declared significant if there is sufficient data or sensitivity. From the point of view of neuroimaging, posterior inference is especially useful because it eschews the multiple-comparisons problem. In classical inference, one tries to ensure that the probability of rejecting the null hypothesis incorrectly is kept small, despite the fact that inferences are made over large volumes of the brain. This induces a multiple-comparisons problem that, for spatially continuous data, requires an adjustment or correction to the *p*-value, using RFT (as mentioned previously). This correction means that classical inference becomes less sensitive or powerful with large search volumes. In contrast, posterior inference does not have to contend with the multiple-comparisons problem because there are no false positives. The probability that activation has occurred at any particular voxel, given the data, is the same, irrespective of whether one has analyzed only that voxel or the entire brain. For this reason, posterior inference using PPMs represents a more powerful approach than classical inference in neuroimaging.

1. *Hierarchical Models and Empirical Bayes*

PPMs require the posterior distribution or conditional distribution of the activation (a contrast of conditional parameter estimates) given the data. This posterior density can be computed, under Gaussian assumptions, using Bayes rule. Bayes rule requires the specification of a likelihood function and the prior density of the model's parameters. The models used to form PPMs and the likelihood functions are exactly the same as in classical SPM analyses (namely the GLM). The only extra bit of information that is required is the prior probability distributions of the parameters. Although it would be possible to specify these using independent data or some plausible physiological constraints, there is an alternative to this fully Bayesian approach. The alternative is *empirical Bayes*, in which the prior distributions are estimated from the data. The empirical Bayes method requires a *hierarchical observation model*, in which the parameters and hyperparameters at any particular level can be treated as priors on the level below. There are numerous examples of hierarchical observation models in neuroimaging. For example, the distinction between fixed- and mixed-effects analyses of multisubject studies relies upon a two-level hierarchical model. However, in neuroimaging there is a natural hierarchical observation model that is common to all brain-mapping experiments. This is the hierarchy induced by the search for the same effects at every voxel within the brain (or grey matter). The first level of the hierarchy corresponds to the experimental effects at any particular voxel, and the second level comprises the effects over voxels. Put simply, the variation in a contrast, over voxels, can be used as the prior variance of that contrast at any particular voxel. Hierarchical linear models have the following form:

$$
\begin{aligned}
y &= X^{(1)}\beta^{(1)} + \varepsilon^{(1)} \\
\beta^{(1)} &= X^{(2)}\beta^{(2)} + \varepsilon^{(2)} \\
\beta^{(2)} &= \ldots
\end{aligned}
\tag{2}
$$

This is exactly the same as Eq. 1, but now the parameters of the first level are generated by a supraordinate linear model, and so on to any hierarchical depth required. These hierarchical observation models are an important extension of the GLM and are usually estimated using expectation maximization (EM) (Dempster *et al.*, 1977). In the present context, the response variables comprise the responses at all voxels, and all $\beta^{(1)}$ are the treatment effects we want to make an inference about. Because we have invoked a second level, the first-level parameters embody random effects and are generated by a second-level linear model. At the second level, $\beta^{(2)}$ is the average effect over voxels and $\varepsilon^{(2)}$ its voxel-to-voxel variation. By estimating the variance of $\varepsilon^{(2)}$, one is implicitly estimating an empirical prior on the first-level parameters at each voxel. This prior can then be used to estimate the posterior probability of $\beta^{(1)}$ being greater than some threshold at each voxel. An example of the ensuing PPM is provided in Fig. 3, along with the classical SPM.

In this section we have seen how the GLM can be used to test hypotheses about brain responses and how, in a hierarchical form, it enables empirical Bayesian or conditional inference. In the next section we will deal with dynamic systems and show how they can be formulated as GLMs. These dynamic models take us closer to an understanding of how brain responses are actually caused by experimental manipulations, and they represent the next step toward causal models of brain responses.

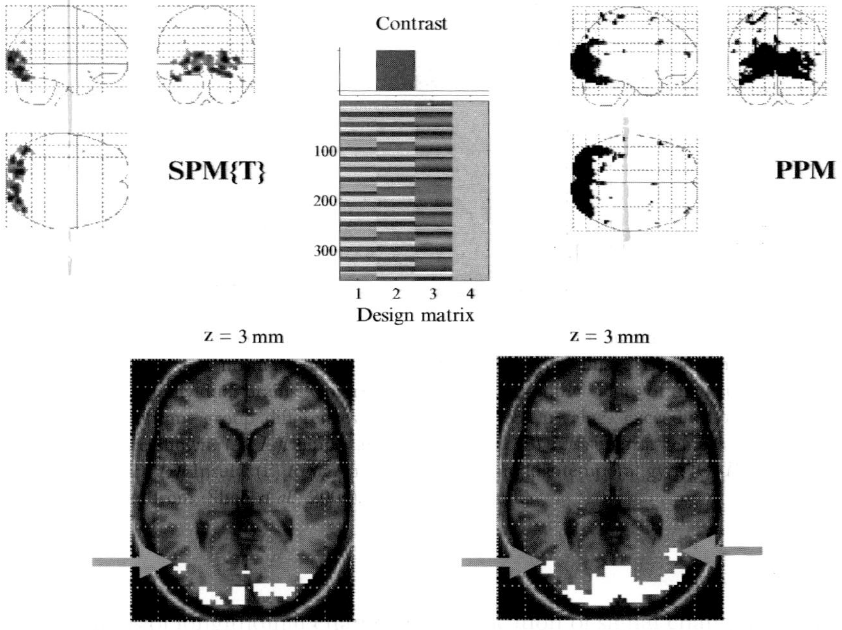

FIG. 3. Statistical parametric map (SPM) and posterior probability map (PPM) for an fMRI study of attention to visual motion. The display format in the lower panel uses an axial slice through extrastriate regions, but the thresholds are the same as employed in maximum-intensity projections (upper panels). Upper right: The activation threshold for the PPM was 0.7 a.u., meaning that all voxels shown had a 90% chance of an activation of 0.7% or more. Upper left: The corresponding SPM using an adjusted threshold at $p = 0.05$. Note the bilateral foci of motion-related responses in the PPM that are not seen in the SPM (grey arrows). As can be imputed from the design matrix (upper middle panel), the statistical model of evoked responses comprised boxcar regressors convolved with a canonical hemodynamic response function. The middle column corresponds to the presentation of moving dots and was the stimulus property tested by the contrast.

E. Dynamic Models

1. *Convolution Models and Temporal Basis Functions*

Friston *et al.* (1994) estimated the form of the hemodynamic impulse-response function (HRF) using a least-squares deconvolution and a time-invariant model, where evoked neuronal responses are *convolved* or smoothed with an HRF to give the measured hemodynamic response (see also Boynton *et al.*, 1996). This simple linear convolution model is the cornerstone of statistical inferences about activations in fMRI with the GLM. An impulse-response function is the response to a single impulse measured at a series of times after the input. It characterizes the input–output behavior of the system (i.e., a voxel) and places important constraints on the sorts of inputs that will excite a response.

Knowing the forms that the HRF can take is important for several reasons, not least because it allows for better statistical models of the data. The HRF may vary from voxel to voxel, and this has to be accommodated in the GLM. To allow for different HRFs in different brain regions, the notion of temporal basis functions to model evoked responses in fMRI was introduced (Friston *et al.*, 1995b) and applied to event-related responses by Josephs *et al.* (1997) (see also Lange and Zeger, 1997). The basic idea behind temporal basis functions is that the hemodynamic response induced by any given trial type can be expressed as the linear combination of several (basis) functions of peristimulus time. The convolution model for fMRI responses takes a stimulus function encoding the neuronal responses and convolves it with an HRF to give a regressor that enters the design matrix. When using basis functions, the stimulus function is convolved with all the basis functions to give a series of regressors (in Fig. 1 we used four stimulus functions and two basis functions to give eight regressors). Mathematically, we can express this model as

$$\begin{array}{rcl} y(t) = X\beta + \varepsilon & & y(t) = u(t) \otimes h(t) \\ X_i = T_i(t) \otimes u(t) & \Leftrightarrow & h(t) = \beta_1 T_1(t) + \beta_2 T_2(t) + \ldots \end{array} \qquad (3)$$

where $\otimes$ means convolution. This equivalence illustrates how temporal basis functions allow one to take any convolution model (right) and convert it into a GLM (left). The parameter estimates β_i are the coefficients or weights that determine the mixture of basis functions of time $T_i(t)$ that best models $h(t)$, the HRF for the trial type and voxel in question. We find the most useful basis set to be a canonical HRF along with its derivatives with respect to the key parameters that determine its form. Temporal basis functions are important because they enable a graceful transition between conventional multilinear regression models with one stimulus function per condition and FIR models with a parameter for each time point following the onset of a condition or trial type. Figure 4 illustrates this graphically. In short, temporal basis functions offer useful constraints on the

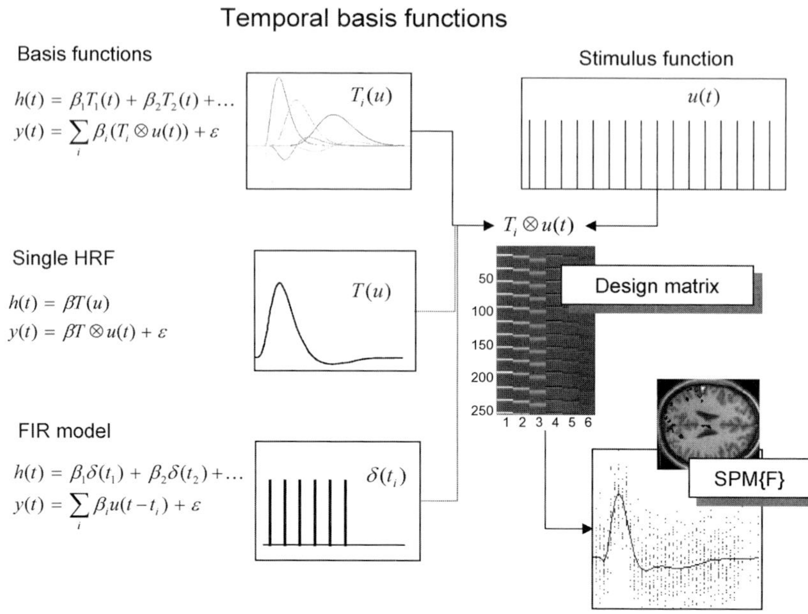

FIG. 4. Temporal basis functions offer useful constraints on the form of the estimated response that retain the flexibility of finite impulse-response (FIR) models and the efficiency of single-regressor models. The specification of these constrained FIR models involves setting up stimulus functions $u(t)$ that model expected neuronal changes (e.g., boxcars of epoch-related responses or spikes [delta functions] at the onset of specific events or trials). These stimulus functions are then convolved with a set of basis functions $T_i(t)$ of peristimulus time that, in some linear combination, model the hemodynamic impulse-response function (HRF). The ensuing regressors are assembled into the design matrix. The basis functions can be as simple as a single canonical HRF (middle) or as complicated as a series of delayed delta functions (bottom). The latter case corresponds to a FIR model, and the coefficients constitute estimates of the impulse-response function at a finite number of discrete sampling times. Selective averaging in event-related fMRI (Dale and Buckner 1997) is mathematically equivalent to this limiting case. (See Color Insert.)

form of the estimated response that retain the flexibility of FIR models and the efficiency of single-regressor models.

F. Biophysical Models

1. *Input-State-Output Systems*

By adopting a convolution model for brain responses in fMRI, we are implicitly positing some underlying dynamic system that converts neuronal responses into observed hemodynamic responses. Our understanding of the biophysical and physiological mechanisms that underpin the HRF has grown considerably in the past few years (Buxton and Frank, 1997; Hoge *et al.*, 1999; Mandeville *et al.*, 1999). Figure 5 shows some simulations based on the hemodynamic model described by Friston *et al.* (2000). Here, neuronal activity induces some autoregulated vasoactive signal that causes transient increases in regional cerebral blood flow (rCBF). The resulting flow increases dilate a venous balloon, increasing its volume (v) and diluting venous blood to decrease deoxyhemoglobin content (q). The blood oxygenation level–dependent (BOLD) signal is roughly proportional to the concentration of deoxyhemoglobin (q/v) and follows the rCBF response after a delay of about a second. The model is framed in terms of differential equations, examples of which are provided in Fig. 5.

Notice that we have introduced variables, like volume and deoxyhemoglobin concentrations, that are not actually observed. These are referred to as the *hidden states* of input-state-output models. The state and output equations of any analytic dynamical system are as follows:

$$\begin{aligned} \dot{x}(t) &= f(x, u, \theta) \\ y(t) &= g(x, u, \theta) + \varepsilon \end{aligned} \tag{4}$$

The first line is an ordinary differential equation and expresses the rate of change of the states as a parameterized function of the states and inputs. Typically the inputs $u(t)$ correspond to designed experimental effects (e.g., the stimulus function in fMRI). There is a fundamental and causal relationship (Fliess *et al.*, 1983) between the outputs and the history of the inputs in Eq. 4. This relationship conforms to a Volterra series, which expresses the output $y(t)$ as a generalized convolution of the input $u(t)$, critically without reference to the hidden states $x(t)$. This series is simply a functional Taylor expansion of the outputs with respect to the inputs (Bendat, 1990). The reason it is a functional expansion is that the inputs are a function of time.

$$\begin{aligned} y(t) &= \sum_i \int_0^t \ldots \int_0^t \kappa_i(\sigma_1,\ldots,\sigma_i) u(t-\sigma_1),\ldots,u(t-\sigma_i) d\sigma_1,\ldots,d\sigma_i \\ \kappa_i(\sigma_1,\ldots,\sigma_i) &= \frac{\partial^i y(t)}{\partial u(t-\sigma_1),\ldots,\partial u(t-\sigma_i)} \end{aligned} \tag{5}$$

where $\kappa_i(\sigma_1,\ldots,\sigma_i)$ is the ith-order kernel. In Eq. 5, the integrals are restricted to the past. This renders Eq. 5, causal. The key thing here is that Eq. 5, is simply a convolution and can be expressed as a GLM as in Eq. 3. This means that we can take a neurophysiologically realistic model of hemodynamic responses and use it as an observation model to estimate parameters using observed data. Here, the model is parameterized in terms of kernels that have a direct analytic relation to the original parameters θ of the biophysical system (through Eq. 5). The

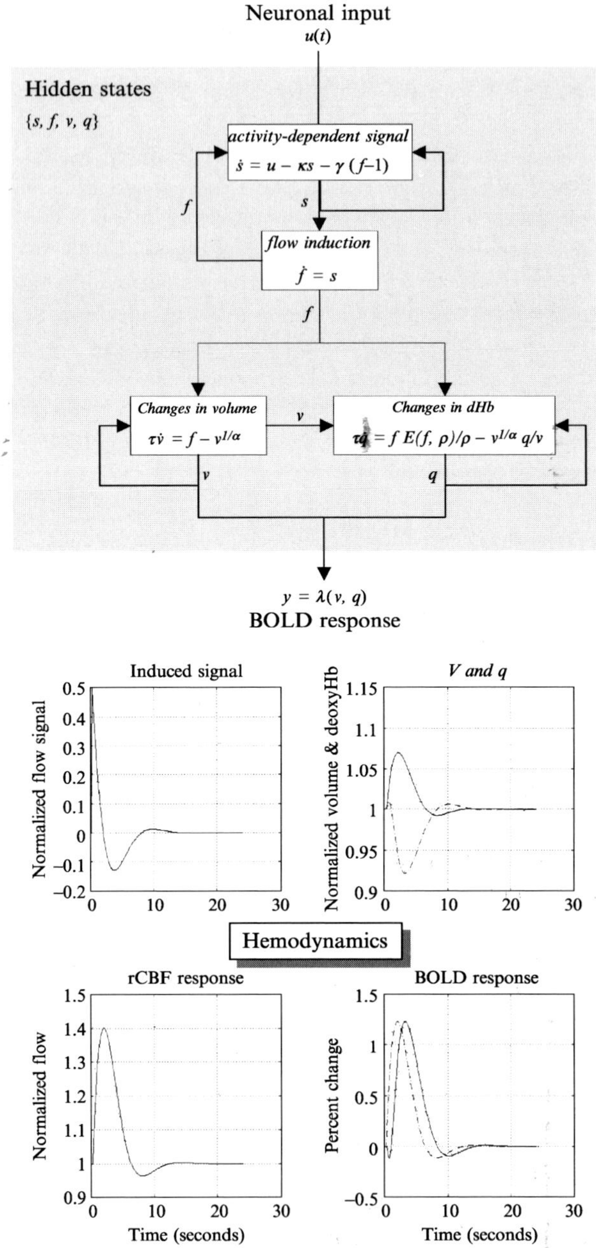

FIG. 5. Right: Hemodynamics elicited by an impulse of neuronal activity as predicted by a dynamical biophysical model (left side). A burst of neuronal activity causes an increase in flow-inducing signal that decays with first-order kinetics and is downregulated by local flow. This

first-order kernel is simply the conventional HRF. High-order kernels correspond to high-order HRFs and can be estimated using basis functions as described previously. In fact, by choosing basis functions according to the

$$A(\sigma)_i = \frac{\partial \kappa(\sigma)_1}{\partial \theta_i} \tag{6}$$

one can estimate the biophysical parameters because, to a first-order approximation, $\beta_i = \theta_i$. The critical step we have taken here is to start with a causal dynamic model of how responses are generated and construct a general linear observation model that allows us to estimate and infer things about the parameters of that model. This contrasts with the conventional use of the GLM with design matrices that are not informed by a forward model of how data are caused. This approach to modeling brain responses has a much more direct connection with underlying physiology and rests upon an understanding of the underlying system.

2. *Nonlinear System Identification*

Once a suitable causal model has been established (e.g., Fig. 5), we can estimate second-order kernels. These kernels represent a nonlinear characterization of the HRF that can model interactions among stimuli in causation of responses. One important manifestation of the nonlinear effects, captured by the second-order kernels, is a modulation of stimulus-specific responses by preceding stimuli that are proximate in time. This means that responses at high stimulus presentation rates saturate and, in some instances, show an "inverted U" behavior. This behavior appears to be specific to BOLD effects (as opposed to evoked changes in cerebral blood flow) and may represent a *hemodynamic refractoriness*. This effect has important implications for event-related fMRI, where one may want to present trials in quick succession.

The results of a typical nonlinear analysis are given in Fig. 6. The results in the right panel represent the average response, integrated over a 32-second train of stimuli as a function of stimulus-onset asynchrony (SOA). These responses are based on the kernel estimates (lefthand panels) using data from a voxel in the left-posterior temporal region of a subject obtained during the presentation of single words at different rates. The solid line represents the estimated response and shows a clear maximum at just less than one second. The dots are responses based on empirical data from the same experiment. The broken line shows the

---

signal increases rCBF, which dilates the venous capillaries, increasing volume (v). Concurrently, venous blood is expelled from the venous pool decreasing deoxyhemoglobin content (q). The resulting fall in deoxyhemoglobin concentration leads to transient increases in blood oxygenation level–dependent (BOLD) signal and a subsequent undershoot. Left: The hemodynamic model on which these simulations were based (Friston *et al.*, 2000).

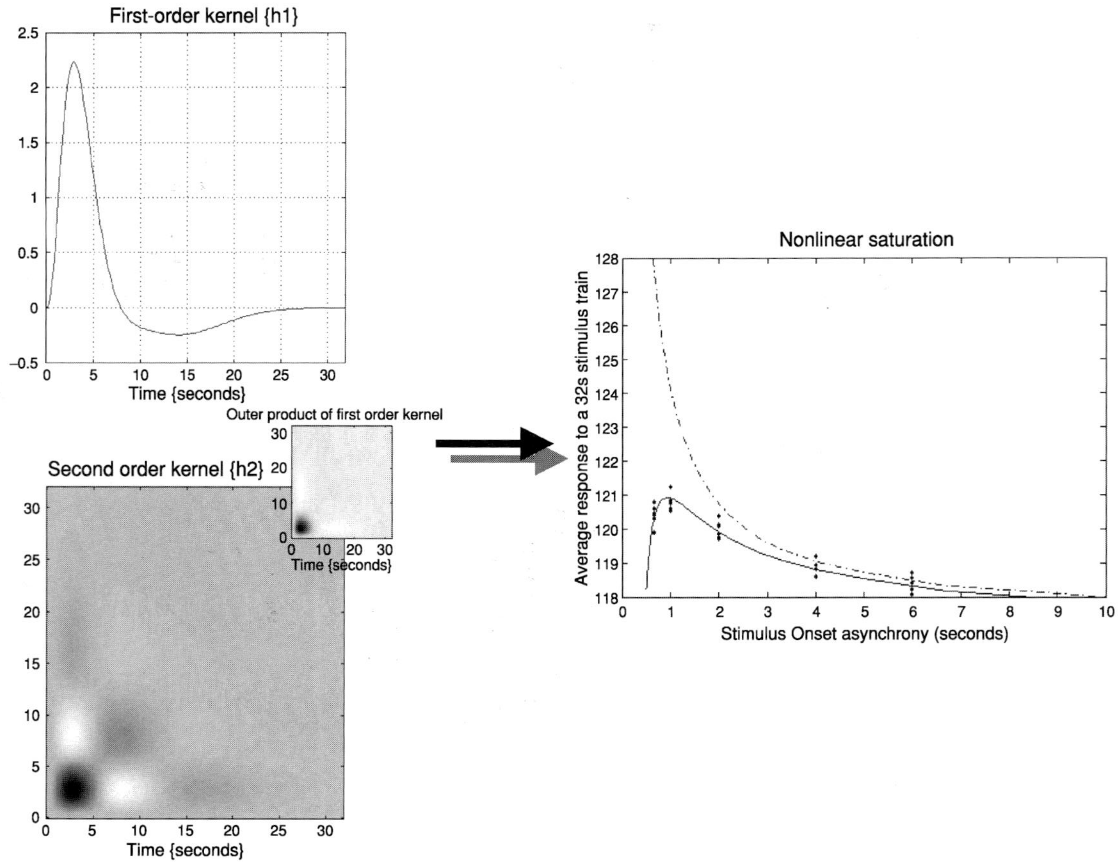
First-order kernel {h1}
Time {seconds}
Outer product of first order kernel
Time {seconds}
Second order kernel {h2}
Time {seconds}
Nonlinear saturation
Average response to a 32s stimulus train
Stimulus Onset asynchrony (seconds)

expected response in the absence of nonlinear effects (i.e., that predicted by setting the second-order kernel to zero). It is clear that nonlinearities become important at around two seconds, leading to an actual diminishment of the integrated response at subsecond SOAs. The implication of this sort of result is that the assumptions of the linear convolution models discussed above are violated with subsecond SOAs (Buckner *et al.*, 1996; Burock *et al.*, 1998).

In summary, we started with models of regionally specific responses, framed in terms of the GLM, in which responses were modeled as linear mixtures of designed changes in explanatory variables. Hierarchical extensions to linear observation models enable random-effects analyses and, in particular, an empirical Bayesian approach. The mechanistic utility of these models is realized though the use of forward models that embody causal dynamics. Some simple variants of these are the linear convolution models used to construct explanatory variables in conventional analyses of fMRI data. These are a special type of generalized convolution models that are mathematically equivalent to input-state-output systems that comprise hidden states. Estimation and inference with these dynamic models tells us something about *how* the response was caused, but only at the level of a single voxel. The next section retains the same perspective on models, but in the context of distributed responses and functional integration.

IV. Models of Functional Integration

A. Functional and Effective Connectivity

Imaging neuroscience has firmly established functional specialization as a principle of brain organization in the human. The integration of specialized areas has proven more difficult to assess. Functional integration is usually inferred on

FIG. 6. Left panels: Volterra kernels from a voxel in the left superior temporal gyrus at −56, −28, and 12 mm. These kernel estimates were based on a single-subject study of aural word presentation at different rates (from 0 to 90 words per minute) using a second-order approximation to a Volterra series expansion modeling the observed hemodynamic response to stimulus input (a delta function for each word). These kernels can be thought of as a characterization of the second-order hemodynamic response function. The first-order kernel (upper panel) represents the (first-order) component usually presented in linear analyses. The second-order kernel (lower panel) is presented in image format. The color scale is arbitrary; white is positive and black is negative. The insert on the right represents $\kappa_1\kappa_1^T$, the second-order kernel predicted by a simple model that involves a linear convolution with κ_1 followed by some static nonlinearity. Right panel: Integrated responses over a 32-second stimulus train as a function of stimulus-onset asynchrony (SOA). Solid line: Estimates based on the nonlinear convolution model parameterized by the kernels on the left. The broken line represents the responses expected in the absence of second-order effects (i.e., in a truly linear system). The dots represent empirical averages based on the presentation of actual stimulus trains.

the basis of correlations among measurements of neuronal activity. The concept of functional connectivity has been defined as statistical dependencies or correlations *among remote neurophysiological events*. However, correlations can arise in a variety of ways. For example, in multiunit electrode recordings, correlations can result from stimulus-locked transients evoked by a common input, or they can reflect stimulus-induced oscillations mediated by synaptic connections (Gerstein and Perkel, 1969). Integration within a distributed system is usually better understood in terms of effective connectivity: Effective connectivity refers explicitly to *the influence that one neural system exerts over another*, at either a synaptic (i.e., synaptic efficacy) or a population level. It has been proposed that "the [electrophysiological] notion of effective connectivity should be understood as the experiment- and time-dependent, simplest possible circuit diagram that would replicate the observed timing relationships between the recorded neurons" (Aertsen and Preißl, 1991). This speaks to two important points: (1) Effective connectivity is dynamic, i.e., dependent on activity and time; and (2) it depends upon a model of the interactions. The estimation procedures employed in functional neuroimaging can be divided into those based on (1) linear regression models (e.g., McIntosh and Gonzalez-Lima, 1994) or (2) nonlinear dynamic causal models.

There is a necessary link between functional integration and multivariate analyses because the latter are necessary in order to model interactions among brain regions. Multivariate approaches can be divided into those that are inferential in nature and those that are data-led or exploratory. We will first consider multivariate approaches that are generally based on functional connectivity or covariance patterns (which are generally exploratory) and then turn to models of effective connectivity (which allow for some form of inference).

1. *Eigenimage Analysis and Related Approaches*

Friston *et al.* (1993) introduced voxel-based principal component analysis (PCA) of neuroimaging time series to characterize distributed brain systems implicated in sensorimotor, perceptual, or cognitive processes. These distributed systems are identified with principal components or *eigenimages* that correspond to spatial modes of coherent brain activity. This approach represents one of the simplest multivariate characterizations of functional neuroimaging time series and falls into the class of exploratory analyses. Principal component or eigenimage analysis generally uses singular value decomposition (SVD) to identify a set of orthogonal spatial modes that capture the greatest amount of variance expressed over time. As such, the ensuing modes embody the most prominent aspects of the variance–covariance structure of a given time series. Noting that covariance among brain regions is equivalent to functional connectivity renders eigenimage analysis particularly interesting, because it was among the first ways of addressing functional integration (i.e., connectivity) with neuroimaging data.

Subsequently, eigenimage analysis has been elaborated in a number of ways. Notable among these are canonical variate analysis (CVA) and multidimensional scaling (Friston *et al.*, 1996b,c). CVA was introduced in the context of MANCOVA (multiple analysis of covariance) and uses the generalized eigenvector solution to maximize the variance that can be explained by some explanatory variables relative to error. CVA can be thought of as an extension of eigenimage analysis that refers explicitly to some explanatory variables and allows for statistical inference.

In fMRI, eigenimage analysis (e.g., Sychra *et al.*, 1994) is generally used as an exploratory device to characterize coherent brain activity. These variance components may or may not be related to experimental design. For example, endogenous coherent dynamics have been observed in the motor system at very low frequencies (Biswal *et al.*, 1995). Despite its exploratory power, eigenimage analysis is fundamentally limited for two reasons. First, it offers only a linear decomposition of any set of neurophysiological measurements, and second, the particular set of eigenimages or spatial modes obtained is uniquely determined by constraints that are biologically implausible. These aspects of PCA confer inherent limitations on the interpretability and usefulness of eigenimage analysis of biological time series and have motivated the exploration of nonlinear-PCA and neural-network approaches.

Two other important approaches deserve mention here. The first is independent component analysis (ICA). ICA uses entropy maximization to find, using iterative schemes, spatial modes or their dynamics that are approximately *independent*. This is a stronger requirement than *orthogonality* in PCA, and it involves removing high-order correlations among the modes (or dynamics). It was initially introduced as *spatial* ICA (McKeown *et al.*, 1998), in which the independence constraint was applied to the modes (with no constraints on their temporal expression). More recent approaches use, by analogy with magnetophysiological and electrophysiological time-series analysis, *temporal* ICA, in which the dynamics are enforced to be independent. This requires an initial dimension reduction (usually using conventional eigenimage analysis). Finally, there has been an interest in cluster analysis (Baumgartner *et al.*, 1997). Conceptually, this can be related to eigenimage analysis through multidimensional scaling and principal coordinate analysis.

All these approaches are interesting, but hardly anyone uses them. This is largely because they tell you nothing about how the brain works and do not allow one to ask specific questions. Simply demonstrating statistical dependencies among regional brain responses (i.e., demonstrating functional connectivity) does not address how these responses were caused. To address this, one needs explicit models of integration, or, more precisely, effective connectivity.

B. Dynamic Causal Modeling with Bilinear Models

This section is about modeling interactions among neuronal populations, at a cortical level, using neuroimaging time series. The aims of these dynamic causal models are to estimate, and make inferences about, the coupling among brain areas and to explore how that coupling is influenced by changes in experimental context (e.g., time or cognitive set). The basic idea is to construct a reasonably realistic neuronal model of interacting cortical regions or nodes. This model is then supplemented with a forward model of how neuronal or synaptic activity translates into a measured response (see previous section). This enables the parameters of the neuronal model (i.e., effective connectivity) to be estimated from observed data.

Intuitively, this approach regards an experiment as a designed perturbation of neuronal dynamics that are promulgated and distributed throughout a system of coupled anatomical nodes to change region-specific neuronal activity. These changes engender, through a measurement-specific forward model, responses that are used to identify the architecture and time constants of the system at a neuronal level. This represents a departure from conventional approaches (e.g., structural equation modeling and autoregression models Büchel and Friston, 1997; Harrison *et al.*, 2003; McIntosh and Gonzalez-Lima, 1994), in which one assumes that the observed responses are driven by endogenous or intrinsic noise (i.e., innovations). In contrast, dynamic causal models assume that the responses are driven by designed changes in inputs. An important conceptual aspect of dynamic causal models pertains to how the experimental inputs enter the model and cause neuronal responses. Experimental variables can elicit responses in one of two ways. First, they can do so through direct influences on specific anatomical nodes. This would be appropriate, for example, in the modeling of sensory evoked responses in early visual cortices. The second class of input exerts its effect vicariously, through a modulation of the coupling among nodes. These sorts of experimental variables would normally be more enduring; consider, for example, changes in attention to a particular attribute or maintenance of some perceptual set. These distinctions are seen most clearly in relation to particular forms of causal models used for estimation—for example the bilinear approximation:

$$
\begin{aligned}
\dot{x} &= f(x, u) \\
&= Ax + uBx = Cu \\
y &= g(x) + \varepsilon \\
A &= \frac{\partial f}{\partial x} \quad B = \frac{\partial^2 f}{\partial x \partial u} \quad C = \frac{\partial f}{\partial u}
\end{aligned}
\tag{7}
$$

where $\dot{x} = \partial x/\partial t$. This is an approximation to any model of how changes in neuronal activity in one region x_i are caused by activity in the other regions. Here, the output function $g(x)$ embodies a hemodynamic model, linking neuronal activity to BOLD, for each region (e.g., that in Fig. 5). The matrix A represents the connectivity among the regions in the absence of input u. Effective connectivity is the influence that one neuronal system exerts over another in terms of inducing a response $\partial \dot{x}/\partial x$. This latent connectivity can be thought of as the intrinsic coupling in the absence of experimental perturbations. The matrix B is effectively the change in latent coupling induced by the input. It encodes the input-sensitive changes in A or, equivalently, the modulation of effective connectivity by experimental manipulations. Because B is a second-order derivative, it is referred to as *bilinear*. Finally, the matrix C embodies the extrinsic influences of inputs on neuronal activity. The parameters $\theta = \{A, B, C\}$ are the connectivity or coupling matrices that we wish to use to identify and define the functional architecture and interactions among brain regions at a neuronal level.

Because Eq. 7 has exactly the same form as Eq. 4, we can express it as a GLM and estimate the parameters using EM in the usual way (Friston *et al.*, 2003). Generally, estimation in the context of highly parameterized models like dynamic causal models requires constraints in the form of priors. These priors enable conditional inferences about the connectivity estimates. The sorts of questions that can be addressed with dynamic causal models are now illustrated by a look at how attentional modulation might be mediated in sensory processing hierarchies in the brain.

1. *Dynamic Causal Models and Attentional Modulation*

It has been established that the superior posterior parietal cortex (SPC) exerts a modulatory role on V5 area responses using Volterra-based regression models (Friston and Büchel, 2000) and that the inferior frontal gyrus (IFG) exerts a similar influence on the SPC using structural equation modeling (Büchel and Friston, 1997). The example here shows that a dynamic causal model leads to the same conclusions, but starting from a completely different construct. The experimental paradigm and data-acquisition parameters are described in the legend to Fig. 7. This figure also shows the location of the regions that entered into the dynamic causal model. These regions were based on maxima from conventional SPMs testing for the effects of photic stimulation, motion, and attention. Regional time courses were taken as the first eigenvariate of 8-mm spherical volumes of interest centered on the maxima shown in the figure. The inputs, in this example, comprise one sensory perturbation and two contextual inputs. The sensory input was simply the presence of photic stimulation, and the first contextual one was presence of motion in the visual field. The second contextual input, encoding attentional set, was unity during attention to speed changes and zero otherwise. The outputs corresponded to the four regional eigenvariates in Fig. 7 (left panel).

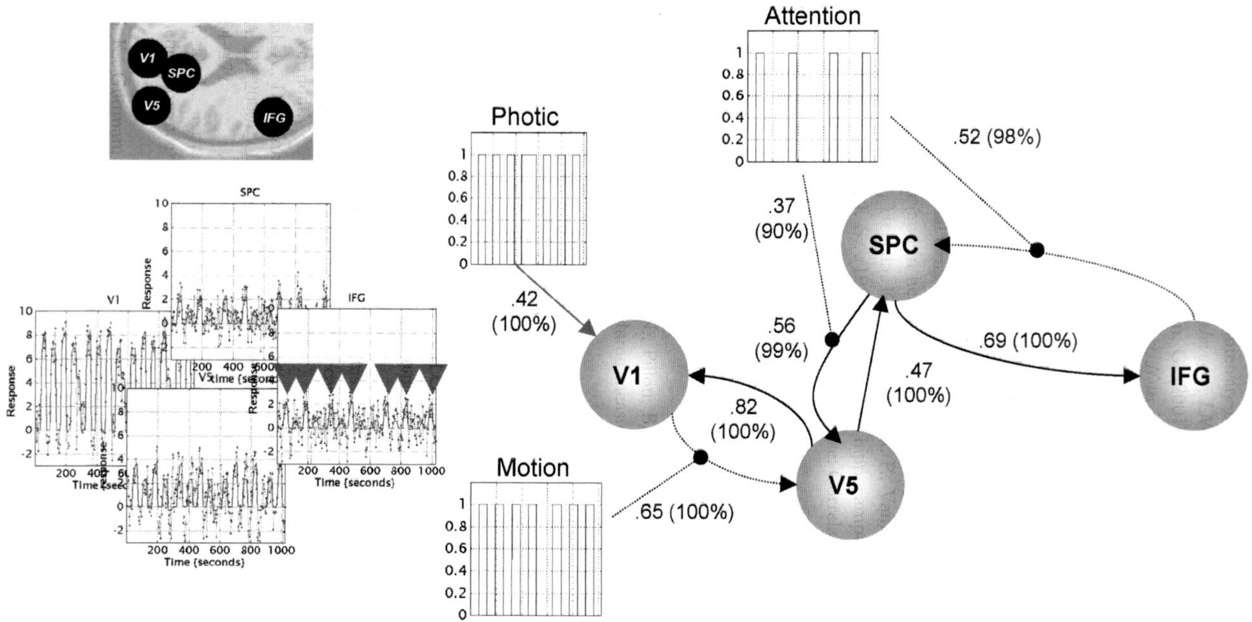

FIG. 7. Results of a dynamic causal model analysis of attention to visual motion with fMRI. Right panel: Functional architecture based upon the conditional estimates shown alongside their connections, with the percent confidence that they exceeded a non-trivial threshold shown in brackets. The most interesting aspects of this architecture involve the role of motion and attention in exerting bilinear effects. Critically, the influence of motion is to enable connections from V1 to the motion-sensitive area V5. The influence of attention is to enable backward connections from the inferior frontal gyrus (IFG) to the superior parietal cortex (SPC). Furthermore, attention increases the influence of the SPC on V5. Dotted arrows connecting regions represent significant bilinear effects in the absence of a significant intrinsic coupling. Left panel: Fitted responses based upon the conditional estimates and the adjusted data are shown for each region in the dynamic causal model. The insert (upper left) shows the location of the regions. Subjects were studied with fMRI under identical stimulus conditions (visual motion subtended by radially moving dots) while the attentional component of the task (detection of velocity changes) was manipulated. The data were acquired from a normal subject at 2 Tesla using

The intrinsic connections were constrained to conform to a hierarchical pattern in which each area was reciprocally connected to its supraordinate area. Photic stimulation entered at, and only at, V1. The effect of motion in the visual field was modeled as a bilinear modulation of the V1-to-V5 connectivity, and attention was allowed to modulate the backward connections from the IFG and the SPC.

The results of this dynamic causal model are shown in Fig. 7 (right panel). Of primary interest here is the modulatory effect of attention that is expressed in terms of the bilinear coupling parameters for this input. As expected, we can be highly confident that attention modulates the backward connections from the IFG to the SPC and from the SPC to V5. Indeed, the influences of the IFG on the SPC are negligible in the absence of attention (dotted connection). It is important to note that the only way that attentional manipulation could affect brain responses was through this bilinear effect. Attention-related responses are seen throughout the system (attention epochs are marked with arrows in the plot of IFG responses in Fig. 7). This attentional modulation is sufficiently accounted for by changes in just two connections. This change is, presumably, instantiated by instructional set at the beginning of each epoch.

The second thing that this analysis illustrates is how functional segregation is modeled in a dynamic causal model. Here one can regard the V1 area as "segregating" motion from other visual information and distributing it to the motion-sensitive area V5. This segregation is modeled as a bilinear "enabling" of V1-to-V5 connections when, and only when, motion is present. Note that in the absence of motion the intrinsic V1-to-V5 connection was trivially small (in fact, the estimate was -0.04). The key advantage of entering motion through a bilinear effect, as opposed to a direct effect on V5, is that we can finesse the inference that V5 shows motion-selective responses with the assertion that these responses are mediated by afferents from V1. The two bilinear effects presented here represent two important aspects of functional integration that a dynamic causal model were designed to characterize.

---

a Magnetom VISION (Siemens, Erlangen, Germany) whole-body MRI system, equipped with a head volume coil. Contiguous multislice T2*-weighted fMRI images were obtained with a gradient echo-planar sequence (TE = 40 ms; TR = 3.22 seconds; matrix size = $64 \times 64 \times 32$; voxel size $3 \times 3 \times 3$ mm). Each subject had four consecutive 100-scan sessions comprising a series of 10-scan blocks under 5 different conditions, D F A F N F A F N S. The first condition, D, was a dummy condition that allowed for magnetic saturation effects. F (fixation) corresponded to a low-level baseline where the subjects viewed a fixation point at the center of a screen. In condition A (attention), subjects viewed 250 dots moving radially from the center at 4.7 degrees per second and were asked to detect changes in radial velocity. In condition N (no attention), the subjects were asked simply to view the moving dots. In condition S (stationary), subjects viewed stationary dots. The order of A and N was swapped for the last two sessions. In all conditions, subjects fixated on the center of the screen. There were no speed changes during scanning. No overt response was required in any condition. (See Color Insert.)

2. *Structural-Equation Modeling*

The central idea behind dynamic causal modeling is to treat the brain as a deterministic nonlinear dynamic system that is subject to inputs and produces outputs. Effective connectivity is parameterized in terms of coupling among unobserved brain states (e.g., neuronal activity in different regions). The objective is to estimate these parameters by perturbing the system and measuring the response. This is in contrast to established methods for estimating effective connectivity from neurophysiological time series, which include structural-equation modeling and models based on multivariate autoregressive processes. In these models, there is no designed perturbation, and the inputs are treated as unknown and stochastic. Furthermore, the inputs are often assumed to express themselves instantaneously such that, at the point of observation, the change in states will be zero. From Eq. 7, in the absence of bilinear effects we have

$$\begin{aligned} \dot{x} &= 0 \\ &= Ax + Cu \\ x &= -A^{-1}Cu \end{aligned} \tag{8}$$

This is the regression equation used in structural-equation modeling, where $A = A' - I$ and A' contains the off-diagonal connections among regions. The key point here is that A is estimated based on the assumption that u is some random innovation with known covariance. This is not really tenable for designed experiments, when a u represents a carefully structured experimental input. Although structural-equation modeling and related autoregressive techniques are useful for establishing dependence among regional responses, they are not surrogates for informed causal models based on the underlying dynamics of these responses.

In this section we have covered multivariate techniques ranging from eigenimage analysis, which does not have an explicit forward or causal model, to dynamic causal modeling, which does. The bilinear approximation of any dynamic causal model has been illustrated through its use with fMRI to study attentional modulation. The parameters of bilinear dynamic causal models include first-order effective connectivity A and its experimentally induced changes B. Although the Bilinear approximation is useful, it is possible to model effective connectivity among neuronal subpopulations explicitly. We will conclude with a dynamic causal model that embraces a number of neurobiological facts and takes us much closer to a mechanistic understanding of how brain responses are generated. This example uses responses measured with an EEG.

C. Dynamic Causal Modeling with Neural Mass Models

ERPs have been used for decades as electrophysiological correlates of perceptual and cognitive operations. However, the exact neurobiological mechanisms underlying their generation are largely unknown. In this section we use neuronally plausible models to understand ERPs. The example used in this section shows that changes in connectivity are sufficient to explain certain ERP components. Specifically, we will look at the P300, a late component associated with rare or unexpected events. If the unexpected nature of rare stimuli depends on learning which stimuli are frequent, then the P300 must be caused by plastic changes in connectivity that mediate perceptual learning. We conclude by showing that recent advances in the modeling of evoked responses now afford measures of connectivity among cortical sources that can be used to quantify the effects of perceptual learning.

1. *Hierarchical Neural Mass Models*

The minimal model we have developed (David *et al.*, 2005) uses the connectivity rules described by Felleman and Van Essen (1991) to assemble a network of coupled sources. These rules are based on a partitioning of the cortical sheet into supragranular and infragranular layers and a granular layer (layer 4). Bottom-up or forward connections originate in agranular layers and terminate in layer 4.

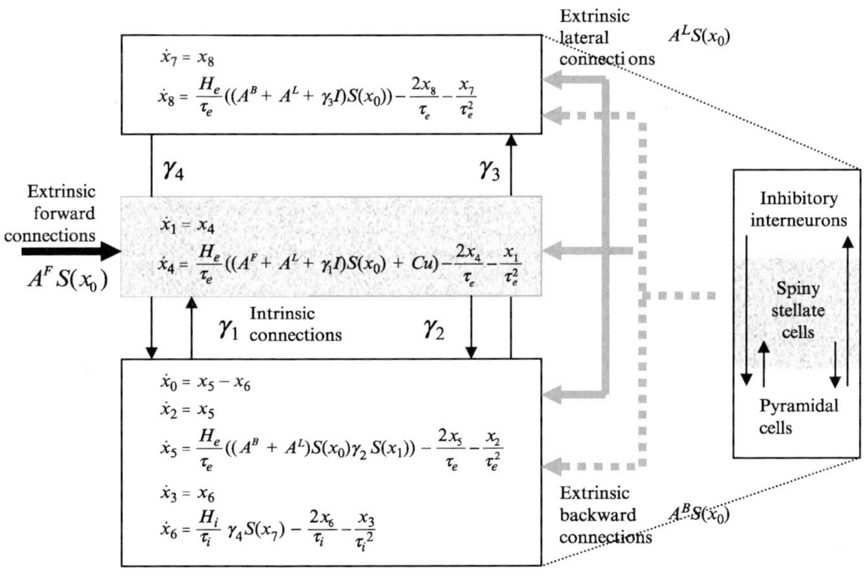

FIG. 8. Schematic of the dynamic causal model used to model electrical responses. This schematic shows the state equation describing the dynamics of sources or regions. Each source is modeled with three subpopulations (pyramidal, spiny stellate, and inhibitory interneurons), as described by Jansen and Rit (1995) and David and Friston (2003). These have been assigned to granular and agranular cortical layers, which receive forward and backward connections, respectively.

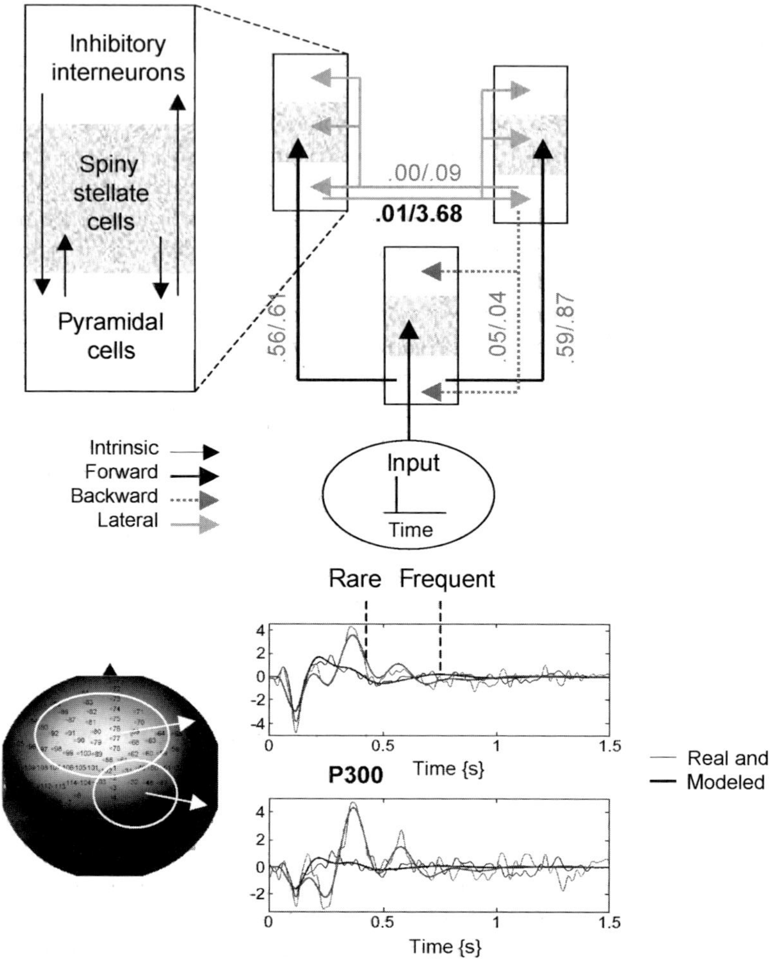

FIG. 9. Summary of a dynamic causal modeling of event-related potentials (ERPs) elicited during an auditory P300 paradigm that employed rare and frequent pure tones. Upper panel: Schematic showing the architecture of the neuronal model used to explain the empirical data. Sources were coupled with extrinsic cortico–cortical connections, following the rules of Felleman and van Essen (1991). The free parameters of this model included intrinsic and extrinsic connection strengths that were adjusted to best explain the observed ERPs. In this example, the lead field was also estimated, with no spatial constraints. The parameters were estimated for ERPs recording during the presentation of rare and frequent tones and are reported beside their corresponding connection (frequent/rare). The most notable finding was that the P300 could be explained by a selective increase in lateral connection strength (highlighted in bold). Lower panel: The channel positions (left) and ERPs (right) averaged over two subsets of channels (circled on the left). Note the correspondence between the measured ERPs and those generated by the model. See David et al. (2004) for details. We modeled ERPs that exhibited a strong modulation of the P300 component, on comparison of

Top-down or backward connections target agranular layers. Lateral connections originate in agranular layers and target all layers. These long-range or extrinsic cortico–cortical connections are excitatory and arise from pyramidal cells.

Each region or source is modeled using a neural mass model described by David and Friston (2003), based on the model of Jansen and Rit (1995). This model emulates the activity of a cortical area using three neuronal subpopulations assigned to granular and agranular layers. A population of excitatory pyramidal (output) cells receives inputs from inhibitory and excitatory populations of interneurons, via intrinsic connections (intrinsic connections are confined to the cortical sheet). Within this model, excitatory interneurons can be regarded as spiny stellate cells found predominantly in layer 4 and in receipt of forward connections. Excitatory pyramidal cells and inhibitory interneurons are considered to occupy agranular layers and receive backward and lateral inputs (see Fig. 8).

To model event-related responses, the network receives inputs via input connections. These connections are exactly the same as forward connections and deliver inputs u to the spiny stellate cells in layer 4. In the present context, inputs u model subcortical auditory inputs. The vector C controls the influence of the input on each source. The lower, upper, and leading diagonal matrices A^F, A^B, A^L encode forward, backward, and lateral connections, respectively. The dynamic causal model here is specified in terms of the state equations shown in Fig. 8 and a linear output equation:

$$\begin{aligned}\dot{x} &= f(x, u) \\ y &= Lx_0 + \varepsilon\end{aligned} \tag{9}$$

where x_0 represents the transmembrane potential of pyramidal cells and L is a lead field matrix coupling electrical sources to the EEG channels. This should be compared to the dynamic causal model above for hemodynamics. Here, the equations governing the evolution of neuronal states are much more complicated and realistic than the bilinear approximation in Eq. 7. Conversely, the output equation is a simple linearity, in contrast to the nonlinear observer used for fMRI. As an example, the state equation for the inhibitory subpopulation is as

responses to frequent and rare stimuli using an auditory oddball paradigm. The auditory stimuli, which were 1000- or 2000-Hz tones with rise and fall times of 5 ms and duration of 80 ms, were presented binaurally. The tones were presented every 2 seconds, for 15 minutes, in a pseudo-random sequence with 2000-Hz tones occurring 20% of the time and 1000-Hz tones occurring 80% of the time. The subject was instructed to keep a mental record of the number of 2000-Hz tones (nonfrequent target tones). Data were acquired using 128 EEG electrodes with 1000-Hz sample frequencies. Before averaging, data were referenced to mean earlobe activity and band-pass filtered between 1 and 30 Hz. Trials that showed ocular artifacts and bad channels were removed from further analysis.

follows (propagation delays on the extrinsic connections have been omitted for clarity here and in):

$$\dot{x}_7 = x_8$$
$$\dot{x}_8 = \frac{H_e}{\tau_e}((A^B + A^L + \gamma_3 I)S(x_0)) - \frac{2x_8}{\tau_e} - \frac{x_7}{\tau_e^2} \tag{10}$$

Within each subpopulation, the evolution of neuronal states rests on two operators. The first transforms the average density of presynaptic inputs into the average postsynaptic membrane potential. This is modeled by a linear transformation with excitatory (e) and inhibitory (i) kernels parameterized by $H_{e,i}$ and $\tau_{e,i}$. $H_{e,i}$ control the maximum postsynaptic potential and $\tau_{e,i}$ represent a lumped rate constant. The second operator S transforms the average potential of each subpopulation into an average firing rate. This is assumed to be instantaneous and is a sigmoid function. Interactions among the subpopulations depend on constants $\gamma_{1,2,3,4}$, which control the strength of intrinsic connections and reflect the total number of synapses expressed by each subpopulation. In Eq. 10, the top line expresses the rate of change of voltage as a function of current. The second line specifies how current changes as a function of voltage, current, and presynaptic input from extrinsic and intrinsic sources. Having specified the dynamic causal model, one can estimate the coupling parameters from empirical data using EM, as described previously.

2. *Perceptual Learning and the P300*

The example shown in Fig. 9 is an attempt to model the P300 in terms of changes in backward and lateral connections among cortical sources. In this example, two (averaged) channels of EEG data were modeled with three cortical sources. Using this generative or forward model, we estimated differences in the strength of these connections for rare and frequent stimuli. As expected, we could account for detailed differences in the ERPs (the P300) by changes in connectivity (see the legend of Fig. 9 for details). Interestingly, these differences were expressed selectively in the lateral connections. If this model is a sufficient approximation of the real sources, these changes are a noninvasive measure of plasticity, mediating perceptual learning in the human brain.

V. Conclusion

In this article we have reviewed some key models that underpin image analysis and have touched briefly on ways of assessing specialization and integration in the brain. These models can be regarded as a succession of modeling endeavors, drawing more and more on our understanding of how brain-imaging signals are generated, in terms of both biophysics and the underlying neuronal interactions. We have seen how hierarchical linear observation models encode the treatment effects elicited by experimental design. GLMs based on convolution models imply an underlying dynamic input-state-output system. The form of these systems can be used to constrain convolution models and explore some of their simpler nonlinear properties. By creating observation models based on explicit forward models of neuronal interactions, we can now start to model and assess interactions among distributed cortical areas and make inferences about coupling at the neuronal level. The coming years will probably see an increasing realism in the dynamic causal models introduced here (see Horwitz *et al.*, 2001). There are already attempts to use plausible models of neuronal ensembles to estimate network parameters of evoked responses in EEGs. These endeavors are likely to encompass fMRI signals in the near future, enabling the conjoint modeling, or fusion, of different modalities and the marriage of computational neuroscience with modeling of brain responses.

Acknowledgments

The Wellcome Trust funded this work. We would like to thank all our colleagues at the Wellcome Department of Imaging Neuroscience for their help with this review.

References

Absher, J. R., and Benson, D. F. (1993). Disconnection syndromes: An overview of Geschwind's contributions. *Neurology* **43,** 862–867.

Adler, R. J. (1981). *In* "The Geometry of Random Fields" Wiley, New York.

Aertsen, A., and Preißl, H. (1991). Dynamics of activity and connectivity in physiological neuronal Networks. *In* "Non Linear Dynamics and Neuronal Networks" (H. G. Schuster, Ed.), VCH Publishers, New York. Ashburner, J., and Friston, K. J. (1991). Voxel-based morphometry—the methods. *NeuroImage* **11,** 805–821.

Bandettini, P. A., Jesmanowicz, A., Wong, E. C., and Hyde, J. S. (1993). Processing strategies for time course data sets in functional MRI of the human brain. *Mag. Res. Med.* **30,** 161–173.

Baumgartner, R., Scarth, G., Teichtmeister, C., Somorjai, R., and Moser, E. (1997). Fuzzy clustering of gradient-echo functional MRI in the human visual cortex, Part 1: reproducibility. *J. Mag. Res. Imaging* **7,** 1094–1101.

Bendat, J. S. (1990). "Nonlinear System Analysis and Identification from Random Data." John Wiley and Sons, New York. Berry, D. A., and Hochberg, Y. (1990). Bayesian perspectives on multiple comparisons. *J. Statistical Planning Inference* **82,** 215–227.

Biswal, B., Yetkin, F. Z., Haughton, V. M., and Hyde, J. S. (1995). Functional connectivity in the motor cortex of resting human brain using echo-planar MRI. *Mag. Res. Med.* **34,** 537–541.

Boynton, G. M., Engel, S. A., Glover, G. H., and Heeger, D. J. (1996). Linear systems analysis of functional magnetic resonance imaging in human V1. *J. Neurosci.* **16,** 4207–4221.

Büchel, C., Wise, R. J. S., Mummery, C. J., Poline, J.-B, and Friston, K. J. (1996). Nonlinear regression in parametric activation studies. *NeuroImage* **4,** 60–66.

Büchel, C., and Friston, K. J. (1997). Modulation of connectivity in visual pathways by attention: Cortical interactions evaluated with structural equation modelling and fMRI. *Cerebral Cortex* **7,** 768–778.

Buckner, R., Bandettini, P., O'Craven, K., Savoy, R., Petersen, S., Raichle, M., and Rosen, B. (1996). Detection of cortical activation during averaged single trials of a cognitive task using functional magnetic resonance imaging. *Proc. Natl. Acad. Sci. USA* **93,** 14878–14883.

Burock, M. A., Buckner, R. L., Woldorff, M. G., Rosen, B. R., and Dale, AM. (1998). Randomized Event-Related Experimental Designs Allow for Extremely Rapid Presentation Rates Using Functional MRI. *NeuroReport* **9,** 3735–3739.

Buxton, R. B., and Frank, L. R. (1997). A model for the coupling between cerebral blood flow and oxygen metabolism during neural stimulation. *J. Cereb. Blood. Flow Metab.* **17,** 64–72.

Dale, A., and Buckner, R. (1997). Selective averaging of rapidly presented individual trials using fMRI. *Hum Brain Mapp.* **5,** 329–340.

David, O., and Friston, K. J. (2003). A neural mass model for MEG/EEG: Coupling and neuronal dynamics. *NeuroImage* **20,** 1743–1755.

David, O., Harrison, L., Kilner, J., Penny, W., and Friston, K. J. (2005). Studying effective connectivity with a neural mass model of evoked MEG/EEG responses. *BioMag.*, in press.

Dempster, A. P., Laird, N. M., and Rubin (1977). Maximum likelihood from incomplete data via the EM algorithm. *J. Roy. Stat. Soc. Series B* **39,** 1–38.

Fliess, M., Lamnabhi, M., and Lamnabhi-Lagarrigue, F. (1983). An algebraic approach to nonlinear functional expansions. *IEEE Trans. Circuits Syst.* **30,** 554–570.

Friston, K. J., Frith, C. D., Liddle, P. F., and Frackowiak, R. S. J. (1991). Comparing functional (PET) images: The assessment of significant change. *J. Cereb. Blood Flow Metab.* **11,** 690–699.

Friston, K. J., Frith, C. D., Passingham, R. E., Liddle, PF, and Frackowiak, R. S. J. (1992). Motor practice and neurophysiological adaptation in the cerebellum: A positron tomography study. *Proc. Roy. Soc. Lon. Series B* **248,** 223–228.

Friston, K. J., Frith, C. D., Liddle, P. F., and Frackowiak, R. S. J. (1993). Functional Connectivity: The principal component analysis of large data sets. *J Cereb. Blood Flow Metab.* **13,** 5–14.

Friston, K. J., Jezzard, P. J., and Turner, R. (1994). Analysis of functional MRI time-*series*. *Hum. Brain Mapp.* **1,** 153–171.

Friston, K. J., Holmes, A. P., Worsley, K. J., Poline, J. B., Frith, C. D., and Frackowiak, R. S. J. (1995a). Statistical Parametric Maps in functional imaging: A general linear approach. *Hum. Brain Mapp.* **2,** 189–210.

Friston, K. J., Frith, C. D., Turner, R., and Frackowiak, R. S. J. (1995b). Characterising evoked hemodynamics with fMRI. *NeuroImage* **2,** 157–165.

Friston, K. J., Price, CJ, Fletcher, P., Moore, C., Frackowiak, R. S. J., and Dolan, R. J. (1996a). The trouble with cognitive subtraction. *NeuroImage* **4,** 97–104.

Friston, K. J., Poline, J.-B., Holmes, A. P., Frith, C. D., and Frackowiak, R. S. J. (1996b). A multivariate analysis of PET activation studies. *Hum. Brain Mapp.* **4,** 140–151.

Friston, K. J., Frith, C. D., Fletcher, P., Liddle, P. F., and Frackowiak, R. S. J. (1996c). Functional topography: Multidimensional scaling and functional connectivity in the brain. *Cerebral Cortex* **6,** 156–164.

Friston, K. J., Mechelli, A., Turner, R., and Price, C. J. (2000). Nonlinear responses in fMRI: The Balloon model, Volterra kernels, and other hemodynamics. *NeuroImage.* **12,** 466–477.

Friston, K. J., and Büchel, C. (2000). Attentional modulation of effective connectivity from V2 to V5/MT in humans. *Proc Natl Acad Sci USA*. **97,** 7591–7596.

Friston, K. J., Harrison, L., and Penny, W. (2003). Dynamic causal modelling. *NeuroImage* **19,** 1273–1302.

Gerstein, G. L., and Perkel, D. H. (1969). Simultaneously recorded traits of action potentials: Analysis and functional interpretation. *Science* **164,** 828–830.

Goltz, F. (1881). *In* "Transactions of the 7th International Medical Congress" (W. MacCormac, Ed.), Vol. I, pp. 218–228. JW Kolkmann, London.

Grafton, S., Mazziotta, J., Presty, S., Friston, K. J., Frackowiak, R. S.J, and Phelps, M. (1992). Functional anatomy of human procedural learning determined with regional cerebral blood flow and PET. *J Neurosci.* **12,** 2542–2548.

Harrison, L. M., Penny, W., and Friston, K. J. (2003). Multivariate autoregressive modelling of fMRI time series. *NeuroImage.* **19,** 1477–1491.

Hoge, R. D., Atkinson, J., Gill, B., Crelier, G. R., Marrett, S., and Pike, G. B. (1999). Linear coupling between cerebral blood flow and oxygen consumption in activated human cortex. *Proc. Natl. Acad. Sci.* **96,** 9403–9408.

Holmes, A., and Ford, I. (1993). A Bayesian approach to significance testing for statistic images from PET. *In* "Quantification of Brain Function, Tracer Kinetics and Image analysis in Brain PET" (K. Uemura, N. A. Lassen, T. Jones, and I. Kanno, Eds.)., *Excerpta Medica, Int. Cong.* Series No. **1030,** 521–534.

Horwitz, B., Friston, K. J., and Taylor, J. G. (2001). Neural modelling and functional brain imaging: An overview. *Neural Networks* **13,** 829–846.

Jansen, B. H., and Rit, V. G. (1995). Electroencephalogram and visual evoked potential generation in a mathematical model of coupled cortical columns. *Biol. Cybern* **73,** 357–366.

Lange, N., and Zeger, S. L. (1997). Non-linear Fourier time series analysis for human brain mapping by functional magnetic resonance imaging (with discussion.). *J. Roy. Stat. Soc. Ser. C.* **46,** 1–29.

Lueck, C. J., Zeki, S., Friston, K. J., Deiber, M. P., Cope, N. O., *et al.* (1989). The colour centre in the cerebral cortex of man. *Nature* **340,** 386–389.

Mandeville, J. B., Marota, J. J., Ayata, C., Zararchuk, G., Moskowitz, M. A., B Rosen, B., and Weisskoff, R. M. (1999). Evidence of a cerebrovascular postarteriole Windkessel with delayed compliance. *J. Cereb. Blood Flow Metab.* **19,** 679–689.

McIntosh, A. R., and Gonzalez-Lima, F. (1994). Structural equation modelling and its application to network analysis in functional brain imaging. *Hum. Brain Mapp.* **2,** 2–22.

McKeown, M., Jung, T.-P., Makeig, S., Brown, G., Kinderman, S., Lee, T.-W., and Sejnowski, T. (1998). Spatially independent activity patterns in functional MRI data during the Stroop colour naming task. *Proc. Natl. Acad. Sci.* **95,** 803–810.

Petersen, S. E., Fox, P. T., Posner, M. I., Mintun, M, and Raichle, ME. (1989). Positron emission tomographic studies of the processing of single words. *J. Cog. Neurosci.* **1,** 153–170.

Phillips, C. G., Zeki, S., and HB Barlow, H. B. (1984). Localisation of function in the cerebral cortex Past present and future. *Brain* **107,** 327–361.

Price, C. J., Wise, R. J. S., Ramsay, S., Friston, K. J., Howard, D, *et al.* (1992). Regional response differences within the human auditory cortex when listening to words. *Neurosci. Lett.* **146,** 179–182.

Price, C. J., and Friston, K. J. (1997). Cognitive Conjunction: A new approach to brain activation experiments. *NeuroImage* **5,** 261–270.

Sychra, J. J., Bandettini, P. A., Bhattacharya, N., and Lin, Q. (1994). Synthetic images by subspace transforms I Principal component images and related filters. *Med. Physics* **21,** 193–201.

Talairach, P., and Tournoux, J. (1988). "A Stereotactic coplanar atlas of the human brain" Stuttgart Thieme.

Worsley, K. J., Evans, AC, Marrett, S., and Neelin, P. (1992). A three-dimensional statistical analysis for rCBF activation studies in human brain. *J. Cereb. Blood Flow Metab.* **12,** 900–918.

Worsley, K. J., Marrett, S., Neelin, P., Vandal, A. C., Friston, K. J., and Evans, A. C. (1996). A unified statistical approach or determining significant signals in images of cerebral activation. *Hum. Brain Mapp.* **4,** 58–73.

Zeki, S. (1990). The motion pathways of the visual cortex. *In* "Vision: Coding and Efficiency" (C. Blakemore, Ed.), pp. 321–345. Cambridge University Press, UK.

Further Reading

Felleman, DJ, and Van Essen, D. C. (1992). Distributed hierarchical processing in the primate cerebral cortex. *Cereb. Cortex* **1,** 1–47.

VOXEL-BASED MORPHOMETRIC ANALYSIS USING SHAPE TRANSFORMATIONS

Christos Davatzikos

Section of Biomedical Image Analysis, Department of Radiology
University of Pennsylvania, Philadelphia, Pennsylvania 19104

I. Introduction

The widespread use of neuroimaging methods in a variety of clinical and basic science fields has created the need for systematic and highly automated image analysis methodologies that extract pertinent information from images in a way that enables comparisons across different studies, laboratories, and image databases. Quantifying morphological characteristics of the brain from tomographic images, most often magnetic resonance images (MRIs), is an invaluable way for researchers to understand the way in which a disease can affect brain anatomy, to construct new diagnostic methods utilizing image information, and to create longitudinal follow-up studies evaluating potential drugs.

The conventional type of morphological analysis of brain images has relied on manual tracings of regions of interest (ROI) (Bobinski *et al.*, 1999; Convit *et al.*, 1997; Cuenod *et al.*, 1993; De Santi *et al.*, 2001; deToledo-Morrell *et al.*, 1997; Dickerson *et al.*, 2001; Du *et al.*, 2001; Frisoni *et al.*, 1996; Jack *et al.*, 1997, 1999; Killiany *et al.*, 1993, 2000; Krasuski *et al.*, 1998; Laakso *et al.*, 1995, 2000; Lehericy *et al.*, 1994; Rosen *et al.*, 2003; Xu *et al.*, 2000). These methods typically require that the reliability and repeatability of manual tracings be established across different raters, but also within the same rater at different times, as a first

INTERNATIONAL REVIEW OF
NEUROBIOLOGY, VOL. 66
DOI: 10.1016/S0074-7742(05)66004-7

step. However, methods based on manually defined ROIs are limited in many ways. First, they rely on the need for *a priori* knowledge of the regions that are affected by a disease, so that respective ROIs can be defined, and therefore they may fail to discover new findings. Although a good hypothesis might be available at the beginning of a morphometric study, one would typically want to discover new knowledge that cannot, by definition, be part of the hypothesis. Take one example selected from the literature on the neuroimaging of dementia: although the role of hippocampal and entorhinal cortical atrophy in early prediction of Alzheimer's disease (AD) is widely accepted, relatively little is known about the potential involvement of other brain regions. Such knowledge could help in the construction of more sensitive methods for detection of and differentiation among different dementias. The complete investigation of the role of all brain structures in a disease, as well as the diagnosis of that disease, would be prohibitively labor intensive for an adequately large sample size if manual methods were employed. Moreover, inter- and intra-rater reliability issues would become crucial limiting factors, particularly in longitudinal studies, in which it is extremely difficult to maintain intra- and inter-rater reliability over time. Second, the spatial specificity of ROI-based methods is limited by the sizes of the ROIs being measured, which are typically rather coarse. A region that might be affected by disease may be only part of one predefined ROI, or it might span two or more ROIs, which inevitably washes out the results and reduces the statistical power of the measurement method. Alternative methods, such as stereology, are limited in a similar way. Although, in principle, one could delimit the ROIs measured to be as small as desired in order to increase spatial specificity, this would decrease rater reliability for measurement methods that are based on human raters. Finally, manual ROI tracing is severely limited in many modern studies, since it is no longer unusual to include over a thousand scans per study.

II. Shape Transformations and Voxel-Based Methods

In order to address these limitations, image analysis methods based on shape analysis have been studied in the literature during the past 15 years (Ashburner *et al.*, 1998; Bookstein, 1989; Christensen *et al.*, 1994; Collins *et al.*, 1992; Davatzikos *et al.*, 1996; Freeborough and Fox, 1998; Golland *et al.*, 1999, 2001; Joshi *et al.*, 2001; Miller *et al.*, 1993; Pizer *et al.*, 1999; Rexilius *et al.*, 1999; Sandor and Leahy, 1997; Styner and Gerig, 2001; Szekely *et al.*, 1996; Thompson *et al.*, 1997). One very promising approach to morphometric analysis has been based on *shape transformations*, and the associated methods are often called unbiased, or hypothesis-free, methods. A shape transformation is a spatial map that adapts one individual's brain anatomy to that of another. The resulting transformation measures the differences between the two anatomies with very high spatial specificity—ultimately, the specificity allowed by the image's voxel size. More generally, a template of anatomy is selected first, and this serves as a measurement unit. The shape transformation that maps other brains to the template is determined via some sort of image analysis algorithm, and it is used as a means of quantifying the individual anatomies. Inter-individual comparisons are then performed through the application of standard statistical methods to the respective shape transformations. Voxels that display significant group differences or longitudinal changes are grouped into regions. Therefore, there is no need to define ROIs in advance. Instead, the ROIs are determined retrospectively via the voxel-wise statistical analysis of the shape transformations. The concept of this approach is shown in Fig. 1a, which is based on earlier work on the corpus callosum (Davatzikos *et al.*, 1996).

Although this approach has gained widespread attention only in the past decade, it has its roots in the seminal work by D'Arcy Thompson (1917), who

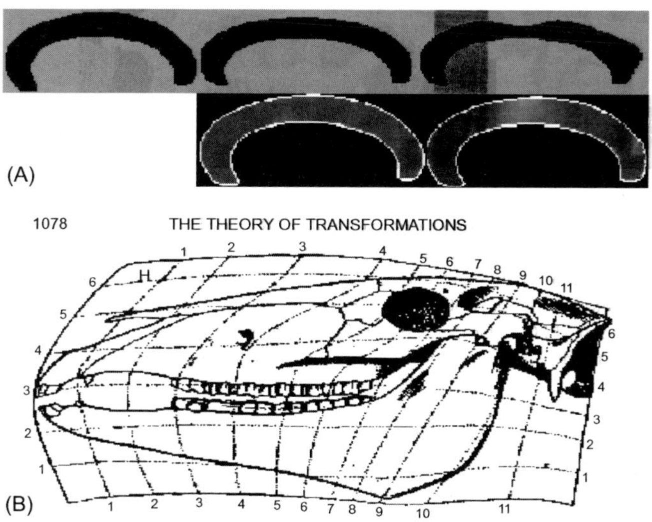

FIG. 1. Using a shape transformation for morphometric measurements. (A) (Top left) A callosal template. (Top middle and right) Two individual shapes. (Bottom) Respective color-coded maps of the determinant of the Jacobian of the shape transformation mapping the template to the two shapes. Contraction is colored green and expansion is colored red. Voxel-wise comparison of these images reveals local shape differences of the respective shapes. (B) Seminal work by D'Arcy Thompson in 1917 using shape transformations to make comparisons among species. (See Color Insert.)

studied differences among species by measuring deformations of coordinate grids from images of one species to images of another (see Fig. 1b). At that time, only very limited manual drawing methods were available to D'Arcy Thompson, which imposed limits on the spatial specificity of this approach. The approach was later adopted by Bookstein in the landmark-based morphometrics literature (Bookstein, 1989) and further extended by Grenander's pattern theory (Grenander, 1985) and Miller's work on diffeomorphisms (Miller *et al.*, 1993), and by P. Thompson's work on tensor mapping (Thompson *et al.*, 1997), among several other investigators. Our group performed one of the first applied studies, focusing on sex differences in the corpus callosum (Davatzikos *et al.*, 1996; Davatzikos and Resnick, 1998). Despite the mathematical subtleties behind this approach, the basic principle is simple and follows three standard measurement steps, similar to the steps one would perform in order to measure, for example, lengths of objects: (1) A measurement unit is selected to be used as a reference. This might be the meter or the yard in standard length measurements; In our work, the measurement unit is a typical brain image in shape measurements and is referred to as the template. (2) The template is stretched over the extent of an individual's brain so that homologous features are mapped to each other; the shape transformation perfectly quantifies this stretching, and the process is analogous to stretching the meter or the yard over the full length of an object, if one seeks to obtain a length measurement. (3) Inter-individual comparisons are performed by comparison of respective shape transformations, exactly as one would compare a 2.3-meter object to a 2.8-meter object. Of course, obtaining and comparing three- or four-dimensional shape transformations is far more complicated than obtaining and comparing scalar length measurements, but the fundamental principle is the same.

Techniques using shape transformations can generally be divided into three main categories, which are described next.

A. Deformation-Based Methods

This is the most popular family of methods, and it derives directly from the principles described above, namely that the shape transformation from a template to an individual anatomy is a quantitative measure of the individual anatomy. Various methods have been presented in the literature. The beginning of the modern phase of this approach in medical imaging can perhaps be dated back to 1989, with Bookstein's work on landmark-based morphometrics (Bookstein, 1989). In this approach, a number of reliably identifiable anatomical landmarks are first defined by a human rater. A Procrustes fit is then used to place the landmarks into a standardized coordinate space, in which statistical analysis is carried out. Shape deformations are visualized by interpolation of the mapping on the landmarks everywhere else in the brain, with a thin plate spline model. This approach has been mainly used in 2D neuroimaging studies, which are often restricted to the corpus callosum since few anatomical structures lend themselves to 2D analysis. Moreover, defining landmarks in 3D with high accuracy and reproducibility is often a very difficult and impractical task, especially in large studies.

Algorithms based on maximizing the similarity between an image treated as a template and other images in the study have been widely used (Christensen and Johnson, 2001; Christensen *et al.*, 1994; Chung *et al.*, 2001; Collins *et al.*, 1994; Ferrant *et al.*, 1999; Friston *et al.*, 1995; Miller *et al.*, 1993; Rueckert *et al.*, 1999; Thirion, 1996). These methods assume that if a shape transformation renders two images similar, it implies anatomical correspondence between the underlying anatomies. This is a reasonable assumption, but it can easily be violated in practice, since two images can be made similar via shape transformations that do not respect the underlying anatomical correspondences. For example, one can simply flow gray matter (GM) into GM, white matter (WM) into WM, and cerebrospinal fluid (CSF) into CSF, thereby creating images that look alike (since these three tissue types have similar intensity distributions throughout the brain) but lack the underlying shape transformations that reflect true anatomical measures. Such a method could, for example, morph the precentral gyrus to the postcentral gyrus.

An important issue with intensity-based transformations is that of inverse consistency. In particular, if we attempt to match Image 1 to Image 2, then Image 2 to Image 1, we should get shape transformations that are inverse to each other. This condition is not necessarily met in practice, especially by image-similarity measures. Therefore, techniques that specifically impose inverse consistency have also been examined in the literature (Christensen, 1999; Christensen and Johnson, 2001; Shen and Davatzikos, 2002).

Somewhat related to image-intensity matching are methods that optimize information-theoretic criteria in order to find appropriate shape transformations. The main advantage of these methods over image-similarity methods is that they can potentially be used across different imaging modalities, for example, when tissue intensities are different in two images to be matched. The most popular criterion of optimality has been mutual information (Rueckert *et al.*, 1999; Viola and Wells III, 1995; Wells *et al.*, 1995), which is maximized when the "predictability" of the warped image based on the template is maximized, and which tends to occur when the different tissue types in two images are well registered.

A different class of algorithms is based on some form of feature matching (Chui, 2003; Davatzikos, 1996; Ferrant *et al.*, 1999; Rangarajan *et al.*, 1997; Sandor and Leahy, 1997; Thompson and Toga, 1996; Wang and Staib, 1999; Wang *et al.*, 2003). A number of features, such as edges or curves or surfaces, are typically extracted from the images via an image analysis algorithm, or simply drawn manually, and are then used to drive a three-dimensional (3D) deformable registration method, which effectively interpolates feature correspondence in the remainder of the brain. Related are medialness models (Pizer *et al.*, 1999), which use the medial axes of anatomical shapes as features instead of using boundaries themselves. Feature-based methods pay more attention to the biological relevance of the shape-matching procedure, since they use only anatomically distinct features to find the transformation, whereas image-matching methods seek transformations that produce images that look alike, with little warranty that the implied correspondences have anatomical meaning. However, the latter approaches take advantage of the full dataset, and not only of a relatively sparse subset of features.

One method that has been developed by our group attempts to bridge between these two extremes by developing attribute vectors that aim at making each voxel a feature (Shen, 2004; Shen and Davatzikos, 2002, 2003), and it was called Hierarchical Attribute Matching Mechanism for Elastic Registration (HAMMER). HAMMER is a hierarchical warping mechanism that has two key characteristics. First, it places emphasis on determining anatomical correspondences, which in turn drive the 3D warping procedure. In particular, we have used feature-extraction methods whose goal is to determine, from the images, a number of parameters that can characterize at least some key anatomical features as distinctively as possible. In our 2002 work (Shen and Davatzikos, 2002), we used geometric moment invariants (GMIs) as a means for achieving this goal. GMIs are quantities that are constructed from images that are first segmented into GM, WM, and CSF, or any other set of tissues of interest. They are determined from the image content around each voxel, and they quantify the anatomy in the vicinity of that voxel. GMIs of different tissues and different orders are collected into an attribute vector for each voxel in an image. Ideally, we would like each voxel to have a distinctive attribute vector; of course, this is not possible in reality. Figure 2 shows a color-coded image of the degree of similarity between the GMI-based attribute vector of a point on the anterior horn of the left ventricle and the attribute vectors of every other point in the image. The GMI attribute vector of this point, as well as those of many other points in the brain, is reasonably distinctive, as Fig. 2 shows. HAMMER was constructed to solve an optimization problem that involves finding a shape transformation that maximizes the similarity of respective attribute vectors, while being smoothed by a standard Laplacian regularization term. We have recently explored more distinctive attribute vectors, aiming at constructing even more distinctive morphological signatures for every image voxel. Toward this goal, we used wavelet-based hierarchical image descriptions of large neighborhoods centered on each image voxel (Xue *et al.*, 2003, 2005).

A second key characteristic of HAMMER addresses a fundamental problem encountered in high-dimensional image matching. This problem is that the cost

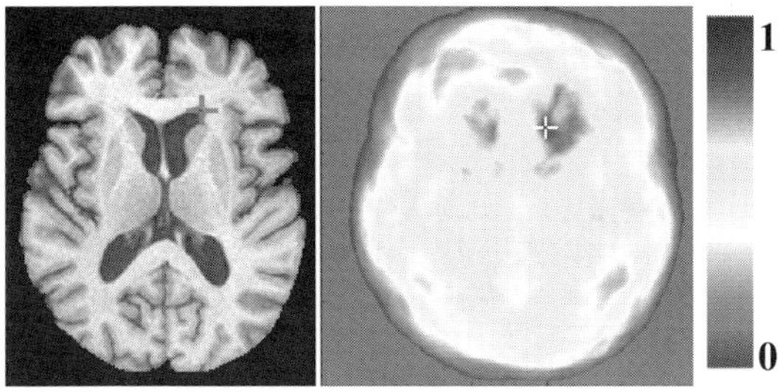

FIG. 2. The point marked by a cross has a relatively distinctive GMI-based attribute vector. The color-coded image on the right shows the degree of similarity between the attribute vector of the marked (by crosses) point and the attribute vector of every other point in the brain. 1 is maximum similarity and 0 is minimum similarity. (See Color Insert.)

function being optimized typically has many local minima, which trap an iterative optimization procedure into solutions that correspond to poor matches between the template and the individual. This is due, in part, to the ambiguity in finding point correspondences. For example, if many candidate points in an individual image have attribute vectors that are similar to that of a particular template voxel, then this introduces an ambiguity that results in local minima of the corresponding energy function. In contrast, consider the situation in which there are a few anchor points for which correspondence (the value of the shape transformation) can be determined rather unambiguously, perhaps because each anchor point's attribute vector is very different from all but its corresponding anchor point. In that case, the shape transformation on all other (nonanchor) points could be determined via some sort of interpolation from the anchor points. This problem would not have local minima. Of course, the cost function being minimized would be only a lower-dimensional approximation, rather than a cost function involving every single voxel in the image. HAMMER is based on this fact, and it forms successive lower-dimensional cost functions that are initially based only on key anchor points and then gradually involve more and more points. More points are considered as a better estimate of the shape transformation is obtained, and potential local minima are avoided. Anchor points are defined based on how distinctive their attribute vectors are.

A third feature of HAMMER is that it is inverse-consistent, in terms of the driving correspondences. This means that if the individual is deformed to the template, instead of the converse, the mapping between any two driving points during this procedure would be identical. This feature is a computationally fast approximation for the problem of finding fully 3D inverse-consistent shape transformations that was originally proposed by Johnson and Christensen (2001). Representative results elucidating HAMMER's performance are shown in Figs. 3 and 4.

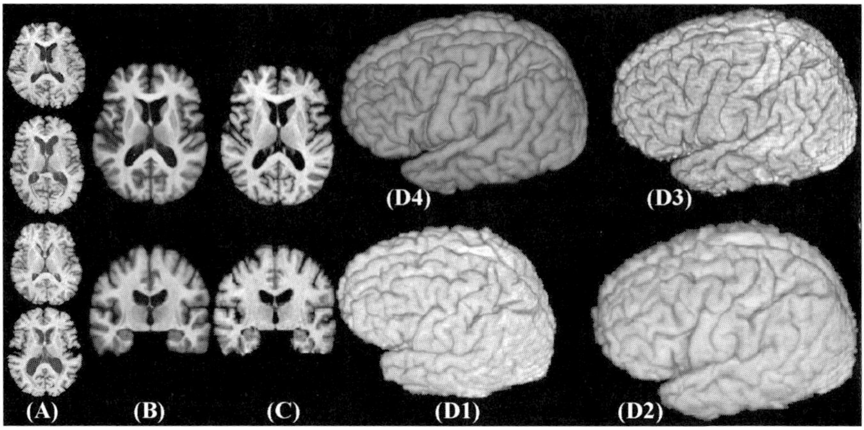

FIG. 3. Results using HAMMER warping algorithm. (A) 4 representative sections from MR images of the BLSA database (B) Representative sections from the image formed by averaging 150 images warped by HAMMER to match the template shown in (C). (D1–D4) 3D renderings of a representative case, its warped configuration using HAMMER, the template, and the average of 150 warped images, respectively. The anatomical detail seen in (B) and (D4) is indicative of the registration accuracy. The red crosses in (D3–D4) are identically placed, in order to allow visualization of point correspondences.

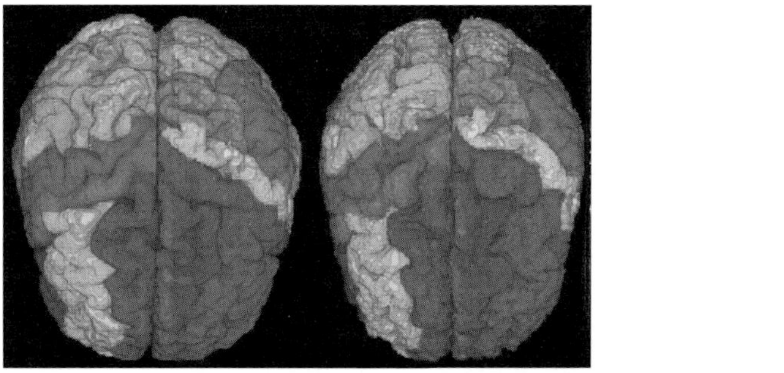

FIG. 4. Representative example of automated definition of regions of interest, by warping a prelabeled atlas (left) to an individual's MR images (the warped atlas is shown on the right as a color-coding of a volume rendering of the target brain). This automated ROI definition makes it possible to apply the method to studies with large sample sizes in a streamlined way. (See Color Insert.)

B. The Voxel-Based Morphometry Method Used by the Statistical Parametric Mapping Software

An alternative to the approach of measuring a deformation field that accurately maps a template to an individual has been adopted by the statistical parametric mapping (SPM) group (Ashburner and Friston, 2000) and used in several studies (e.g., Good *et al.*, 2002). The main rationale behind that method is that one does not necessarily need to accurately match all images in a study to the template. A coarse match, which is one that removes some but not all variability across subjects, is used to spatially normalize images into a stereotaxic coordinate system. Then, residual differences in tissue types encountered in the vicinity of an image voxel are assumed to reflect underlying morphological differences across individuals and groups. For example, if a particular brain disease causes a selective loss of tissue in a particular region of the brain, the residual density, after spatial normalization, of brain tissue in a diseased population will be lower than that in a population of healthy controls. This approach is reasonably effective when the tissue atrophy is spread over a large region of the brain, in which case accurate registration should not be critical. However, for diseases that affect brain tissue relatively focally, this approach is limited. Moreover, some of the morphological characteristics of interest are removed by the spatial normalization transformation, and therefore they are never measured by the residual density. This is unlike deformation-based methods. Even more importantly, the amount of information that is lost during spatial normalization is not controlled in any way; it might actually vary from one image to another, depending on the template's similarity to the respective individuals and the limitations of the spatial normalization algorithm. Despite these issues, voxel-based morphometry (VBM) has been freely available via the SPM software and has been used quite widely and successfully, especially in studies that do not seek subtle and spatially localized brain abnormalities.

C. The Mass-Preserving Principle for Regional Volumetric Analysis

Morphometric analysis based solely on the shape transformation that maps a template to an individual anatomy is affected by errors in determination of the shape transformation. If the warping mechanism used by a particular method is not able to perfectly match the anatomy of each individual with the anatomy of the template, then subtle structural characteristics are lost (and are never recovered in subsequent stages of the analysis). These errors can be significant obstacles to the exploration of subtle differences between two or more individuals or time points. In order to address this problem, we developed a mass-preserving framework for shape transformations, which is more robust, for reasons that are explained below. Our approach is shown schematically in Fig. 5.

In the mass-preserving framework of Regional Analysis of Volumes Examined in Normalized Space (RAVENS) (Davatzikos, 1998; Davatzikos *et al.*, 2001; Goldszal *et al.*, 1998), if the shape transformation applies an expansion to a structure, the density of the structure decreases accordingly to ensure that the total amount of tissue is preserved. Conversely, tissue density increases during contraction. Consequently, tissue density in the template's (stereotaxic) space is directly proportional to the volume of the respective structure in its original form.

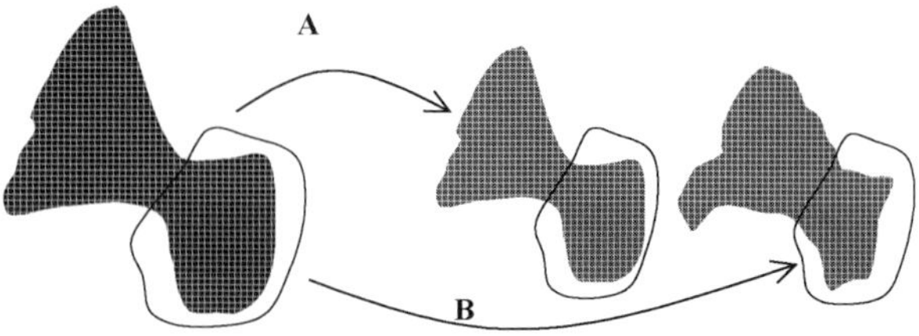

FIG. 5. Schematic representation of the mass-preserving framework of the RAVENS analysis. A shape transformation A that causes contraction of the structure as it maps it to a stereotaxic space increases the tissue density within the structure, so that the total amount of tissue is preserved. The transformation B is different (e.g., it might correspond to a result with greater error in the shape transformation). However, the total amount of tissue is preserved under both transformations, A and B. For example, integrating the tissue density within the outlined regions gives exactly the same result, and equal to the area of the outlined region in the original shape. This property is lacking in direct measurements of the shape transformation.

Therefore, regional volumetric measurements and comparisons are performed via measurements and comparisons of the respective RAVENS density maps. One RAVENS map is generated for each tissue of interest (these tissues are typically GM, WM, and CSF). We validated RAVENS on 24 MRIs that showed synthetic atrophy (Davatzikos *et al.*, 2001). Specifically, we randomly selected standard "spoiled gradient-recalled acquisition in the steady state" (SPGR) images of 12 Baltimore Longitudinal Study of Aging (BLSA) subjects, and we outlined the precentral and superior temporal gyri in all of them. We then introduced a uniform 30% volumetric contraction in these two outlined gyri, thereby generating another 12 images with synthesized atrophy in them. Figure 6 (top) shows cross-sections of a typical image before and after contraction of the precentral gyrus (segmented images are shown).

We then used RAVENS to determine the 24 respective brain-tissue density maps, and we applied a point-wise statistical analysis to them via paired t-tests. Regions of statistically significant differences between the two sets of 12 are shown in Fig. 6 (bottom), overlaid on the average WM RAVENS map of the 24 subjects (used for reference). The highlighting of the two regions in which atrophy was introduced shows the spatial specificity of RAVENS. We also (Davatzikos *et al.*, 2001) compared the sensitivity of RAVENS with the widely used VBM approach of the SPM package (Ashburner and Friston, 2000), and we found that RAVENS performed significantly better in this validation study.

III. Longitudinal Stability

There is a growing interest in longitudinal studies, which are important in the study of development, normal aging, early markers of AD, and responses to various treatments. Therefore, securing the longitudinal stability of measurements

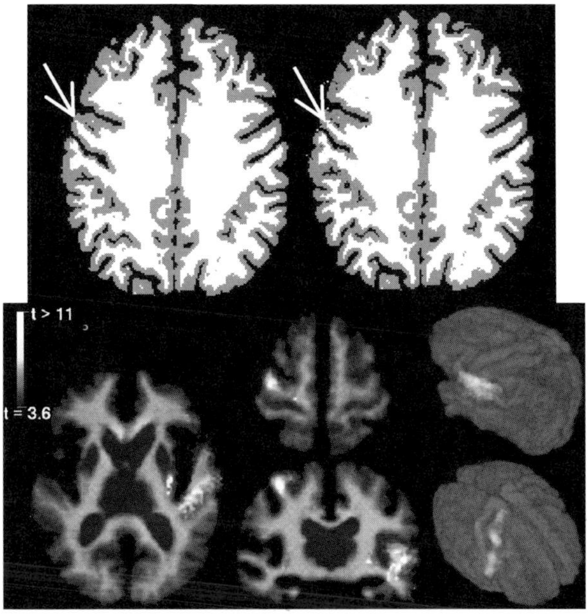

FIG. 6. (Top) Representative slices from the level of the precentral gyrus, with simulated atrophy indicated by the arrows (left is before and right is after uniform 30% atrophy within the gyrus was applied). (Bottom) Regions detected by the RAVENS analysis, overlaid on the average WM RAVENS maps of the 24 individuals. The two detected regions were exactly where atrophy was simulated. Reprinted from (Davatzikos *et al.*, 2001). (See Color Insert.)

is of paramount importance. However, in a longitudinal morphometric study, we would like to measure the shape transformation during each time point, and then examine longitudinal changes in the shape transformation. This approach is valid in theory, but limited in practice. This is because small error measurements are dramatically amplified when we calculate temporal differences. Although temporal smoothing can be applied retrospectively to shape measurements, it is far better if we incorporate temporal smoothing into the procedure for finding the shape transformation, when the image information is available to the algorithm, rather than trying to retrospectively adjust a noisy shape transformation. The issue of longitudinal measurements' robustness stands out when we undertake to measure the progression of a normal older adult into mild cognitive impairment (MCI), a situation in which it is important to have the ability to detect subtle morphological changes well before severe cognitive decline appears. To further illustrate the difficulties faced by the current 3D method, in Fig. 7 we have shown some representative longitudinal volumetric measurements from single subjects as well as from averages obtained from 90 BLSA subjects over 6 years.

In order to address this issue and be able to obtain longitudinally stable measurements, we have developed an approach to finding the shape transformation in four

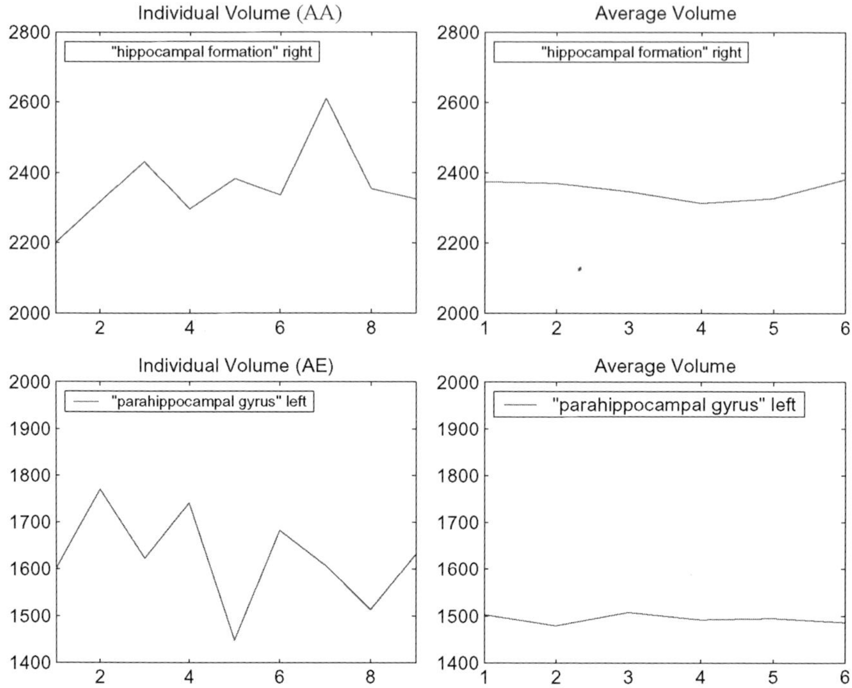

FIG. 7. Example illustrating the problems faced when applying a 3D atlas warping method independently to each time-point in a longitudinal study. (Left) Plots of volumetric measurements from two representative BLSA participants and 2 structures, using 3D HAMMER (right hippocampal formation and left parahippocampal gyrus). (Right) Analogous plots showing average volumes of these two structures, obtained by averaging the volumetric measurements of 90 BLSA participants for each of 6 years. Considerable variation is apparent. For example, the standard deviation around the baseline is about 5% for the left hippocampus of subject AD. Although a difference of 5% cannot be appreciated by visual inspection (see Fig. 8, below), it can adversely affect the accuracy of longitudinal measurements. As should be expected, variation of the average hippocampal volume is much lower (less than 1%), because of the averaging of over 90 individuals.

dimensions, where the 4th dimension is time (Shen and Davatzikos, 2004). The formulation can readily be reduced to a 3D problem if only cross-sectional data is available. We should note that a step in this direction was proposed by Freeborough and Fox (1998), who used the image at one time point as the template for shape reconstruction in another frame. However, that approach still measures longitudinal differences independently for different time points, and therefore it does not apply temporal smoothing in any other way than by using the same anatomy of a different time point as the template.

Our 4D warping approach (Shen and Davatzikos, 2004) simultaneously establishes longitudinal correspondences in the individual and correspondences

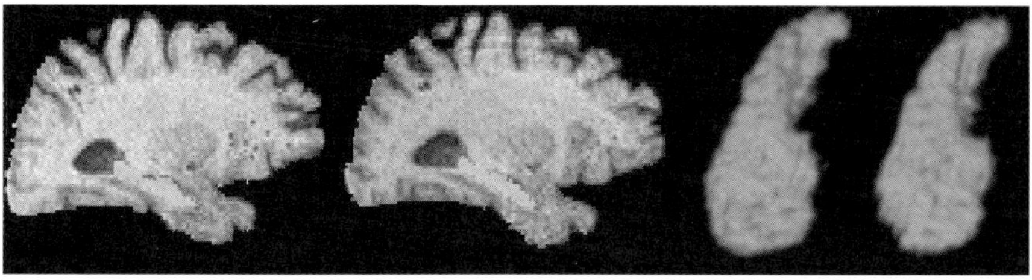

FIG. 8. Automated segmentation results using 3D HAMMER for subject AD, years 3 and 4. The 5% difference in volumes, in this case, is not readily appreciated visually from these images. (The sections are only approximately corresponding, since the scans were at slightly different orientations. 3D renderings are shown on the right.) (See Color Insert.)

between the template and the individual. This is different from the 3D warping methods, which aim at establishing only the inter-subject correspondences between the template and the individual in a single time point. Specifically, 4D HAMMER uses a fully automatic four-dimensional atlas-matching method that constrains the smoothness in both the spatial and the temporal domain during the hierarchical atlas-matching procedure, thereby producing smooth and accurate estimations of longitudinal changes. Most importantly, morphological features and matches guiding this deformation process are determined via 4D image analysis, which significantly reduces noise and improves robustness in the detection of anatomical correspondence. Put simply, image features that are consistently recognized in all time points guide the warping procedure, whereas spurious features, such as noisy edges, appear inconsistently at different time points and are eliminated. We found that this 4D approach yielded longitudinal measurements that were both more stable and more accurate than those obtained from to 3D warping (Shen and Davatzikos, 2004).

IV. Diagnosis: Putting it All Together

The VBM analysis methods described in Section II have enjoyed widespread acceptance in the past decade, since they do not rely on any *a priori* hypotheses regarding the structures to be measured but rather apply unbiased analyses of the entire set of data on a voxel-by-voxel basis. Accordingly, they highlight regions in which there are statistically significant differences between two groups, for example. However, the existence of significant differences in certain brain regions does not necessarily imply that volumetric measurements of those regions are sufficient for the diagnosis of disease. For example, say that normal control older subjects differ from patients developing MCI in the volumes of the hippocampus and the entorhinal cortex (ERC), but volumes of normal and MCI individuals are highly overlapping. In this case, diagnosis based solely on volumes of the hippocampus and the ERC could be unreliable. In the recent years, interest in integrating voxel-wise morphometric measurements into tools that can be used for diagnosis has grown quite a bit (Gerig *et al.*, 2001; Golland *et al.*, 2001; Lao *et al.*, 2003). One of the motivating factors behind these developments is the complex and spatio-temporally distributed nature of the changes that most diseases cause, particularly in the brain. For example, the anatomical structures that carry the most discriminative power are likely to depend on the stage of the disease, as the disease progressively spreads throughout various brain regions (Braak *et al.*, 1998), but they also depend on age and other demographic and genetic factors (Moffat *et al.*, 2000), since disease is to be distinguished from complex and progressively changing background (and normal) variations in anatomy and function that may depend on demographic and/or genetic background. Moreover, disease might cause changes of the image characteristics beyond those measured by volumetrics; such changes might include, for example, brightening or darkening of an MR image due to demyelination, deposition of minerals, or other macrostructural or microstructural changes caused by disease. Vascular disease also causes well-known MR signal changes, for example, brightening of the T_2-weighted signal in the WM of the brain. It is thus becoming clear that multiple modalities and multiple anatomical regions must be considered jointly in a multivariate classification fashion, in order for the desirable diagnostic power to be achieved. Moreover, regions that are less affected by disease should be considered alongside known affected regions (which, in the example of AD, might include primarily temporal-lobe structures in relatively early disease stages), since differential atrophy or image-intensity changes between these regions are likely to further amplify diagnostic accuracy and discrimination from a background of normal variation.

The approach described by Lao *et al.* (2003) is based on the RAVENS mass-preserving morphological representation described earlier in this chapter. It hierarchically decomposes a RAVENS map into images of different scales, each of which captures the morphology of the anatomy of interest at a different degree of spatial resolution. The most important morphological parameters are then selected and used in conjunction with a nonlinear pattern-classification technique to form a hypersurface, the high-dimensional analogue to a surface, which is constructed in such a way that it optimally separates two groups of interest, for example, normal controls and patients of a particular disease. Effectively, that approach defines a nonlinear combination of a large number of volumetric measurements from the entire brain, each taken at a different scale. The scale typically depends on the size of the respective anatomical structure and the size of the region that is most affected by the disease. This nonlinear combination of volumetric measurements is the best way to distinguish between the two groups, and therefore the best way to perform diagnosis via classification of a new scan into patients or normal controls.

V. Neuroimaging Studies of Aging, Schizophrenia, and Genetic Influences on Brain Development

Voxel-based morphometric analysis has been adopted in a variety of studies. We will briefly summarize three studies in which we have applied these techniques, in order to illustrate their use.

Baltimore Longitudinal Study of Aging. The neuroimaging arm of the BLSA was initiated in 1993, and it is now in its 12th year (Resnick *et al.*, 2000, 2003).

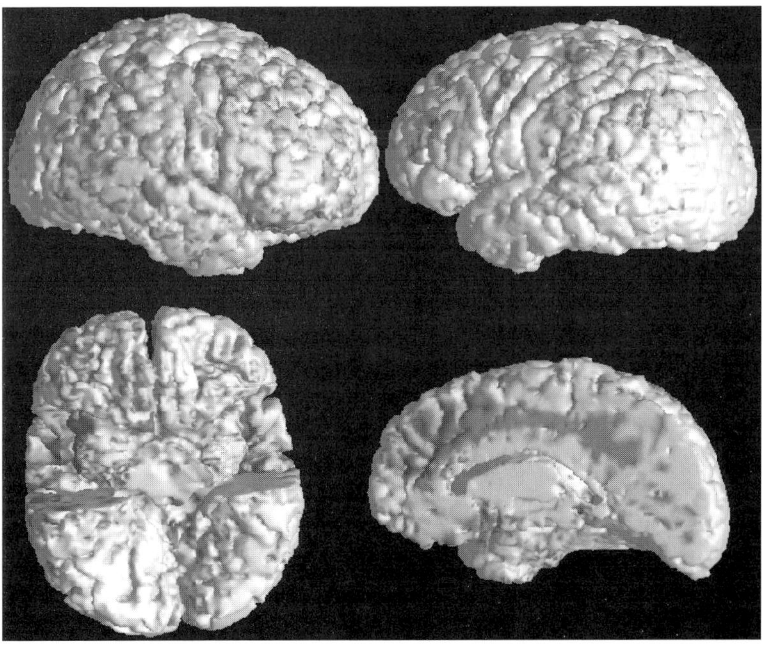

FIG. 9. Regions displaying significant longitudinal grey matter atrophy over a 4-year period. Estimates of longitudinal atrophy were determined by segmentation into GM, WM and CSF, then applying the mass-preserving RAVENS methodology described in the text, which deforms each individual's brain into alignment with a template brain, while preserving tissue mass by converting it to density. Voxel-based analysis of the resultant tissue density maps is equivalent to voxel-based volumetric analysis and therefore of atrophy quantification. Reprinted from (Resnick *et al.*, 2003). (See Color Insert.)

Approximately 150 healthy older adults have been followed annually over this period with structural, functional (PET-O^{15}), and neuropsychological evaluations. Analysis and integration of these data aim at determining early markers of AD that occur in a background of normal structural and functional changes. The RAVENS methodology described above has been applied to structural images from this study in order to measure patterns of significant longitudinal atrophy in normal healthy adults. Figure 9 displays a 3D rendering of the regions that displayed significant longitudinal atrophy over a 4-year period.

In addition to atrophy and other shape changes, signal changes are very pronounced even in normal aging. Most importantly, the WM tends to get darker in standard T1-SPGR images, perhaps due to underlying vascular disease, demyelination, mineral deposition, or other degenerative processes. The methods described in this chapter can also be used in voxel-based analysis of signal changes. In 2002 (Davatzikos and Resnick, 2002) we studied WM darkening in the BLSA sample. Figure 10 displays a statistical parametric map of voxel-wise

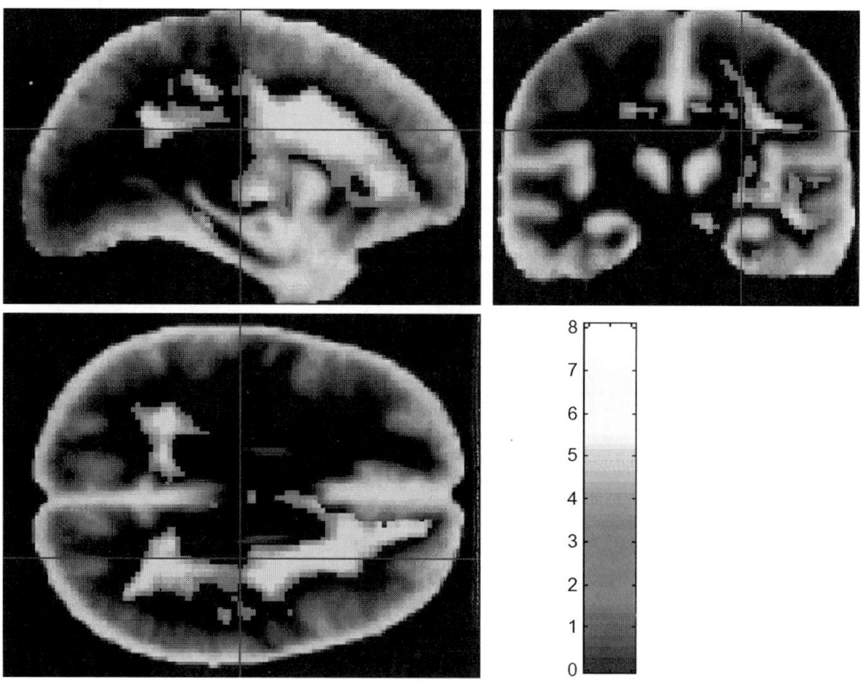

FIG. 10. Statistical parametric maps of T1 signal darkening with normal aging, obtained using the RAVENS methodology for elastic registration. T1 darkening might be due to a variety of factors, such as vascular disease, demyelination, mineral deposition or other degenerative processes. Reprinted from (Davatzikos and Resnick, 2002). (See Color Insert.)

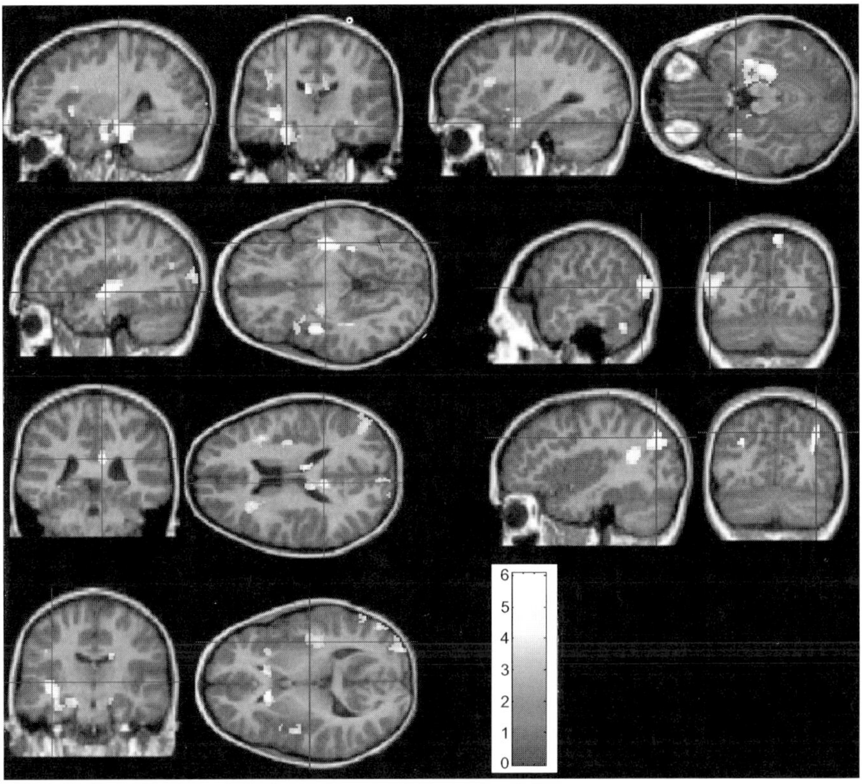

FIG. 11. Visual summary of several detected regions with significant group differences between XXY brains and normal control brains. The underlying image is the template that we used to normalize individual brains. Color-coding was based on the values of the t-statistic. Only the voxels with significant group differences, i.e., the corrected p-values exceeding a significance threshold 0.005, are shown. (A) *Left* hippocampal formation; (B) *Left* superior temporal gyrus; (C) Cingulate region; (D) *Left* Insula; (E) *Right* Amygdala; (F) *Left* middle temporal gyri; (G) *Right* parietal lobe WM. Reprinted from (Shen *et al.*, 2004). (See Color Insert.).

analysis of longitudinal change. Quantification of signal changes is just as important as measurement of atrophy, for many reasons. First, many degenerative changes can be first manifested as signal changes, prior to tissue atrophy. Second, in early AD stages, vascular disease or demyelination of WM can have an effect on cognitive decline that is additive to that of AD pathology. Therefore, structural and functional changes caused by vascular disease, which most often change MR signal characteristics, must be well characterized in order for the imaging signatures that are specific to AD to be isolated. Moreover, our previous work suggests that, from certain perspectives, signal changes carry information orthogonal to atrophy (Davatzikos and Resnick, 2002).

Neuroimaging of schizophrenia. We have performed voxel-based analysis of the RAVENS tissue-density maps in a study of 158 normal controls and schizophrenia patients. Using voxel-based analysis of the RAVENS maps, we have identified spatially complex patterns of tissue-volume differences between healthy controls and patients. Moreover, using multivariate nonlinear classification methods (described in Section IV), we have identified morphological signatures that are unique to the disease, which can potentially be used for early diagnostic purposes.

Neuroimaging of XXY children. We recently examined brain morphometry variation associated with XXY males (Klinefelter's syndrome) by using an automated whole-brain volumetric analysis method (Shen *et al.*, 2004). The application to 34 XXY males and 62 normal male controls revealed that, relative to the brains of normal controls, there was pronounced volume reduction in the brains of XXY males, localized at the insula, temporal gyri, amygdala, hippocampus, cingulate gyri, and occipital gyri. Most of these statistically significant regions are located in the GM structures, with the exception of one cluster of atrophy involved in WM structure, i.e., WM of the right parietal lobe. Figure 11 shows some representative results from the statistical analysis of the RAVENS tissue-density maps obtained in this study.

VI. Conclusion

Voxel-based morphometric analysis provides researchers with an unbiased way of examining high-dimensionality image data, inasmuch as it does not rely on any *a priori* hypotheses regarding the regions of interest to be examined and instead analyzes the entire image dataset and identifies regions in which morphological measurements are of interest (e.g., regions in which volumetric measurements differ between normal controls and patients or between two serial scans). There is a plethora of methods for voxel-based morphometric analysis (Ashburner *et al.*, 2003), and each of these methods has merits and limitations. We described a technique based on high-dimensional elastic warping of brain images, formulated in a mass-preserving framework so that tissue volumes are properly preserved and measured in this process. We also discussed advantages of high-dimensional pattern classification and multivariate analysis over voxel-based (mass-univariate) methods, since the former capture complex associations among image measurements in different parts of the brain. Finally, we presented a few representative studies in which tissue-density statistical analysis was used as a means for volumetric analysis. Future work in this field is certain to focus on the pursuit of better methods for accurately warping one anatomical image to another, as well as powerful methods for multivariate analysis, which is a particularly difficult task in view of the huge dimensionality of spatio-temporal image measurements.

References

Ashburner, J., Hutton, C., Frackowiak, R. S. J., Johnsrude, I., Price, C., and Friston, K. J. (1998). Identifying global anatomical differences: Deformation-based morphometry. *Hum. Brain Mapp.* **6,** 348–357.

Ashburner, J., and Friston, K. J. (2000). Voxel-based morphometry: The methods. *Neuroimage* **11,** 805–821.

Ashburner, J., Csernansky, J. G., Davatzikos, C., Fox, N. C., Frisoni, G. B., and Thompson, P. M. (2003). Computer-assisted imaging to assess brain structure in healthy and diseased brains. *Lancet (Neurology)* **2,** 79–88.

Bobinski, M., de Leon, M. J., Convit, A., De Santi, S., Wegiel, J., Tarshish, C. Y., Saint Louis, L. A., and Wisniewski, H. M. (1999). MRI of entorhinal cortex in mild Alzheimer's disease. *Lancet* **353,** 38–40.

Bookstein, F. L. (1989). Principal Warps: Thin-Plate Splines and the Decomposition of Deformations. *IEEE Trans. Pattern Anal. Mach. Intell.* **11,** 567–585.

Braak, H., Braak, E., Bohl, J., and Bratzke, H. (1998). Evolution of Alzheimer's disease–related cortical lesions. *J. Neural. Transm.* **54**(Suppl.), 97–106.

Christensen, G., Rabbitt, R. D., and Miller, R. I. (1994). 3D brain mapping using a deformable neuroanatomy. *Phys. Med. Biol.* **39,** 609–618.

Christensen, G. E. (1999). Consistent linear–elastic transformations for image matching. *Lec. Notes in Comp. Sci.: Proc. MICCAI '98* **1613,** 224–237.

Christensen, G. E., and Johnson, H. J. (2001). Consistent image registration. *IEEE Trans. Med. Imaging* **20,** 568–582.

Chui, H., and Rangarajan, A. (2003). A new point matching algorithm for non-rigid registration. *Comput. Vis. Image Underst.* **89,** 114–141.

Chung, M. K., Worsley, K. J., Paus, T., Cherif, C., Collins, D. L., Giedd, J. N., Rapoport, J. L., and Evanst, A. C. (2001). A unified statistical approach to deformation-based morphometry. *Neuroimage* **14,** 595–606.

Collins, D. L., Neelin, P., Peters, T. M., and Evans, A. C. (1994). Automatic 3D intersubject registration of MR volumetric data in standardized Talairach space. *J. Comput. Assist. Tomogr.* **18,** 192–205.

Collins, L., Peters, T. M., Dai, W., and Evans, A. C. (1992). Model-Based Segmentation of Individual Brain Structures from MRI Data. *Proc. SPIE Conf. Visualization Biomed. Comp.* **1808,** 10–23.

Convit, A., De Leon, M. J., Tarshish, C., De Santi, S., Tsui, W., Rusinek, H., and George, A. (1997). Specific hippocampal volume reductions in individuals at risk for Alzheimer's disease. *Neurobiol. Aging* **18,** 131–138.

Cuenod, C. A., Denys, A., Michot, J. L., Jehenson, P., Forette, F., Kaplan, D., Syrota, A., and Boller, F. (1993). Amygdala atrophy in Alzheimer's disease. An *in vivo* magnetic resonance imaging study. *Arch. Neurol.* **50,** 941–945.

Davatzikos, C. (1996). Spatial normalization of 3D images using deformable models. *J. Comput. Assist. Tomogr.* **20,** 656–665.

Davatzikos, C., Vaillant, M., Resnick, S., Prince, J. L., Letovsky, S., and Bryan, R. N. (1996). A computerized approach for morphological analysis of the corpus callosum. *J. Comput. Assist. Tomogr.* **20,** 88–97.

Davatzikos, C. (1998). Mapping of image data to stereotaxic spaces. *Hum. Brain Mapp.* **6,** 334–338.

Davatzikos, C., and Resnick, S. M. (1998). Sex differences in anatomic measures of interhemispheric connectivity: Correlations with cognition in men but not in women. *Cereb. Cortex* **8,** 635–640.

Davatzikos, C., Genc, A., Xu, D., and Resnick, S. M. (2001). Voxel-Based Morphometry Using the RAVENS Maps: Methods and Validation Using Simulated Longitudinal Atrophy. *Neuroimage* **14,** 1361–1369.

Davatzikos, C., and Resnick, S. M. (2002). Degenerative age changes in white matter connectivity visualized *in vivo* using magnetic resonance imaging. *Cereb. Cortex* **12,** 767–771.

De Santi, S., de Leon, M. J., Rusinek, H., Convit, A., Tarshish, C. Y., Roche, A., Tsui, W. H., Kandil, E., Boppana, M., Daisley, K., Wang, G. J., Schlyer, D., and Fowler, J. (2001). Hippocampal formation glucose metabolism and volume losses in MCI and AD. *Neurobiol. Aging* **22,** 529–539.

deToledo-Morrell, L., Sullivan, M. P., Morrell, F., Wilson, R. S., Bennett, D. A., and Spencer, S. (1997). Alzheimer's disease: *In vivo* detection of differential vulnerability of brain regions. *Neurobiol. Aging* **18,** 463–438.

Dickerson, B. C., Goncharova, I., Sullivan, M. P., Forchetti, C., Wilson, R. S., Bennett, D. A., Beckett, L. A., and deToledo-Morrell, L. (2001). MRI-derived entorhinal and hippocampal atrophy in incipient and very mild Alzheimer's disease. *Neurobiol. Aging* **22,** 747–754.

Du, A. T., Schuff, N., Amend, D., Laakso, M. P., Hsu, Y. Y., Jagust, W. J., Yaffe, K., Kramer, J. H., Reed, B., Norman, D., Chui, H. C., and Weiner, M. W. (2001). Magnetic resonance imaging of the entorhinal cortex and hippocampus in mild cognitive impairment and Alzheimer's disease. *J. Neurol. Neurosurg. Psychiatry* **71,** 441–447.

Ferrant, M., Warfield, S., Guttman, C. R. G., Mulkern, R. V., Jolesz, F., and Kikinis, R. (1999). 3D image matching using a finite element based elastic deformation model. *Proc. MICCAI '98,* 202–209.

Freeborough, P. A., and Fox, N. C. (1998). Modeling Brain Deformations in Alzheimer's Disease by Fluid Registration of Serial 3D MR Images. *J. Comput. Assist. Tomogr.* **22,** 838–843.

Frisoni, G. B., Beltramello, A., Weiss, C., Geroldi, C., Bianchetti, A., and Trabucchi, M. (1996). Linear measures of atrophy in mild Alzheimer disease. *Am. J. Neuroradiol.* **17,** 913–923.

Friston, K. J., Ashburner, J., Frith, C. D., Poline, J. B., Heather, J. D., and Frackowiak, R. S. J. (1995). Spatial registration and normalization of images. *Hum. Brain Mapp.* **2,** 165–189.

Gerig, G., Styner, M., and Lieberman, J. (2001). Shape versus size: Improved understanding of the morphology of brain structures. *In* "MICCAI 2001," (W. J. Niessen and M. A. Viergever, Eds.), pp. 24–32. Kluver, Utrecht, the Netherlands.

Goldszal, A. F., Davatzikos, C., Pham, D., Yan, M., Bryan, R. N., and Resnick, S. M. (1998). An image processing protocol for the analysis of MR images from an elderly population. *J. Comput. Assist. Tomogr.* **22,** 827–837.

Golland, P., Grimson, W. E. L., and Kikinis, R. (1999). Statistical shape analysis using fixed topology skeletons: Corpus callosum study. *Lect. Notes Comp. Sci.* **1613,** 382–387.

Golland, P., Grimson, W. E. L., Shenton, M. E., and Kikinis, R. (2001). Deformation Analysis for Shape Based Classification. *Lect. Notes Comp. Sci.* **2082,** 517–530.

Good, C. D., Scahill, R. I., Fox, N. C., Ashburner, J., Friston, K. J., Chan, D., Crum, W. R., Rossor, M. N., and Frackowiak, R. S. J. (2002). Automatic differentiation of anatomical patterns in the human brain: Validation with studies of degenerative dementias. *Neuroimage* **17,** 29–46.

Grenander, U. (1985). Advances in pattern theory: The 1985 Rietz lecture Annals of Statitics, 17: 1–30.

Jack, C. R., Petersen, R. C., Y.C., X., O'Brien, P. C., Smith, G. E., Ivnik, R. J., Boeve, B. F., Waring, S. C., Tangalos, E., and Kokmen, E. (1999). Prediction of AD with MRI-based hippocampal volume in mild cognitive impairment. *Neurology* **52,** 1397–1403.

Jack, C. R., Jr., Petersen, R. C., Xu, Y. C., Waring, S. C., O'Brien, P. C., Tangalos, E. G., Smith, G. E., Ivnik, R. J., and Kokmen, E. (1997). Medial temporal atrophy on MRI in normal aging and very mild Alzheimer's disease. *Neurology* **49,** 786–794.

Johnson, H. J., and Christensen, G. (2001). Landmark and intensity-based consistent thin-plate spline image registration. *Proc. Conf. Informat. Processing Med. Imaging, LNCS* **2081,** 329–343.

Joshi, S., Pizer, S., Fletcher, P. T., Thall, A., and Tracton, G. (2001). Multi-scale 3-D Deformable Model Segmentation Based on Medial Description. *Lect. Notes in Comp. Sci.* **2082,** 64–77.

Killiany, R. J., Moss, M. B., Albert, M. S., Sandor, T., Tieman, J., and Jolesz, F. (1993). Temporal lobe regions on magnetic resonance imaging identify patients with early Alzheimer's disease. *Arch. Neurol.* **50,** 949–954.

Killiany, R. J., Gomez-Isla, T., Moss, M., Kikinis, R., Sauder, T., Jolesz, F., Tauzi, R., Jones, K., Hyman, B. T., and Albert, M. S. (2000). Use of structural magnetic resonance imaging to predict who will get Alzheimer's disease. *Ann. Neurol.* **47,** 430–439.

Krasuski, J. S., Alexander, G. E., Horwitz, B., Daly, E. M., Murphy, D. G., Rapoport, S. I., and Schapiro, M. B. (1998). Volumes of medial temporal lobe structures in patients with Alzheimer's disease and mild cognitive impairment (and in healthy controls). *Biol. Psychiatry* **43,** 60–68.

Laakso, M. P., Soininen, H., Partanen, K., Helkala, E. L., Hartikainen, P., Vainio, P., Hallikainen, M., Hanninen, T., and Riekkinen, P. J. Sr. (1995). Volumes of hippocampus, amygdala and frontal lobes in the MRI-based diagnosis of early Alzheimer's disease: Correlation with memory functions. *J. Neural Transm. Park. Dis. Dement. Sect.* **9,** 73–86.

Laakso, M. P., Hallikainen, M., Hanninen, T., Partanen, K., and Soininen, H. (2000). Diagnosis of Alzheimer's disease: MRI of the hippocampus vs delayed recall. *Neuropsychologia* **38,** 579–584.

Lao, Z., Shen, D., Xue, Z., Karacali, B., Resnick, S. M., and Davatzikos, C. (2003). Morphological classification of brains via high-dimensional shape transformations and machine learning methods. *Neuroimage* **21,** 46–57.

Lehericy, S., Baulac, M., Chiras, J., Pierot, L., Martin, N., Pillon, B., Deweer, B., Dubois, B., and Marsault, C. (1994). Amygdalohippocampal MR volume measurements in the early stages of Alzheimer disease. *Am. J. Neuroradiol.* **15,** 929–937.

Miller, M. I., Christensen, G., Amit, Y., and Grenander, U. (1993). Mathematical textbook of deformable neuroanatomies. *Proc. Nat. Acad. Sci.* **90,** 11944–11948.

Moffat, S. D., Szekely, C. A., Zonderman, A. B., Kabani, N. J., and Resnick, S. M. (2000). Longitudinal change in hippocampal volume as a function of apolipoprotein E genotype. *Neurology* **55,** 134–136.

Pizer, S., Fritsch, D. S., Yushkevich, P. A., Johnson, V. E., and Chaney, E. L. (1999). Segmentation, registration and measurement of shape variation via image object shape. *IEEE Trans. Med. Imaging* **18,** 851–865.

Rangarajan, A., Chui, H., and Bookstein, F. L. (1997). The softassign procrustes matching algorithm. *In* "Information Processing in Medical Imaging" (J. Duncan and G. Gindi, Eds.), pp. 29–70. Springer, Poultney, Vermont.

Resnick, S. M., Goldszal, A., Davatzikos, C., Golski, S., Kraut, M. A., Metter, E. J., Bryan, R. N., and Zonderman, A. B. (2000). One-year age changes in MRI brain volumes in older adults. *Cereb. Cortex* **10,** 464–472.

Resnick, S. M., Pham, D. L., Kraut, M. A., Zonderman, A. B., and Davatzikos, C. (2003). Longitudinal magnetic resonance imaging studies of older adults: A shrinking brain. *J. Neurosci.* **23,** 3295–3301.

Rexilius, J., Warfield, S. K., Guttman, C. R. G., Wei, X., Benson, R., Wolfson, L., Shenton, M. E., Handels, H., and Kikinis, R. (1999). A novel nonrigid registration algorithm and applications. *Proc. MICCAI '98,* 202–209.

Rosen, A. C., Prull, M. W., Gabrieli, J. D., Stoub, T., O'Hara, R., Friedman, L., Yesavage, J. A., and deToledo-Morrell, L. (2003). Differential associations between entorhinal and hippocampal volumes and memory performance in older adults. *Behav. Neurosci.* **117,** 1150–1160.

Rueckert, D., Sonoda, L. I., Hayes, C., Hill, D. L. G., Leach, M. O., and Hawkes, D. J. (1999). Non-rigid registration using free-form deformations: Application to breast MR images. *IEEE Trans. Med. Imaging* **18,** 712–721.

Sandor, S., and Leahy, R. (1997). Surface based labelling of cortical anatomy using a deformable atlas. *IEEE Trans. Med. Imaging* **16,** 41–54.

Shen, D., and Davatzikos, C. (2002). HAMMER: Hierarchical attribute matching mechanism for elastic registration. *IEEE Trans. Med. Imaging* **21,** 1421–1439.

Shen, D. (2004). 4D image warping for measurement of longitudinal brain changes. *In* "Proceedings of the IEEE International Symposium on Biomedical Imaging." Accepted for publication IEEE, Arlington, Va.

Shen, D., and Davatzikos, C. (2004). Measuring temporal morphological changes robustly in brain MR images via 4-dimensional template warping. *Neuroimage* **21,** 1508–1517.

Shen, D. G., and Davatzikos, C. (2003). Very high resolution morphometry using mass-preserving deformations and HAMMER elastic registration. *Neuroimage* **18,** 28–41.

Shen, D., Liu, D., Hong Liu Clasen, L., Giedd, J., and Davatzikos, C. (2004). Automated Morphometric Study of Brain Variation in XXY Males. *Neuroimage* **23,** 648–653.

Styner, M., and Gerig, G. (2001). Medial models incorporating object variability for 3D shape analysis. *Lect. Notes Comp. Sci.* **2082,** 502–516.

Szekely, G., Kelemen, A., Brechbuhler, C., and Gerig, G. (1996). Segmentation of 2-D and 3-D objects from MRI volume data using constrained deformations of flexible Fourier contour and surface models. *Med. Image Anal.* **1,** 19–34.

Thirion, J. P. (1998). Image matching as a diffusion process: An analogy with Maxwell's demons. *Medical Image Analysis* **2,** 243–260.

Thompson, D. W. (1917). "On Growth and Form." Cambridge University Press, Cambridge.

Thompson, P., and Toga, A. W. (1996). A surface-based technique for warping three-dimensional images of the brain. *IEEE Trans. Med. Imaging* **15,** 402–417.

Thompson, P. M., Mac Donald, D., Mega, M. S., Holmes, C. J., Evans, A., and Toga, A. W. (1997). Detection and mapping of abnormal brain structure with a probabilistic atlas of cortical surfaces. *J. Comput. Assist. Tomogr.* **21,** 567–581.

Viola, P., and Wells, W. M., III (1995). Alignment by maximization of mutual information. *In* "Proceedings of the International Conference on Computer Vision" (E. Grimson, S. Shafer, A. Blake, and K. Sugihara, Eds.), p. 16–23. IEEE, Los Alamitos, California.

Wang, Y., and Staib, L. H. (1999). Elastic model-based non-rigid registration incorporating statistical shape information. *Lect. Notes Comp. Sci., MICCAI '98* **1496,** 1162–1173.

Wang, Y., Peterson, B. S., and Staib, L. H. (2003). 3D brain surface matching based on geodesics and local geometry. *Comput. Vis. Image Underst.* **89,** 252–271.

Wells, I., William, M., Viola, P., and Kikinis, R. (1995). Multi-modal volume registration by maximization of mutual information. *In* "Medical Robotics and Computer Assisted Surgery," p. 55–62. Wiley, New York.

Xu, Y., Jack, C. R., Jr., O'Brien, P. C., Kokmen, E., Smith, G. E., Ivnik, R. J., Boeve, B. F., Tangalos, R. G., and Petersen, R. C. (2000). Usefulness of MRI measures of entorhinal cortex versus hippocampus in AD. *Neurology* **54,** 1760–1767.

Xue, Z., Shen, D., and Davatzikos, C. (2003). Correspondence detection using wavelet-based attribute vectors. *In* "Medical Image Computing and Computer-Assisted Intervention (MICCAI)," pp. 762–770. Springer-Verlag, Montreal.

Xue, Z., Shen, D., and Davatzikos, C. (2004). Determining correspondence in 3D MR brain images using attribute vectors as morphological signatures of voxels. *IEEE Trans. Med. Imaging.* **23,** 1276–1291.

THE CUTTING EDGE OF fMRI AND HIGH-FIELD fMRI

Dae-Shik Kim

Center for Biomedical Imaging, Boston University School of Medicine
Boston, Massachusetts 02118

I. Brain Activities are Compartmentalized

One of the most characteristic features of the brain's information processing is its functional compartmentalization. Cytoarchitectonically and myloarchitectonically distinct areas (Brodmann, 1909) of the mammalian brain perform different functions, and, within such brain areas, neurons with similar receptive and response properties are clustered into isofunctional domains. Such cortical parcellation has been especially well studied in the visual systems of primates and felines. In these species, visual functions are localized at the scale of centimeters to millimeters in multiple cortical areas, each processing a separate aspect of the visual world (Felleman and Van Essen, 1991; Payne and Peters, 2002). Furthermore, within each visual area, there is a topographic representation of the receptive field properties in a "cortical map." For example, neurons with similar response properties, such as ocular dominance, orientation, and direction preferences, are clustered into "columns" at a submillimeter scale, spanning the entire cortical plate from the pia to the white matter.

Since the pioneering studies by Hubel and Wiesel (1962), the layout, function, and plasticity of cortical columns and maps have been studied extensively in mammalian brains by a variety of techniques. Each of the traditional mapping

INTERNATIONAL REVIEW OF
NEUROBIOLOGY, VOL. 66
DOI: 10.1016/S0074-7742(05)66005-9

0074-7742/05 $35.00

techniques, however, suffers from a fundamental limitation: While single and multiunit recording have set the gold standard for definition of neuronal receptive field properties, it is very difficult to use these techniques to study the overall layout of the receptive field properties in a given area. This problem can be partly overcome by the use of metabolic activity labels such as 2-deoxyglucose (2DG) (Sokoloff *et al.*, 1977). However, this method cannot be used for *in vivo* studies, as the experimental animal has to be sacrificed before the radioactive labeling pattern can be visualized. Furthermore, this method allows the cortical pattern of activity in response to a maximum of two different stimuli to be labeled, and this significantly limits the number of independent map properties that can be studied (Lowel *et al.*, 1987). The more recently developed "optical imaging of intrinsic signals" method allows the simultaneous recording of neuronal activity over large areas of cortex (Frostig *et al.*, 1990). Utilizing the fact that gray-matter tissues dominated by deoxyhemoglobin absorb relatively more light at near-infrared wavelengths (605 $\pm$ 10 nm), one can discern the "active" areas of the cortex by shining light on the exposed cortical surface. While this optical imaging technique has been extensively utilized to examine the tangential organization of cortical maps (Hubener *et al.*, 1997; Kim *et al.*, 1999), unfortunately, it provides only very limited depth information, because only signals from the superficial 300–500 μm can be accessed (Bonhoeffer and Grinvald, 1993; Stetter and Obermayer, 1999; Stetter *et al.*, 2000; Yae *et al.*, 1992). Furthermore, optical imaging at this spatial resolution is an invasive technique, since a craniotomy has to be performed in order for the cortical surface to be exposed. The routine use of optical imaging to assess human cognitive functions is therefore not possible at this point except at a very coarse spatial resolution, which would be unsuitable for cognitive studies.

II. Blood Oxygen Level–Dependent Functional MRI

These limitations are the reasons why, in the potpourri of neuroimaging techniques used for understanding brain function, functional magnetic resonance imaging (fMRI) has come to play a dominant role in both human-model and animal-model studies. The most frequently employed fMRI method is based on blood oxygen level–dependent (BOLD) contrast and was first described by Ogawa *et al.* (1990) in rodent-brain studies and was subsequently used to generate noninvasive functional images in the human brain (Bandettini *et al.*, 1992a; Kwong *et al.*, 1992; Ogawa *et al.*, 1992). The great utility and conceptual fascination of BOLD is based on the fact that, in it, the paramagnetic deoxyhemoglobin (doxyHb) sequestered in red blood cells is employed as nature's own contrast agent, thus eliminating the need for external (and usually invasive) contrast agents.

To this end, BOLD contrast originates from the intravoxel magnetic-field inhomogeneity induced by deoxyhemoglobin sequestered in compartmentalized red blood cells. Magnetic susceptibility differences between the deoxyHb-containing compartments and the surrounding space devoid of this strongly paramagnetic molecule generate magnetic-field gradients across and near the boundaries of these compartments. Therefore, signal intensities in MR images sensitized to BOLD contrast are altered if the regional deoxyHb content is perturbed. This occurs in the brain because of spatially specific metabolic and hemodynamic responses to enhanced neuronal activity. Increases in neuronal and metabolic activity result in a *reduction* of the relative deoxyhemoglobin concentration due to an increase in blood flow (and hence an increased supply of fresh oxyhemoglobin) that is not matched by a commensurate level of increase in oxygen consumption. Consequently, in conventional BOLD fMRI, brain "activity" can be measured as an *increase* in T_2- or T_2*-weighted MR signals (Bandettini *et al.*, 1992b; Kwong *et al.*, 1992; Ogawa *et al.*, 1992). Since its introduction little more than 10 years ago, BOLD fMRI has been successfully applied—among numerous other examples—to precisely localize the cognitive (Wagner *et al.*, 1998), motor (Kim *et al.*, 1993), and perceptual (Engel *et al.*, 1997; Sereno *et al.*, 1995; Tootell *et al.*, 1997) functions of the human cerebral cortex.

III. Neural Correlates of BOLD fMRI

While BOLD neuroimaging studies have provided researchers with unprecedented amounts of insight into the workings of the human brain *in vivo*, the explanatory power of BOLD fMRI is currently limited, as there is a fundamental gap in our understanding of the linkage between the observable BOLD contrast and the underlying neuronal events. While the question of "the missing link" has many facets, the most pressing question is that of *spatial correspondence*. In other words: does an increase in BOLD contrast in a given area accurately reflect the location of the underlying neural activity changes? This is naturally a question of great potential concern. Just as an illustration, consider a recent study in which the topographic representations of the cortical fields in the somatosensory cortex of anesthetized monkeys, as revealed using functional MRI, were suggested to be in concordance with the more traditional electrophysiological maps for only about 55% of the cases (Disbrow *et al.*, 2000). If this holds true, then results such as these would greatly diminish the usefulness of cognitive neuroimaging *in toto*. To this end, recent studies from several labs suggest that there may be an intriguing relationship between the ability of BOLD to pinpoint the underlying neuronal events and the fundamental hemodynamic processes that give rise to BOLD. This possibility stems from the fact that the BOLD signal is not a direct measure of neuronal activity *per se.* Rather, it is a complex convolution of changes in cerebral hemodynamic balance following a focal neural activity. Therefore, for a proper understanding of the BOLD signals and their fundamental spatial specificity, the size, geometry, and location of the cortical vasculature surrounding the neural tissue must be taken into consideration. Here, for the sake of further discussion, we will define a microvascular/tissue component as an area of rapid exchange between blood water and tissue water (i.e., capillaries and adjacent tissues). In the parenchymal tissue of the cat cerebral cortex, such capillaries form a dense network with an average spacing of about 20 μm (Pawlik *et al.*, 1981). On the other hand, the macrovascular component comprises of arteries, small arterioles, small venules and veins where such an exchange does not exist. The vascular origin of the BOLD effect depends heavily on *magnetic field strength* (Ogawa *et al.*, 1998; Weisskoff *et al.*, 1994):

$$R_2{}^* = \alpha\{\Delta\chi_o\omega_o(1-Y)\}b_{vl} \text{ (large blood vessels)} \tag{1}$$

$$R_2{}^* = \eta\{\Delta\chi_o\omega_o(1-Y)\}^2\, b_{vs}\, p\text{(small blood vessels)} \tag{2}$$

where $R_2{}^* = 1/T_2{}^*$, α and η are constants, ω_o is the strength of the external magnetic field in frequency units (rad/sec), $\Delta\chi_o$ is the maximum susceptibility difference expected in the presence of fully deoxygenated blood, Y is the fraction of oxygenated blood present, $\{\Delta\chi_o\omega_o(1-Y)\}$ is the frequency shift due to the susceptibility difference between the deoxyhemoglobin-containing blood vessels, b_{vl} is the blood volume for large blood vessels and b_{vs} is the small-vessel blood volume, and p is the fraction of active small vessels (i.e., filled with deoxyhemoglobin-containing red blood cells). The most important feature of these equations is that the BOLD contrast grows as the *square* of the external magnetic field for small blood vessels, where the effect is dominated by dynamic averaging of magnetic-field inhomogeneities surrounding deoxyhemoglobin-containing blood vessels (Eq. 2). In contrast, the dependence on the external magnetic field is *linear* for large blood vessels (Eq. 1). This, in turn, suggests that the use of a high-field scanner *per se* will not only increase the overall signal-to-noise ratio (SNR) but will specifically increase the functional signal components from more neurophysiologically relevant parenchymal tissue.

The neurophysiological relevance of BOLD will further depend on the following questions.

A. Do BOLD Signal Changes Reflect the Magnitude of Neural Activity Change Linearly?

The amplitude of the fMRI signal-intensity change has been used to obtain information beyond simple identification of spatial compartmentalization of brain function by correlating variations in amplitude with behavioral measures, or the

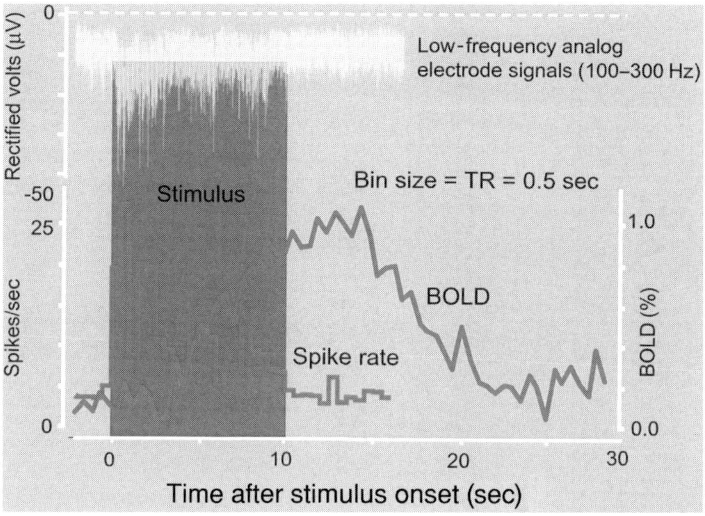

FIG. 1. The time course of BOLD and single-unit recordings from the same cortical location. Identical visual stimuli were used for fMRI and subsequent single-unit recording sessions. Blue trace: peristimulus histogram of the spike activity. Bin size for the histogram = 0.5 sec = TR (scan repeat time) for fMRI. Red trace: BOLD percent changes during visual stimulation. X-axis: time after stimulus onset. Left Y-axis: Spikes per second. Right Y-axis: BOLD percent changes. Gray box: stimulus duration. The black trace above indicates the original low-frequency analog signals (100–300 Hz) underlying the depicted spike counts. (See Color Insert.)

electroencephalograph (EEG) response. A small but increasing body of results seems to indicate a predominantly linear coupling between BOLD and neuronal activity. For example, cross-species comparison between single-unit data obtained in primates and fMRI data obtained in humans during similar perceptual tasks suggested that fMRI responses might be proportional to the underlying firing rate of the neurons (Heeger *et al.*, 2000; Rees *et al.*, 2000). Such a linear relationship was also observed in anesthetized monkeys by Logothetis *et al.* (2001) in a recent technical tour de force, in which the single-unit responses were acquired together with BOLD fMRI signals *inside* the MRI scanner. In this study, local field potentials (LFPs) and spiking activity were recorded for the first time simultaneously with T_2* BOLD fMRI signals in the monkey cortex, showing a linear relationship between LFPs and spiking rate, but displaying better correlation with the former. In a recent study (Kim *et al.*, 2004; see Fig. 1,2) in which recording was done from multiple sites for the first time, spiking activity was shown to be linearly correlated with the T_2* BOLD response in the cat visual cortex using a single orientation of a moving grid but with different spatial frequencies of the grid lines. However, the correlation varied from point to point on the cortical surface and was generally

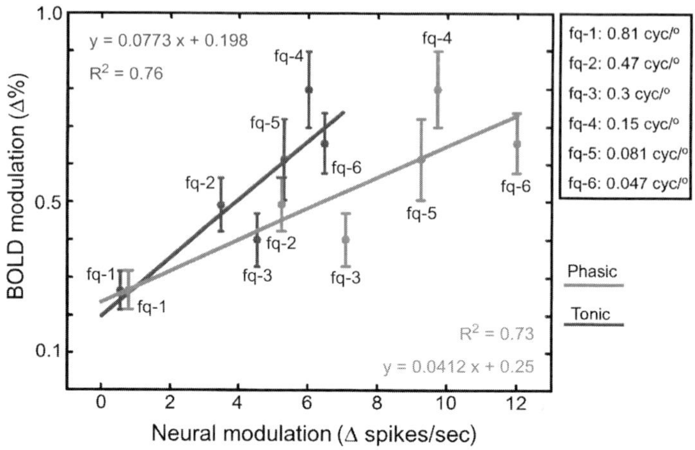

FIG. 2. Results of direct comparison between BOLD and single-unit recordings across all sites (n = 58). X-axis: neural modulation in spikes per second. Y-axis: % BOLD modulation. The data points fq-1 through fq-6 indicate the six spatial frequencies used to elicit neural responses. Coefficient of determination of the regression line $R^2 = 0.85$. Correlation data are plotted for tonic (blue) and phasic (orange) data separately. (See Color Insert.)

valid only when the data were averaged at least over 4–5 mm spatial scale, demonstrating the fact that T_2* BOLD responses are *not* spatially accurate at the level of orientation columns in the visual system. A detailed set of studies was performed asking the same type of questions and using laser Doppler techniques to measure CBF; these studies concluded that linear domains exist between CBF increases and aspects of electrical activity as well as that hemodynamic changes evoked by neuronal activity depend on the afferent input function but that they do not necessarily reflect the output level of activity of a region.

While these results are comforting for the existing body of cognitive neuroimaging data, it is not clear over which spatial scale the linearity between BOLD and neuronal responses holds true (if at all). Is it really invariant across the different spatial scales of the cortical architecture? Can we assume a universal linearity from the scale of entire cortical areas (several millimeters to centimeters) to individual columns (submillimeter)?

B. Small Versus Large Number?

Given the nominal voxel size of most fMRI scans (several millimeters at best), it is safe to conclude that BOLD reflects the activity of many neurons (around 105 neurons for a voxel of 1–2 mm^3 [Braitenberg, 2001]). The overwhelming body of existing electrophysiological data, however, is based on electrode recordings from single-unit activity recording (SUA) or a handful of neurons (multiunit recording activity [MUA]). The real question is hence to ask how accurately the responses of single cells (our gold standard, given the existing body of data) are reflected by a population response, such as in BOLD fMRI. Theoretically, if each neuron "fired" independently of its neighbor's behavior, this would be an ill-posed problem, as fMRI cannot distinguish small activity changes in a large cellular population from large changes in a small population. Fortunately, however, neurons are embedded in tight local circuitries, forming functional clusters with similar receptive field properties ranging from "microcolumns" to "columns" to "hypercolumns." Both the neuronal firing rate and the phase are correlated between neighboring neurons (Singer, 1999), and in most sensory areas there is a good correlation between LFPs, which are assumed to reflect the average activity of a large number of neurons and the responses of individual spiking neurons. In fact, it is difficult to imagine how BOLD contrast could be detectable at all, if it were sensitized to the behavior of uncorrelated individual neurons, as the metabolic demand of a single neuron would hardly be sufficient to initiate the chain of hemodynamic events giving rise to BOLD.

C. What is the Relationship Between Voxel Size and Neural Correspondence?

The MRI voxel size is key in determining the spatial dependence of the correlation between the BOLD and electrode data. A large voxel will improve the relationship to the neuronal event, because a voxel that displays BOLD signal changes will have a much higher probability of including the site of the electrically active column when its size increases, for example, to sizes that are often used in human studies (e.g., $3 \times 3 \times 3$ mm^3). However, such a large voxel can provide only limited information about the pattern of activation, due to its low spatial resolution. Smaller voxels (i.e., voxels the size of individual SUA sites), which could potentially yield a much better spatial resolution, will result in a large variability in neuronal correspondence and the BOLD signal. Furthermore, a large number of "active" voxels will actually originate from positions beyond the site of electrical activity (see Kim *et al.*, 2004 for further details).

D. Spiking or Subthreshold?

According to the standard "integrate-and-fire" model of neuron behavior, an action potential is generated when the membrane potential reaches threshold by depolarization, which in turn is determined by the integration of incoming excitatory postsynaptic potentials (EPSPs) and inhibitory postsynaptic potentials (IPSPs). Action potentials are usually generated only around the axon hillock, whereas synaptic potentials can be generated all across the dendritic tree (mostly on dendritic spines) and the cell soma. The threshold-dependent action potential firing means that much more subthreshold and suprathreshold synaptic activity than action-potential activity is likely at any one time. And the much larger neural surface area associated with synaptic activity means that the total metabolic demand (i.e., the number of Na+/K+ pumps involved, etc.) for synaptic activity ought to be significantly higher than that required for generating action potentials. It therefore seems likely that BOLD contrast—like other methods based on cortical metabolism, such as 2DG and optical imaging—is dominated by the synaptic subthreshold activity. Consequently, BOLD fMRI seems to over-represent the dendritic input activity to a cortical area, rather than the output activity of that same area. This important distinction will have to be taken into consideration when cognitive fMRI data are interpreted.

E. Does BOLD fMRI Represent Neural Synchronicity?

Because of their noninvasive and comprehensive nature, BOLD imaging methods are powerful tools in the investigation of brain function, and they are particularly well suited for studies on the human brain. These methods faithfully monitor the activation of brain regions, but it is still under debate which aspects of neuronal activity are reflected by the amplitude of hemodynamic responses. There is evidence that BOLD responses are positively correlated with neuronal firing and the amplitude of evoked potentials. Because the latter depends not only on the number and discharge rates of activated neurons but also on the temporal coherence of responses, the possibility needs to be considered that the BOLD response is influenced by the amplitude as well as by the temporal structure of neuronal responses. This possibility is supported by a recent study that showed that LFP oscillations, which reflect mainly coherent synaptic activity, correlate better with BOLD signals than with discharge rates. LFP oscillations vary substantially in amplitude and frequency as a function of stimulus configurations and central states. During sleep and relaxed wakefulness, oscillations occur at low frequencies (delta, 0–3 Hz; theta, 4–8 Hz; alpha, 9–14 Hz) while high-frequency oscillations (beta, 15–21 Hz; gamma, 22–90 Hz) prevail during states of arousal and focused attention. Moreover, the analysis of LFP oscillations revealed that stronger stimuli shifted the frequency of LFP oscillations to higher values and that there is a particularly tight correlation between the hemodynamic response and LFP oscillations in the high gamma frequency range (R. Galuske, personal communication; Niessing *et al.*, 2005). Considering the rather loose correlation between discharge rates and hemodynamic responses at constant stimulus intensities, these findings closely link hemodynamic responses to the processes leading to LFP oscillations in the gamma frequency range. This is of particular importance, as gamma frequency ranges have traditionally been associated with cognitive and executive functions such as figure–ground segmentation, expectancy, sensory–motor coordination, short-term memory, and movement preparation. There is thus the exciting possibility that BOLD fMRI may provide information about the neural correlates of cognitive and executive functions in a much more accurate and functionally relevant way than is generally assumed possible.

F. Excitatory or Inhibitory Activity?

To the best of our knowledge, there is currently no evidence that the metabolic demands of excitatory synapses differ greatly from those of inhibitory synapses. Therefore, fundamentally, both the EPSPs the IPSPs can be expected to cause similar metabolic and hemodynamic events, ultimately giving rise to similar BOLD contrasts. On the site of the spiking output activity, however, they have opposite effects: accumulation of EPSPs will increase the probability for spike generation (and therefore also the metabolic demand), while IPSPs will decrease it. Assuming that the BOLD response predominantly reflects changes in synaptic subthreshold activity, it remains unclear whether excitatory and inhibitory cortical events can be differentiated using the BOLD response in any single region. Recently, one group proposed that inhibition, unlike excitation, elicits no measurable change in the BOLD signal (Waldvogel *et al.*, 2000). They hypothesized that because of inhibitory synapses' lower numbers (Beaulieu and Colonnier, 1985), their strategically superior location (inhibitory receptors are located in the basal cell body, whereas excitatory receptors are in the distal dendrites), and their increased efficiency (Koos and Tepper, 1999), there could be lower metabolic demand during inhibition than during excitation. The truth or falsity of this claim notwithstanding, both empirical and theoretical studies suggest that excitatory and inhibitory neurons in the cortex are so tightly interconnected in local circuits (see, e.g., Payne and Peters, 2002 for details of the local circuitry in the cat primary visual cortex; see Kisvarday *et al.*, 1994 for the anatomy of local inhibitory circuits in cats) that one is unlikely to observe an increase in excitation without an increase in inhibition. After all, for an inhibitory neuron to increase its firing rate, it must be receiving more excitatory input, and most of the excitatory input comes from the local cortical neighborhood (Payne and Peters, 2002; Sherpherd, 1990). Naturally, excitation and inhibition would not occur in temporal unison, as no cell would reach threshold if they did. In the several hundred milliseconds to a few seconds during which BOLD contrast emerges (Ogawa *et al.*, 1992), however, such potential temporal differences would most likely be rendered indistinguishable. One viable hypothesis, therefore, is that BOLD contrast reflects a steady-state balance of local excitation and inhibition, particularly if BOLD is more sensitive to subthreshold activity than to spiking activity.

IV. Nonconventional fMRI

BOLD fMRI at the conventional low magnetic field of 1.5 Tesla (T) can commonly achieve a spatial resolution of up to 3–5 millimeters. This is sufficient for labeling cortical organization at scales of hypercolumns (several millimeters) or areas (several centimeters). But functional images at this resolution fail to accurately label the columnar organization of the brain (Fig. 3). Studies at higher magnetic fields (such as 3 or 7 T) can produce significant enhancements of the spatial resolution and specificity of fMRIs. Theoretical and experimental studies have shown at least a linear increase in SNR with magnetic field strength. The increase of the static MR signal can be used to reduce the volume needed for

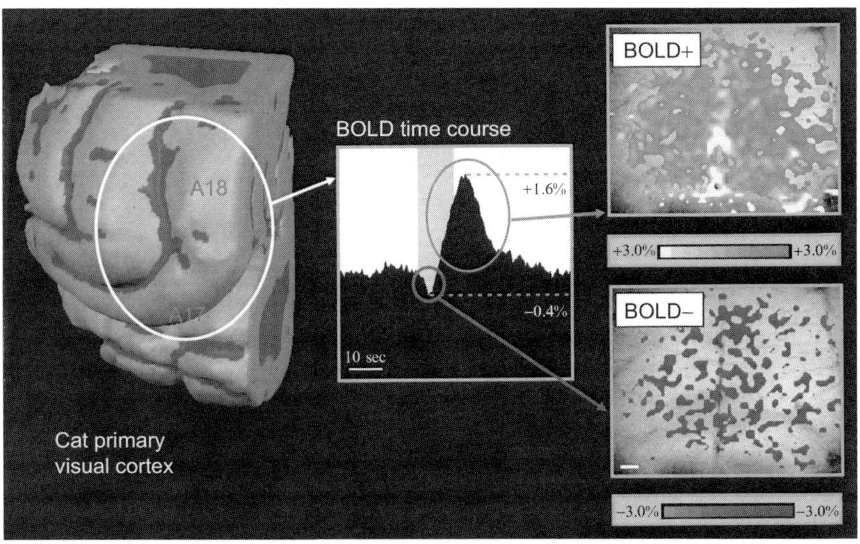

FIG. 3. Improvement of BOLD spatial specificity by use of nonconventional functional MRI signals. Left side: three-dimensional rendering of the cat visual cortex from which the fMRI signals were acquired. The time course on the right side shows biphasic evolution of MR signals, resulting in the early deoxygenation contrast. If used, such deoxygenation signals produce high-resolution images of exceedingly high functional specificity (termed BOLD−) that contrast with conventional BOLD fMRI signals (termed BOLD+).

signal averaging. Furthermore, as the field strength increases, the field gradient around the capillaries becomes larger and extends further into the parenchyma, thus increasing the participation of the brain tissue in functional signal. Concurrently, the shortened T_2^* of the blood at high B_0 reduces the relative contribution from the large veins.

While these results suggest that a stronger magnetic field *per se* will specifically enhance the signal components originating from parenchymal capillary tissue, recent optical spectroscopy and functional MRI data (Kim *et al.*, 2000a; Malonek and Grinvald, 1996, 1997; Vanzetta and Grinvald, 1999) suggest that the spatial specificity of BOLD could be further and more dramatically improved if a (hypothesized) initial *decrease* of MR signals can be utilized for functional imaging formation. To this end, it is suggested that the first event following focal neuronal activity is a prolonged increase in oxygen consumption, caused by an elevation in oxidative metabolism of active neurons. Based on 2DG data (Sokoloff *et al.*, 1977), one can assume that the increase in oxidative metabolism in the mammalian cortex is collocalized with the site of electrical activity. The increase in oxidative metabolism will naturally elevate the local deoxyhemoglobin content in the parenchyma of active neurons, assuming there is no immediate commensurate change in cerebral blood flow (CBF) (Fox and Raichle, 1986). In T_2-or T_2^*-weighted BOLD fMRI images, such an increase in paramagnetic deoxyhemoglobin should therefore be detectable as a transient *decrease* in observable MR signals. Such an initial deoxygenation of the local cortical tissue will last for only a brief period, as fresh blood (fresh oxyhemoglobin) will rush into capillaries in response to the increased metabolism, thus reversing the local ratio of hemoglobin in favor of oxyhemoglobin and hence resulting in a delayed *increase* in observable MR signals (i.e., the conventional BOLD signal). The crucial question here is the location of the above described "biphasic" hemodynamic processes. Grinvald and coauthors (Malonek *et al.*, 1997; Malonek and Grinvald, 1996) hypothesized a fundamentally distinct functional specificity for these two events: The initial deoxygenation, as a consequence of an increase in oxidative metabolism, should be coregistered with the site of electrical activity up to the level of individual cortical columns (in fact, the well-established "optical imaging of intrinsic signals" method [Frostig *et al.*, 1990; Grinvald *et al.*, 1986], which has been cross-validated with single-unit techniques [Crair *et al.*, 1998; Shmuel and Grinvald, 1996], is similarly based on measuring the local transient increase of deoxyhemoglobin). The delayed oxygenation of the cortical tissue, on the other hand, is suggested to be far less specific, due to the spread of hemodynamic activity beyond the site of original neural activity. Both the existence of "biphasic" BOLD response *per se* and the suggested differences in functional specificity have been the subject of heated controversies in recent years (see Buxton, 2001 for a comprehensive update of this saga). Although the initial deoxygenation signal in fMRI (termed "initial dip") has been reported in awake behaving humans

(Hu *et al.*, 1997; Menon *et al.*, 1995) and anesthetized monkeys (Logothetis *et al.*, 1999), studies in rodents failed to detect any significant initial decrease in the BOLD signal following sensory stimulation (Lindauer *et al.*, 2001; Marota *et al.*, 1999; Silva and Kim, 1999; but see Mayhew *et al.*, 2001). The question of whether the use of the initial dip would indeed improve the spatial specificity of BOLD has been far more difficult to address experimentally. This is largely because most fMRI studies done to examine this phenomenon have been conducted in humans (e.g., Hu *et al.*, 1997; Yacoub *et al.*, 1999) and therefore, by necessity, have used relatively coarse nominal spatial resolution above the level of the individual cortical columns. In animal studies using ultra-high magnetic fields (e.g., 9.4 T), in which functional images at submillimeter scales can be acquired, the results of our own group (Kim *et al.*, 2000a) suggest that indeed the use of the "initial dip" can significantly improve the spatial specificity of BOLD. This result has, however, been questioned (Logothetis, 2000; see Kim *et al.*, 2000b for our reply). On the other hand, in a recent pioneering study, preoperative functional MRI and intraoperative optical imaging were performed in the same human subject (Cannestra *et al.*, 2001). Though the spatial overlap between optical imaging and conventional (positive) fMRI was poor, there was a dramatic improvement in the spatial correspondence between the two datasets when the initial dip portion of the MRI signal was used. Furthermore, combined single-unit and oxygen-tension-probe measurements (Thompson *et al.*, 2003) convincingly demonstrated both the presence and the functional significance of the initial deoxygenation-signal component.

As an alternative to the initial deoxygenation signals, the spatial specificity of T2*-based fMRI can be further improved if only the arterial contribution and/or (to attenuate) the draining-vessel artifacts are utilized for functional image construction. For example, perfusion-weighted images based on arterial spin labeling can be made sensitive to the CBF changes that occur on the way from upstream arterial networks to the capillaries, thus providing better spatial localization abilities (Duong *et al.*, 2001; Luh *et al.*, 2000) than do T_2* BOLD imaging methods.

Finally, as an alternative to the initial deoxygenation and CBF signals, and in particular at high magnetic fields, blood-vessel artifacts can be diminished using the so-called Hahn spin echo (HSE) BOLD fMRI method. HSE fMRI signals can originate from extravascular space around the microvasculature (Boxerman *et al.*, 1993; Ogawa *et al.*, 1993) as well as from blood itself within the blood vessels (Ugurbil *et al.*, 2000, 1999; van Zijl *et al.*, 1998). The former location provides spatial specificity in the hundred-micron spatial scale because capillaries are separated, on the average, by 25 μm (Pawlik *et al.*, 1981). The blood effect, however, can be associated with both large and small blood vessels and hence degrades the spatial specificity of fMRI. However, the apparent T_2 of venous blood decreases quadratically with the increase of magnetic-field magnitude

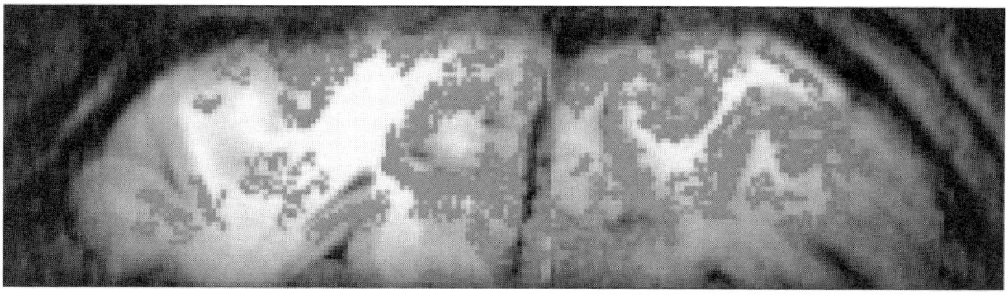

FIG. 4. 0.5- × 0.5- × 2-mm^3 Hahn spin echo (HSE) images of the visual activation in the human visual cortex obtained at 7 Tesla (T), demonstrating that high-spatial-resolution images can be obtained robustly in the human brain at 7 T by the normally weak but more accurate HSE imaging signals single-shot spin echo planar imaging with field-of-view reduction to 3.2 cm × 12.8 cm (64 × 256) (partial Fourier) TE/TR: 70 msec/2 s. Adapted from Ugurbil *et al.* (2003). (See Color Insert.)

(Thulborn *et al.*, 1992) and is diminished from ~180 msec at 1.5 T (Barth and Moser, 1997) to ~6 msec at 9.4 T, significantly smaller than brain tissue T_2 and the time-to-echo (TE) values that would be used at such field strengths. HSE BOLD signals associated with capillaries are inherently small and undetectable at low magnetic fields. However, they are expected to increase supralinearly with increases in magnetic-field strength. They approach potentially useful dimensions at 3 T (Mildner *et al.*, 2001) and are large enough to yield microvasculature-based high-resolution maps, even in the human brain, at 7 T (see Fig. 4).

Notwithstanding the proposed techniques for diminishing vasculature artifacts, in principle, it should be possible to more effectively bypass the degrading effect of large draining vessels *per se*, if one could map the changes in cerebral oxygen ($CMRO_2$) consumption more directly. $CMRO_2$ increases must occur as a result of increased activity of the mitochondria in the "activated" neurons, and high mitochondrial density in the synapse-rich regions, predominantly in the dendrites (Wong-Riley *et al.*, 1998). Such a direct $CMRO_2$ mapping might be possible using the magnetic isotope of oxygen ^{17}O (Ronen *et al.*, 1997; Zhu *et al.*, 2001).

V. Conclusions and Future Prospects for fMRI

In the decade since the first noninvasive measurements of functional blood oxygenation level signals from the human brain, fMRI (both standard and nonstandard) has developed into an indispensable neuroimaging tool that is ubiquitous in both clinical and basic neuroscience settings. The explanatory power of fMRI, however, is currently limited due to the presence of major theoretical and

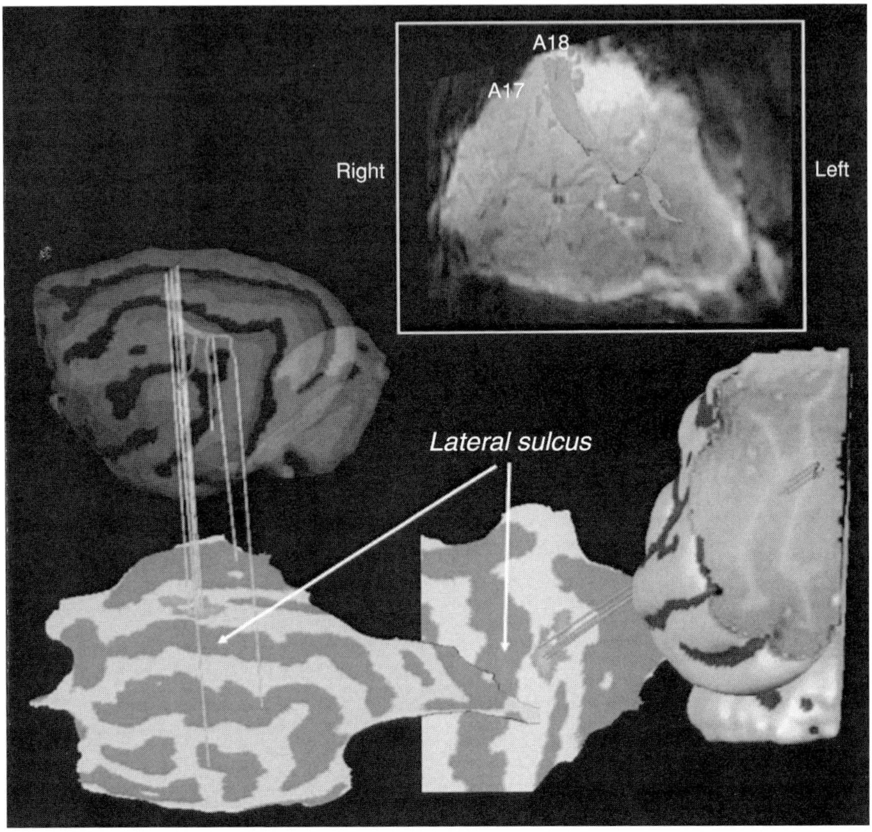

FIG. 5. Combining fMRI with high-resolution diffusion-tensor imaging (DTI)-based fiber tractography. Here, 3D volume reconstruction of the cat cortex is depicted as a translucent "glass brain" in which the BOLD-based DTI fibers were embedded, thus preserving the true 3D trajectories of the reconstructed fibers. Lower panels: "flattened" representation of cortex. Light- and dark-gray areas code for gyri and sulci, respectively. Corresponding points between reconstructed and flattened cortices were connected through the reconstructed DTI fibers. Right panel: same data as in the left panel, viewed from a different perspective. (See Color Insert.)

practical shortcomings. These include (but are not limited to) (1) a lack of detailed understanding of its neural correlate; (2) limited spatial resolution; and (3) the difficulty of combining fMRI with other imaging and measurement techniques. Furthermore, while fMRI noninvasively provides detailed information about the "where" of the brain's functional architecture, such localization information alone must leave unanswered pivotal questions about the brain's information processing (the "how" of the processing), as long as the underlying pattern of

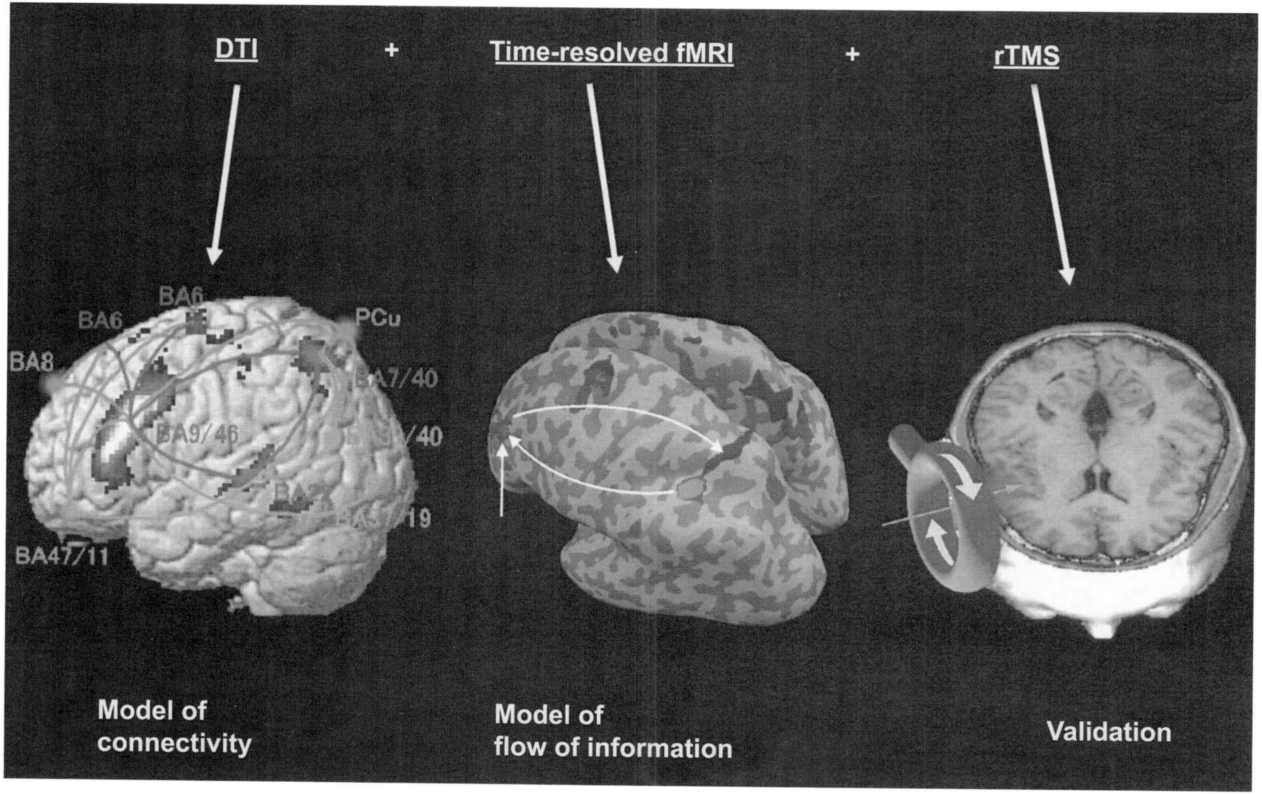

FIG. 6. Cartoon diagram of the multimodal approach that is increasingly being employed for functional MRI studies. If combined properly, such multimodal studies could potentially provide information on localization (by fMRI), anatomical connectivity (by DTI), flow of information (by effective connectivity), and localized disruption (by rTMS) of cortical functionalities from the same brain. (See Color Insert.)

neuronal connectivity cannot be mapped in an equally noninvasive manner. Future fMRI studies in cognitive neuroimaging will have to embrace a significantly more multimodal approach. For example, combining fMRI with diffusion-tensor imaging (Kim *et al.*, 2003) will label the pattern of structural connectivity between functionally active areas. The direction of the flow of functional information within this mesh of neural networks could then be elucidated by time-resolved fMRI, effective connectivities, and possibly also repetitive transcranial magnetic stimulations (rTMS), together with high-resolution fMRI experiments (see Figs. 5, 6).

Acknowledgments

We thank Drs. Kamil Ugurbil, Seong-Gi Kim, Tim Duong, Louis J. Toth, Rainer Goebel, Susumu Mori, Alvaro Pascual-Leone, and Itamar Ronen for their continuing support of this work. This work was supported by grants from NIH (RR08079, MH61937, and MH67530), the MIND institute, the Keck Foundation, the Human Frontiers Science Program, the Whitaker Foundation, and the National Alliance for Research on Schizophrenia and Depression (NARSAD).

References

Bandettini, P. A., Wang, E. C., Hinks, R. S., Rikofsky, R. S., and Hyde, J. S. (1992). Time course EPI of human brain function during task activation. *Magn. Reson. Med.* **25,** 390–397.

Barth, M., and Moser, E. (1997). Proton NMR relaxation times of human blood samples at 1.5 T and implications for functional MRI. *Cell. Mol. Biol. (Noisy-le-grand)* **43,** 783–791.

Beaulieu, C., and Colonnier, M. (1985). A laminar analysis of the number of round-asymmetrical and flat-symmetrical synapses on spines, dendritic trunks, and cell bodies in area 17 of the cat. *J. Comp. Neurol.* **231,** 180–189.

Bonhoeffer, T., and Grinvald, A. (1993). The layout of iso-orientation domains in area 18 of cat visual cortex: Optical imaging reveals a pinwheel-like organization. *J. Neurosci.* **13,** 4157–4180.

Boxerman, J. L., Weisskoff, R. M., Hoppel, B. E., and Rosen, B. R. (1993). MR contrast due to microscopically heterogeneous magnetic susceptibilty: Cylindrical geometry. *In* "12th Annual Meeting of the Society of Magnetic Resonance in Medicine," Vol. 1, p. 389. *Int. Soc. Magn. Reson. Med.*, New York, NY.

Braitenberg, V. (2001). Brain size and number of neurons: An exercise in synthetic neuroanatomy. *J. Comput. Neurosci.* **10,** 71–77.

Brodmann, K. (1909). "Vergleichende Lokalisationslehre der Grosshirnrinde." Verlag von Johann Ambrosius Barth, Leipzip.

Buxton, R. B. (2001). The elusive initial dip. *Neuroimage* **13,** 953–958.

Cannestra, A. F., Pouratian, N., Bookheimer, S. Y., Martin, N. A., Becker, D. P., and Toga, A. (2001). Temporal spatial differences observed by functional MRI and human intraoperative optical imaging. *Cereb. Cortex* 11.

Crair, M. C., Gillespie, D. C., and Stryker, M. P. (1998). The role of visual experience in the development of columns in cat visual cortex. *Science* **279,** 566–570.

Disbrow, E. A., Slutsky, D. A., Roberts, T. P., and Krubitzer, L. A. (2000). Functional MRI at 1.5 tesla: A comparison of the blood oxygenation level-dependent signal and electrophysiology. *Proc. Natl. Acad. Sci. USA* **97,** 9718–9723.

Duong, T. Q., Kim, D. S., Ugurbil, K., and Kim, S. G. (2001). Localized cerebral blood flow response at submillimeter columnar resolution. *Proc. Natl. Acad. Sci. USA* **98,** 10904–10909.

Engel, S. A., Glover, G. H., and Wandell, B. A. (1997). Retinotopic organization in human visual cortex and the spatial precision of functional MRI. *Cereb. Cortex* **7,** 181–192.

Felleman, D. J., and Van Essen, D. C. (1991). Distributed hierarchical processing in the primate cerebral cortex. *Cereb. Cortex* **1,** 1–47.

Fox, P. T., and Raichle, M. E. (1986). Focal physiological uncoupling of cerebral blood flow and oxidative metabolism during somatosensory stimulation in human subjects. *Proc. Natl. Acad. Sci. USA* **83,** 1140–1144.

Frostig, R. D., Lieke, E. E., Ts'o, D. Y., and Grinvald, A. (1990). Cortical functional architecture and local coupling between neuronal activity and the microcirculation revealed by *in vivo* high-resolution optical imaging of intrinsic signals. *Proc. Natl. Acad. Sci. USA* **87,** 6082–6086.

Grinvald, A., Lieke, E., Frostig, R. D., Gilbert, C. D., and Wiesel, T. N. (1986). Functional architecture of cortex revealed by optical imaging of intrinsic signals. *Nature* **324,** 361–364.

Heeger, D. J., Huk, A. C., Geisler, W. S., and Albrecht, D. G. (2000). Spikes versus BOLD: What does neuroimaging tell us about neuronal activity? *Nat. Neurosci.* **3,** 631–633.

Hu, X., Le, T. H., and Ugurbil, K. (1997). Evaluation of the early response in fMRI in individual subjects using short stimulus duration. *Magn. Reson. Med.* **37,** 877–884.

Hubel, D., and Wiesel, T. N. (1962). Receptive field, binocular interaction and functional architecture in the cat's visual cortex. *J. Physiol. (London)* **160,** 106–154.

Hubener, M., Shoham, D., Grinvald, A., and Bonhoeffer, T. (1997). Spatial relationships among three columnar systems in cat area 17. *J. Neurosci.* **17,** 9270–9284.

Kim, D. S., Duong, T. Q., and Kim, S. G. (2000a). High-resolution mapping of iso-orientation columns by fMRI. *Nat. Neurosci.* **3,** 164–169.

Kim, D. S., Duong, T. Q., and Kim, S. G. (2000b). Reply to "Can current fMRI techniques reveal the micro-architecture of cortex?" *Nat. Neurosci.* **3,** 414.

Kim, D. S., Kim, M., Ronen, I., Formisano, E., Kim, K. H., Ugurbil, K., Mori, S., and Goebel, R. (2003). *In vivo* mapping of functional domains and axonal connectivity in cat visual cortex using magnetic resonance imaging. *Magn. Reson. Imaging* **21,** 1131–1140.

Kim, D. S., Matsuda, Y., Ohki, K., Ajima, A., and Tanaka, S. (1999). Geometrical and topological relationships between multiple functional maps in cat primary visual cortex. *Neuroreport* **10,** 2515–2522.

Kim, D. S., Ronen, I., Olman, C., Kim, S. G., Ugurbil, K., and Toth, L. J. (2004). Spatial relationship between neuronal activity and BOLD functional MRI. *Neuroimage* **21,** 876–885.

Kim, S. G., Ashe, J., Hendrich, K., Ellermann, J. M., Merkle, H., Ugurbil, K., and Georgopoulos, A. P. (1993). Functional magnetic resonance imaging of motor cortex: Hemispheric asymmetry and handedness. *Science* **261,** 615–617.

Kisvarday, Z. F., Kim, D. S., Eysel, U. T., and Bonhoeffer, T. (1994). Relationship between lateral inhibitory connections and the topography of the orientation map in cat visual cortex. *Eur. J. Neurosci.* **6,** 1619–1632.

Koos, T., and Tepper, J. M. (1999). Inhibitory control of neostriatal projection neurons by GABAergic interneurons. *Nat. Neurosci.* **2,** 467–472.

Kwong, K. K., Belliveau, J., Chesler, D. A., Goldberg, I. E., Weisskoff, R. M., Poncelet, B. P., Kennedy, D. N., Hoppel, B. E., Cohen, M. S., Turner, R., Cheng, H., Brady, T. J., and Rosen, B. R. (1992). Dynamic magnetic resonance imaging of human brain acrivity during primary sensory stimulation. *Proc. Natl. Acad. Sci. USA* **89,** 5675–5679.

Lindauer, U., Royl, G., Leithner, C., Kuhl, M., Gold, L., Gethmann, J., Kohl-Bareis, M., Villringer, A., and Dirnagl, U. (2001). No evidence for early decrease in blood oxygenation in rat whisker cortex in response to functional activation. *Neuroimage* **13,** 988–1001.

Logothetis, N. (2000). Can current fMRI techniques reveal the micro-architecture of cortex? *Nat. Neurosci.* **3,** 413–414.

Logothetis, N., Pauls, J., Augath, M., Trinath, T., and Oeltermann, A. (2001). Neurophysiological investigation of the basis of the fMRI signal. *Nature* **412,** 150–157.

Logothetis, N. K., Guggenberger, H., Peled, S., and Pauls, J. (1999). Functional imaging of the monkey brain. *Nat. Neurosci.* **2,** 555–562.

Lowel, S., Freeman, B., and Singer, W. (1987). Topographic organization of the orientation column system in large flat-mounts of the cat visual cortex: A 2-deoxyglucose study. *J. Comp. Neurol.* **255,** 401–415.

Luh, W. M., Wong, E. C., Bandettini, P. A., Ward, B. D., and Hyde, J. S. (2000). Comparison of simultaneously measured perfusion and BOLD signal increases during brain activation with T (1)-based tissue identification. *Magn. Reson. Med.* **44,** 137–143.

Malonek, D., Dirnagl, U., Lindauer, U., Yamada, K., Kanno, I., and Grinvald, A. (1997). Vascular imprints of neuronal activity: Relationships between the dynamics of cortical blood flow, oxygenation, and volume changes following sensory stimulation. *Proc. Natl. Acad. Sci. USA* **94,** 14826–14831.

Malonek, D., and Grinvald, A. (1996). Interactions between electrical activity and cortical microcirculation revealed by imaging spectroscopy: Implications for functional brain mapping. *Science* **272,** 551–554.

Malonek, D., and Grinvald, A. (1997). Vascular regulation at sub millimeter range. Sources of intrinsic signals for high resolution optical imaging. *Adv. Exp. Med. Biol.* **413,** 215–220.

Marota, J. J., Ayata, C., Moskowitz, M. A., Weisskoff, R. M., Rosen, B. R., and Mandeville, J. B. (1999). Investigation of the early response to rat forepaw stimulation. *Magn. Reson. Med.* **41,** 247–252.

Mayhew, J., Johnston, D., Martindale, J., Jones, M., Berwick, J., and Zheng, Y. (2001). Increased oxygen consumption following activation of brain: Theoretical footnotes using spectroscopic data from barrel cortex. *Neuroimage* **13,** 975–987.

Menon, R. S., Ogawa, S., Hu, X., Strupp, J. P., Anderson, P., and Ugurbil, K. (1995). BOLD based functional MRI at 4 Tesla includes a capillary bed contribution: Echo-planar imaging correlates with previous optical imaging using intrinsic signals. *Magn. Reson. Med.* **33,** 453–459.

Mildner, T., Norris, D. G., Schwarzbauer, C., and Wiggins, C. J. (2001). A qualitative test of the balloon model for BOLD-based MR signal changes at 3T. *Magn. Reson. Med.* **46,** 891–899.

Niessing, J., Ebisch, B., Schmidt, K. E., Niessing, M., Singer, W., and Galuske, R. (2005). Hemodynamic signals correlate tightly with synchronized gamma oscillations. *Science* **309,** 948–951.

Ogawa, S., Lee, T. M., Nayak, A. S., and Glynn, P. (1990). Oxygenation-sensitive contrast in magnetic resonance image of rodent brain at high magnetic fields. *Magn. Reson. Med.* **14,** 68–78.

Ogawa, S., Menon, R. S., Kim, S. G., and Ugurbil, K. (1998). On the characteristics of functional magnetic resonance imaging of the brain. *Annu. Rev. Biophys. Biomol. Struct.* **27,** 447–474.

Ogawa, S., Menon, R. S., Tank, D. W., Kim, S.-G., Merkle, H., Ellermann, J. M., and Ugurbil, K. (1993). Functional Brain Mapping by Blood Oxygenation Level-Dependent Contrast Magnetic Resonance Imaging. *Biophys. J.* **64,** 800–812.

Ogawa, S., Tank, D. W., Menon, R., Ellermann, J. M., Kim, S. G., Merkle, H., and Ugurbil, K. (1992). Intrinsic signal changes accompanying sensory stimulation: Functional brain mapping with magnetic resonance imaging. *Proc. Natl. Acad. Sci. USA* **89,** 5951–5955.

Pawlik, G., Rackl, A., and Bing, R. J. (1981). Quantitative capillary topography and blood flow in the cerebral cortex of cats: An *in vivo* microscopic study. *Brain Res.* **208,** 35–58.

Payne, B. R., and Peters, A. (2002). "The Cat Primary Visual Cortex." Academic Press, San Diego.

Rees, G., Friston, K., and Koch, C. (2000). A direct quantitative relationship between the functional properties of human and macaque V5. *Nat. Neurosci.* **3,** 716–723.

Ronen, I., Lee, J. H., Merkle, H., Ugurbil, K., and Navon, G. (1997). Imaging H2(17)O distribution in a phantom and measurement of metabolically produced H2(17)O in live mice by proton NMR. *NMR Biomed.* **10,** 333–340.

Sereno, M. I., Dale, A. M., Reppas, J. B., Kwong, K. K., Belliveau, J. W., Brady, T. J., Rosen, B. R., and Tootell, R. B. (1995). Borders of multiple visual areas in humans revealed by functional magnetic resonance imaging. *Science* **268,** 889–893.

Sherpherd, G. M. (1990). "The Synaptic Organization of the Brain." Oxford University Press, Oxford.

Shmuel, A., and Grinvald, A. (1996). Functional organization for direction of motion and its relationship to orientation maps in cat area 18. *J. Neurosci.* **16,** 6945–6964.

Silva, A. C., and Kim, S. G. (1999). Pseudo-continuous arterial spin labeling technique for measuring CBF dynamics with high temporal resolution. *Magn. Reson. Med.* **42,** 425–429.

Singer, W. (1999). Time as coding space? *Curr. Opin. Neurobiol.* **9,** 189–194.

Sokoloff, L., Reivich, M., Kennedy, C., Des Rosiers, M. H., Patlak, C. S., Pettigrew, K. D., Sakurada, O., and Shinohara, M. (1977). The [14C]deoxyglucose method for the measurement of local cerebral glucose utilization: Theory, procedure, and normal values in the conscious and anesthetized albino rat. *J. Neurochem.* **28,** 897–916.

Stetter, M., and Obermayer, K. (1999). Simulation of scanning laser technique for optical imaging o blood-related intrinsic signals. *J. Opt. Soc. Am.* **16,** 58–70.

Stetter, M., Schiessl, I., Otto, T., Sengpiel, F., Hubener, M., Bonhoeffer, T., and Obermayer, K. (2000). Principal component analysis and blind separation of sources for optical imaging of intrinsic signals. *Neuroimage* **11,** 482–490.

Thompson, J. K., Peterson, M. R., and Freeman, R. D. (2003). Single-neuron activity and tissue oxygenation in the cerebral cortex. *Science* **299,** 1070–1072.

Thulborn, K. R., Waterton, J. C., and Matthews, P. M. (1992). Dependence of the transverse relaxation time of water protons in whole blood at high field. *Biochem. Biophys. Acta* **714,** 265–272.

Tootell, R. B., Mendola, J. D., Hadjikhani, N. K., Ledden, P. J., Liu, A. K., Reppas, J. B., Sereno, M. I., and Dale, A. M. (1997). Functional analysis of V3A and related areas in human visual cortex. *J. Neurosci.* **17,** 7060–7078.

Ugurbil, K., Adriany, G., Andersen, P., Chen, W., Gruetter, R., Hu, X., Merkle, H., Kim, D.-S., Kim, S.-G., Strupp, J., Zhu, X.-H., and Ogawa, S. (2000). Magnetic Resonance Studies of Brain Function and Neurochemistry. *Annu. Rev. Biomed. Eng.* **2,** 633–660.

Ugurbil, K., Hu, X., Chen, W., Zhu, X.-H., Kim, S.-G., and Georgopoulos, A. (1999). Functional Mapping in the human brain using high magnetic fields. *Philos. Trans. R. Soc. Lond. B. Biol. Sci.* **354,** 1195–1213.

Ugurbil, K., Toth, L., and Kim, D. S. (2003). How accurate is magnetic resonance imaging of brain function? *Trends. Neurosci.* **26,** 108–114.

van Zijl, P. C., Eleff, S. M., Ulatowski, J. A., Oja, J. M., Ulug, A. M., Traystman, R. J., and Kauppinen, R. A. (1998). Quantitative assessment of blood flow, blood volume and blood oxygenation effects in functional magnetic resonance imaging [see comments]. *Nat. Med.* **4,** 159–167.

Vanzetta, I., and Grinvald, A. (1999). Increased cortical oxidative metabolism due to sensory stimulation: Implications for functional brain imaging. *Science* **286,** 1555–1558.

Wagner, A. D., Schacter, D. L., Rotte, M., Koutstaal, W., Maril, A., Dale, A. M., Rosen, B. R., and Buckner, R. L. (1998). Building memories: Remembering and forgetting of verbal experiences as predicted by brain activity. *Science* **281,** 1188–1191.

Waldvogel, D., van Gelderen, P., Muellbacher, W., Ziemann, U., Immisch, I., and Hallett, M. (2000). The relative metabolic demand of inhibition and excitation. *Nature* **406,** 995–998.

Weisskoff, R. M., Zuo, C. S., Boxerman, J. L., and Rosen, B. R. (1994). Microscopic susceptibility variation and transverse relaxation: Theory and experiment. *Magn. Reson. Med.* **31,** 601–110.

Wong-Riley, M., Anderson, B., Liebl, W., and Huang, Z. (1998). Neurochemical organization of the macaque striate cortex: Correlation of cytochrome oxidase with Na+K+ATPase, NADPH-diaphorase, nitric oxide synthase, and N-methyl-D-aspartate receptor subunit 1. *Neuroscience* **83,** 1025–1045.

Yacoub, E., Le, T. H., Ugurbil, K., and Hu, X. (1999). Further evaluation of the initial negative response in functional magnetic resonance imaging. *Magn. Reson. Med.* **41,** 436–441.

Yae, H., Elias, S. A., and Ebner, T. J. (1992). Deblurring of 3-dimensional patterns of evoked rat cerebellar cortical activity: A study using voltage-sensitive dyes and optical sectioning. *J. Neurosci. Methods* **42,** 195–209.

Zhu, X. H., Merkle, H., Kwag, J. H., Ugurbil, K., and Chen, W. (2001). ^{17}O relaxation time and NMR sensitivity of cerebral water and their field dependence. *Magn. Reson. Med.* **45,** 543–549.

QUANTIFICATION OF WHITE MATTER USING DIFFUSION-TENSOR IMAGING

Hae-Jeong Park

Department of Diagnostic Radiology, Yonsei University, College of Medicine
Seoul 120-749, Korea

INTERNATIONAL REVIEW OF
NEUROBIOLOGY, VOL. 66
DOI: 10.1016/S0074-7742(05)66006-0

0074-7742/05 $35.00

I. Introduction

Diffusion-tensor magnetic resonance imaging (DT-MRI), also known simply as diffusion-tensor imaging (DTI), has rapidly evolved as a new *in vivo* approach to the investigation of white-matter abnormalities or tissue damages: quantifying the diffusivity of the water molecules in brain. The quantification of water diffusion *in vivo* is based on the characteristic movement of water molecules, which varies depending upon the tissue. For example, in pure liquids, such as cerebrospinal fluid, the motion of individual water molecules is random, meaning it has equal probability in all directions. However, the movement of water molecules within myelinated fibers is substantially restricted along the direction perpendicular to the orientation of the axons. Consequently, in white-matter fiber tracts, the principal direction of the water diffusion represents the direction of the fiber bundles. Thus, connecting points along the principal direction of the diffusion makes it possible to appreciate white-matter tracts within the brain. Such fiber-tracking schemes, often referred to collectively as fiber tractography, provide important information about the connectivity between brain regions. Therefore, DTI provides a quantitative assessment of the tissue-specific diffusivity and also provides information on anatomical connection. This chapter will focus on the basics and current advances of DTI for the quantification of white matter.

II. Diffusion in the Brain

A. Introduction to Diffusion in the Brain

Diffusion is the random motion of an ensemble of particles. This is easily observed when a drop of ink is added to water, or when smoke dissipates into the air. The phenomenon of diffusion is named Brownian motion after the Scottish botanist Robert Brown, who in 1827 observed the random motion of minute pollen grains suspended in water. Diffusion is affected by the properties of the medium in which it occurs. Equal, or isotropic, diffusion occurs when a medium does not restrict molecular motion, as is the case with water. Skewed, or anisotropic, diffusion, as is seen within crystals and polymer films, is not equal in all directions.

This diffusion phenomenon has been successfully adopted as a very important marker for the tissue architecture in the brain. Diffusion in the brain, mainly of water molecules, has tissue-specific characteristics. In the cerebrospinal fluid, which has no specific directional restrictions, the diffusion is isotropic (i.e., there is equal probability of motion in all directions), as shown in Fig. 1a. In white matter,

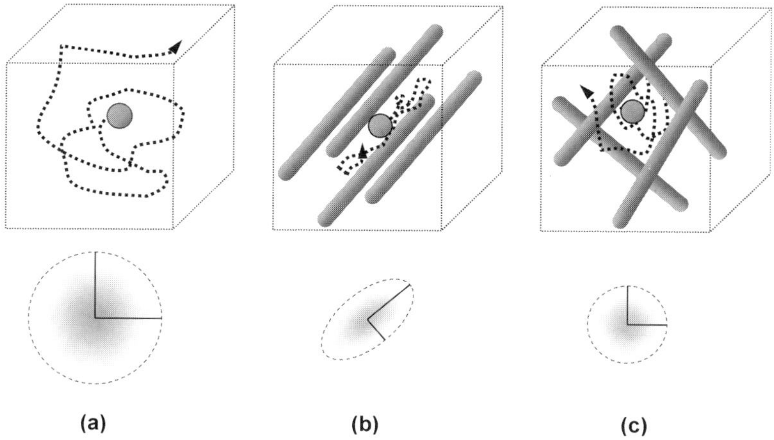

FIG. 1. Diffusion of the water molecule. Unrestricted isotropic self-diffusion (a), restricted anisotropic diffusion in well-arranged structures (b), and restricted isotropic diffusion in complex structures (c). (See Color Insert.)

well-organized tissue structures such as neuronal fibers and myelin sheaths guide the water molecules along the direction of the fiber tracts, and the net diffusion process appears anisotropic (Fig. 1b). The mixture of tissue structures in gray matter—including axons, interconnecting dendrites along the layer, and cell bodies—restricts diffusion, thereby making the net diffusion isotropic (Fig. 1c).

B. Diffusion Coefficient

A typical approach to characterizing the diffusive transport of particles begins with Fick's first law of diffusion, which states that the diffusion flux **j** is equal to the multiplication of a diffusion coefficient D and the negative gradient of concentration c:

$$\mathbf{j} = -D\nabla c \tag{1}$$

Measurement of this diffusion coefficient D makes it possible to characterize the diffusion process under particular structures.

Another approach to deriving the diffusion coefficient is to measure the root-mean-squared distance of the particles. Einstein showed that a diffusion coefficient D could be derived from the variance of the conditional probability distribution at a specific location for a specific duration. The root-mean-squared distance under unrestricted diffusion in d-dimensional space is described by

$$r_{rms} = \sqrt{2dDt} \tag{2}$$

i.e., Einstein's equation, where the diffusion coefficient D has dimensions of distance squared divided by time (e.g., mm^2/s).

If we can measure the change of magnetic resonance (MR) signal intensity according to the root-mean-squared distance of the water particles, we can derive the diffusion coefficient in the brain tissue. By introducing a diffusion-weighting magnetic gradient to a given sample, we can estimate the diffusion coefficient from the MR signal.

III. Basics of DTI

A. Diffusion and Nuclear Magnetic Resonance Signals

This section assumes that the reader has a basic knowledge of MRI principles and so will cover only the rudimentary basics of NMR to explain the relationship between diffusion and nuclear magnetic resonance (NMR) signaling. Further details can be found in books that explain the principles of MRI. Under a magnetic field B_o, spinning protons undergo precession around the axis parallel to the magnetic field B_o with the rate of precession given by the Lamor equation:

$$f = \gamma B_0 \tag{3}$$

in which γ is the gyromagnetic ratio in MHz/Tesla for the spin under consideration. An external 90° radiofrequency (RF) pulse alternating with Lamor frequency excites the net magnetization of the spins to lie in the transverse plane perpendicular to B_o and renders all spins coherent in phase (the excitation process). After the RF signal is turned off, the transverse component of the net magnetization decays exponentially as individual protons dephase with a relaxation time constant, T2, which varies depending on the tissue characteristics (the relaxation process). To generate the T2-weighted image, we detect the difference in the magnitude of tissue-specific free-induction decay of the transverse magnetization. This T2-weighted imaging, as are most other structural MRI techniques, is based on the assumption that the excitation and relaxation processes of a spin occur at stationary positions, changing only in the orientation of the spin. In contrast, the diffusion-weighted imaging (DWI) utilizes the spin translation under the gradient magnetic field. The gradient magnetic field has both strength and direction that can be represented by a vector **G** (i.e., $d\mathbf{B}(\boldsymbol{x})/dx$, where $\mathbf{B}(\boldsymbol{x})$ is the magnetic field at the location $\boldsymbol{x}$) (Fig. 2a). The phase accumulation of a spin for time duration δ under a constant magnet field gradient **G** can be written as a function of the spin location $\mathbf{x}(t)$, as

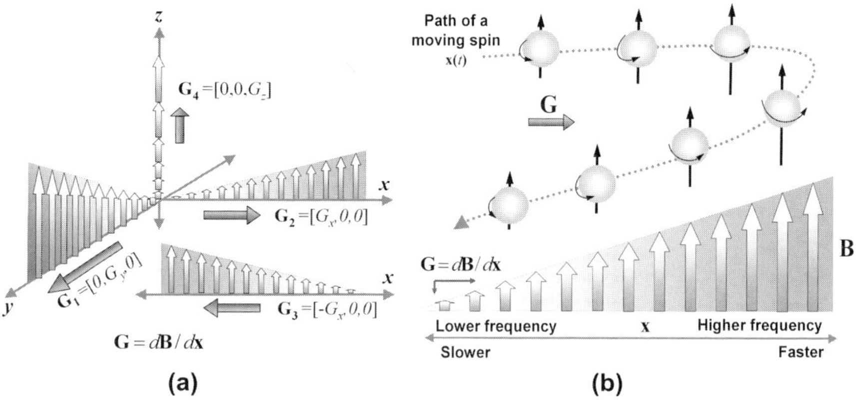

FIG. 2. Nonstationary spin in the magnetic-field gradient. The gradient of the magnetic field (**B**) can be represented with a vector (**G**) that has x, y, and z components in three-dimensional space. Among the four examples of gradients in (a), $\mathbf{G_1}$ is highest in strength, i.e., ($G_y > G_x > G_z$), and it has the direction along the y-axis. $\mathbf{G_2}$ and $\mathbf{G_3}$ have only x-components, and $\mathbf{G_2}$ is in the opposite direction to $\mathbf{G_3}$. Under the magnetic-field gradient, a spin rotates with a different precession frequency corresponding to the magnetic-field strength defined proportionally to the location of the spin (b). When a spin moves in the direction of the stronger magnetic field, the spin rotates faster, and vice versa.

$$\phi = \gamma \int_0^{\delta} \mathbf{G} \cdot \mathbf{x}(t) dt \tag{4}$$

in which a dot operation implies the location $\mathbf{x}(t)$ projected onto the gradient direction **G**. Note that $\mathbf{G} \cdot \mathbf{x}(t)$ is the strength of the magnetic field at position $\mathbf{x}(t)$ and is equal to the precession frequency of the spin when multiplied by γ, the gyromagnetic ratio, i.e., $f = \gamma B_0$. When $\mathbf{x}(t)$ is stationary for the time duration δ, the phase accumulation is simply proportional to the multiplication of the magnitude of **G** and position x_G, a projected position of **x** along the direction of **G**, i.e., $\phi = \gamma |\mathbf{G}| x_G \delta = \gamma |\mathbf{B}(x_G)| \delta = f\delta$.

When a spin moves along the direction of the magnetic-field gradient, phase accumulation becomes more complex, depending on the path **x** on which the spin moves. As illustrated in Fig. 2b, translation along the gradient magnetic-field direction will lead a spin through a varying magnetic field. The spin will have a higher or lower precession frequency when it moves along a higher or lower field. When multitudes of spinning protons move randomly along the magnetic-field gradient, the accumulated phases of spins for a given duration will shift incoherently. The phase incoherence due to the random motion (i.e., the diffusion) of spinning protons is detected as a decrease in the transverse magnetization signal.

B. DWI

1. *Basics of DWI*

DWI was developed to render the MRI sensitive to diffusion. There are various methods of DWI, but we shall discuss two basic methods: gradient-echo DWI and spin-echo DWI. The gradient-echo DWI, illustrated in Fig. 3, utilizes two magnetic fields of identical strength and duration but of opposite directional

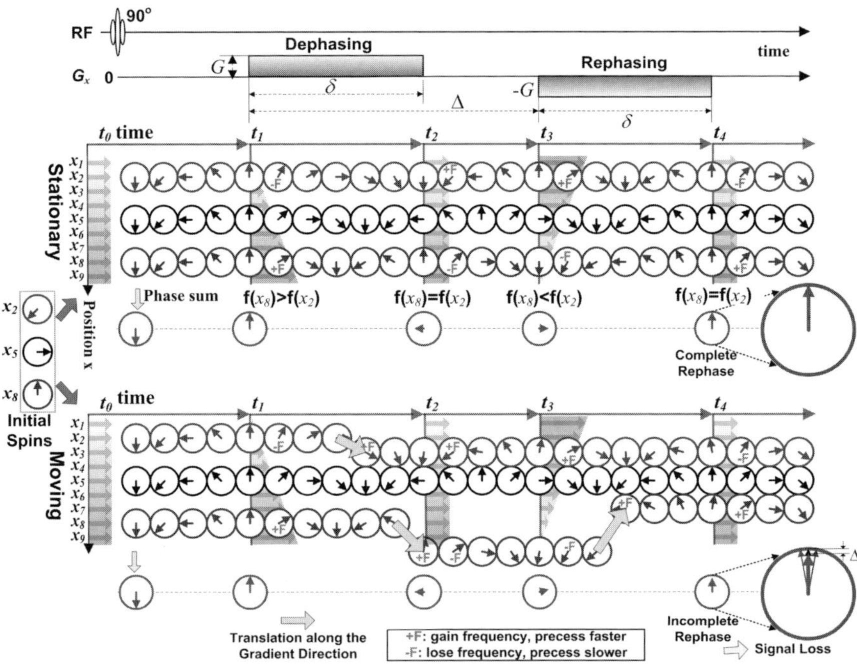

FIG. 3. Gradient-echo diffusion-weighted imaging (DWI). The phase accumulation of three spins initially located at x_2, x_5, and x_8 in the stationary situation (upper panel) and the moving situation (lower panel) during the gradient echo DWI sequence is explained as time elapses. A 90° excitation radiofrequency pulse renders all spins in phase (t_0), and under the uniform magnetic field, this condition is sustained. When a first gradient magnetic pulse is applied (t_1), the strength of the magnetic field increases along the x, i.e., $B(x_2) < B(x_5) < B(x_8)$, and so does the precession frequency, i.e., $f(x_2) < f(x_5) < f(x_8)$. The phase shifts accumulated due to the frequency difference will be sustained after the application of the uniform magnetic field, i.e., $\mathbf{G} = 0$ (t_2). The second magnetic-field gradient will cause the precession frequency at x_8 to be lower than that at x_2. The passing of the same amount of accumulation time (δ) as for the first gradient magnetic pulse will lead to a complete phase realignment of all spins that are stationary (upper panel). For moving situations, the first spin (initially located at x_2) moves to x_3, which causes the spin to rotate faster than at the previous position rotating with $f(x_2)$. Under the second gradient, the compensating phase accumulates proportionally to the frequency $f(x_3)$, which results in incomplete rephase of the phase accumulation with rotation frequency of both $f(x_2)$ and $f(x_3)$ during the first gradient (lower panel). The spin at x_8 also undergoes similar incomplete rephase at time t_4. (See Color Insert.)

gradients. For simplicity, we shall ignore the dephasing due to the spin and spin interaction, which causes free-induction decay of the T2 signal. After a 90° excitation RF pulse, all randomly phased stationary spins align and stay in phase under the homogeneous magnetic field. When the first gradient magnetic pulse is applied, the precession frequencies of protons depend on their location, as the strength of the magnetic field is proportional to the location of the spin. Spins in a weak magnetic field rotate slowly, whereas precession is faster in a strong magnetic field. During the time period of Δ, the phase difference between spins accumulates. When the first gradient is turned off, all spins run with the same frequency, and thus the phase difference is sustained until they experience a new gradient pulse. When the second magnetic-field gradient pulse (with the same strength and duration but in the opposite gradient direction from the first one) is applied, it will cause the spins that were rotating faster at the previous gradient field to rotate more slowly, and vice versa. At the end of the second gradient, all stationary spins will have a coherent phase, which leads to a peak signal amplitude in the receiver coil. The upper panel of Fig. 3 shows the phase changes of the three stationary spins at different locations as time elapses. When spins under the gradient magnetic field move along the gradient direction (lower panel of Fig. 3), the spins will gain or lose the frequency according to their path. Therefore, the second gradient will not completely rephase the translating spins. We detect an attenuated signal of the net magnetization due to their incomplete rephase.

Instead of using two opposite gradient magnetic fields, the spin-echo pulse sequence (Stejskal and Tanner, 1965), as shown in Fig. 4, uses one strong identical diffusion-weighting gradient pulse on each side of the 180° refocusing pulse. The first pulsed gradient accumulates phase shifts for all spins, depending on their position and motion during the application of the first gradient. The 180° pulse inverts the phase of the spins, and the second gradient will induce another phase shift, thus canceling out the phase shifts for the stationary spins (the upper six spins in Fig. 4). Spins that have changed positions during the time period Δ (the lower six spins in Fig. 4) will experience different phase shifts due to the two gradient pulses, and thus they will not become completely refocused. This incomplete phase realignment will consequently lead to signal attenuation.

Signal attenuation due to water diffusion, initially derived by Stejskal and Tanner (1965) based on spin-echo sequence, can be defined by the relationship between the diffusion-weighted signal, S, and the non-diffusion-weighted signal, S_0, as below:

$$\begin{aligned} S &= S_0 \exp(-bD) \\ b &\equiv \gamma^2 |\mathbf{G}|^2 \delta^2 (\Delta - \delta/3) \end{aligned} \qquad (5)$$

where the gradient factor, b, is determined by the diffusion-sensitizing gradient having strength $|\mathbf{G}|$ for the duration δ, with the time delay Δ between the two diffusion-gradient pulses. γ is the gyromagnetic ratio of proton of the water

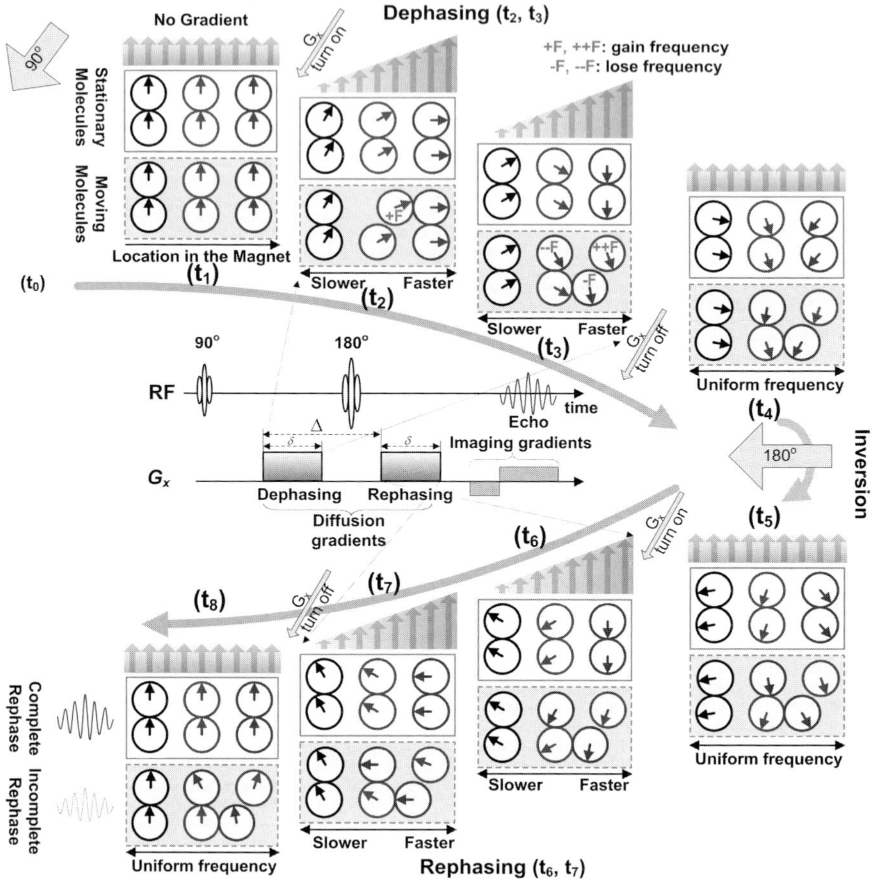

FIG. 4. Spin-echo diffusion-weighted imaging. The time courses of the phase accumulations of six stationary molecules (upper part) and of six moving molecules (lower part) are explained for the spin-echo sequence. Under the uniform magnetic field with no gradient (t_1), the phase alignment of all individual spins caused by a 90° radiofrequency excitation is sustained. When the first gradient magnetic pulse is on (t_2 and t_3), the location-dependent magnetic fields induce spins to rotate with different frequencies. If a spin moves along the direction of the magnetic-field gradient, the phase of the spin changes. A 180° radiofrequency pulse reverses the phase of all spins (t_4 and t_5). The second gradient (t_6 and t_7) will lead to realignment for a stationary spin (t_8), whereas moving molecules will be rendered out of phase. This out-of-phase condition leads to signal attenuation. (See Color Insert.)

molecule (42 MHz/Tesla). The non-diffusion-weighted signal, S_0, is a T2-weighted signal at the echo time (TE), i.e., the time from RF excitation pulse to the center of the echo being received. The diffusion coefficient D reflects molecular "diffusivity" along the direction of the magnetic-field gradient **G**. Since the diffusion coefficient D measured at a voxel can also be affected by many things—tissue perfusion, the partial volume averaging effects, and other experimental errors—it is often called the "apparent" diffusion coefficient (ADC). By measuring the voxel-level signal attenuation under a single magnetic-field gradient, we can derive the ADC D at all voxels in the entire brain composing a diffusivity map (Le Bihan *et al.*, 1986).

With the previously determined b value for NMR acquisition and the measured diffusion weighted signal S, Eq. 5 can be reduced to the relationship between two unknown variables, the ADCs D and S_0. In order to take out S_0 and derive the ADC D, it is necessary to acquire the diffusion-weighted images with at least two different b-values. The ADC D can be derived from two DWIs as follows:

$$\begin{aligned} S(b_0) &= S_0 \exp(-b_0 D) \\ S(b_1) &= S_0 \exp(-b_1 D) \\ D &= -\frac{1}{b_1 - b_0} \ln\left(\frac{S(b_1)}{S(b_0)}\right) \end{aligned} \tag{6}$$

The ADC D is the slope representing the logarithmic difference of attenuation according to the different b-values, as in Fig. 5.

In practice, instead of acquiring two different b-values, researchers often choose b_0 to be zero, which yields $S(b_0) = S_0$, a T2-weighted image without diffusion weighting. Therefore, we can derive the ADC D by acquiring a DWI at a single b-value and a non-diffusion-weighted T2 reference image. One must remember that the signal attenuation due to phase incoherence is caused by both

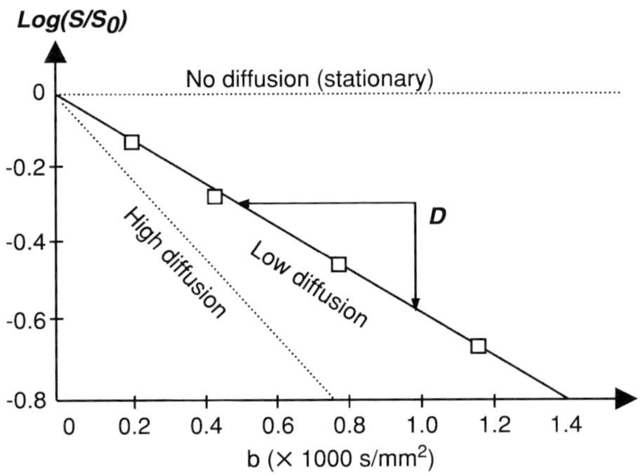

FIG. 5. Diffusion coefficient and diffusion factor b-values. We can derive diffusion coefficient D by calculating the slope of log(S/So) at two or more b-values. Most diffusion-imaging sequences use a b-value on the order of 1000 s/mm^2.

the diffusion process ($\propto \exp(-bD)$) and the T2 relaxation process ($\propto \exp(-TE/T2)$), but the relaxation effects are nicely cancelled out by the division term of Eq. 6 in the spin-echo sequence.

C. DTI

1. *Tensor Representation of Three-Dimensional Diffusion*

We have discussed DWI that measures the attenuation of the NMR signal when multitudes of spins randomly move along the direction of a given magnetic-field gradient. In other words, a single magnetic-field gradient can describe the motion along a single direction. In fact, the random motion of water molecules in the brain tissue occurs in three-dimensional space, and the mobility may not be the same in all directions. In order to quantify diffusion, traditional DWI has measured NMR attenuation due to diffusion along the three laboratory coordinates (x-axis, y-axis, and z-axis) and has derived three diffusion coefficients, or diffusivities, D_{xx}, D_{yy}, and D_{zz}. D_{xx}, D_{yy}, and D_{zz} indicate the diffusivities along the x-axis, y-axis, and z-axis of the laboratory coordinate system, respectively. Using these diffusivities measured along the three orthogonal axes, an isotropic diffusion, as in Fig. 6a, and an anisotropic diffusion aligned precisely along the laboratory coordinates, as in Fig. 6b, can be described. However, three diffusion coefficients cannot fully describe the three-dimensional anisotropic diffusion, which is not aligned with the laboratory coordinates, as shown in Fig. 6c.

Therefore, it is an essential requirement that diffusivities be measured at additional directions for the complete description of any diffusion process. By including the diffusion components along the x and y direction, the x and z

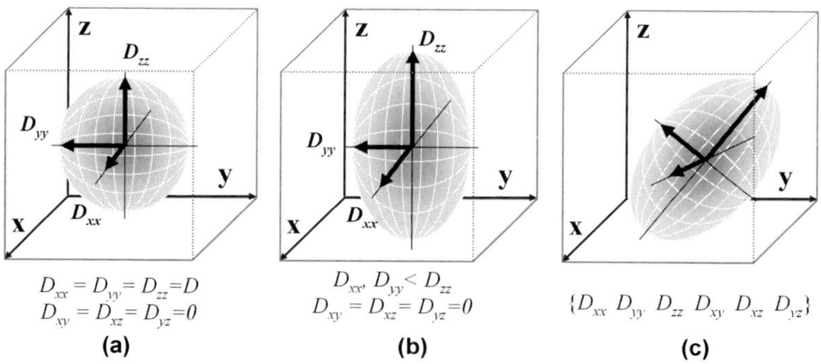

FIG. 6. Representation of isotropic and anisotropic diffusion. Isotropic diffusion, which has the same diffusivity in all directions, and anisotropic diffusion, in which the principal directions of diffusivity align with three MRI coordinates, can be described using only three diffusion coefficients, D_{xx}, D_{yy}, and D_{zz}. However, six diffusion coefficients are necessary to fully describe the three-dimensional diffusion.

direction, and the y and z direction (i.e., D_{xy}, D_{xz}, and D_{yz}) in addition to the traditional measurements of diffusivities along the x-, y-, and z-axes (i.e., D_{xx}, D_{yy}, D_{zz}), it is possible to document the three-dimensional diffusion process. These six diffusivities can be mathematically represented as a symmetric matrix, called the "tensor" as shown in Eq. 7:

$$\mathbf{D} = \begin{pmatrix} D_{xx} & D_{xy} & D_{xz} \\ D_{xy} & D_{yy} & D_{yz} \\ D_{xz} & D_{yz} & D_{zz} \end{pmatrix} \tag{7}$$

The tensor comprises the multiple-directional geometry of diffusion in a three-dimensional space. This diffusion-tensor matrix can be related to a covariance matrix of the three-dimensional Gaussian displacement distribution that explains the probability distribution of the displacement of a single spin for a unit time. The basic properties of the tensor will be discussed in detail in Section IV.

2. *Derivation of Diffusion Tensors*

Diffusion-tensor imaging was originally developed by Basser *et al.* (1994) for three-dimensional assessment of diffusion data *in vivo*, achieved by measurement of DWIs along more than six directional gradients. In order to document the diffusion phenomena, three-dimensional DTI utilizes the diffusion tensor instead of a single diffusion coefficient in the DWI in Eq. 5. Signal attenuation due to the diffusion under the magnetic-field gradient $\mathbf{G} = [\,G_x \quad G_y \quad G_z\,]$ becomes:

$$S = S_0 \exp(-\mathbf{b}\cdot\mathbf{D}) = S_0 \exp\left(-\sum_{i=1}^{3}\sum_{j=1}^{3} b_{ij}D_{ij}\right), \quad i,j = x,y,z \tag{8}$$

where $b_{ij} = -\gamma^2\delta^2(\Delta - \delta/3)G_iG_j, \quad i,j = x,y,z$ for the spin-echo sequences is an element of the gradient factor matrix $\mathbf{b}$. Equation 8 is the more generalized form of Eq. 5, which can be applicable to either anisotropic or isotropic diffusion. An isotropic diffusion can be derived simply by replacing the tensor matrix $\mathbf{D}$ with $D\mathbf{I}$, i.e., the identity matrix multiplied by diffusivity D. Equation 8 can be rewritten as

$$\begin{aligned} \ln\left[\frac{S}{S_0}\right] &= -\mathbf{b}\cdot\mathbf{D} \\ &= -\left[b_{xx}D_{xx} + b_{yy}D_{yy} + b_{zz}D_{zz} + 2b_{xy}D_{xy} + 2b_{xz}D_{xz} + 2b_{yz}D_{yz}\right] \\ &= -\overline{\mathbf{b}}\cdot\overline{\mathbf{D}} \end{aligned} \tag{9}$$

where

$$\begin{aligned} \overline{\mathbf{b}} &= [\,b_{xx} \quad b_{yy} \quad b_{zz} \quad 2b_{xy} \quad 2b_{xz} \quad 2b_{yz}\,] \\ \overline{\mathbf{D}} &= [\,D_{xx} \quad D_{yy} \quad D_{zz} \quad D_{xy} \quad D_{xz} \quad D_{yz}\,] \end{aligned}$$

For the derivation of the diffusion tensor matrix **D** with six independent variables from Eq. 9, at least six equations are required. Therefore, six diffusion measurements along the noncollinear noncoplanar gradient directions, plus one non-diffusion-weighted measurement as the reference datum, are required in order to obtain the complete solution.

Figure 7 shows examples of these seven images, where S_0 is the non-diffusion-weighted image and S_1, S_2, S_3, S_4, S_5 , and S_6 are the DWIs derived from the six different diffusion-sensitizing magnetic gradients, **G**. From these data sets, a complete set of equations are illustrated below:

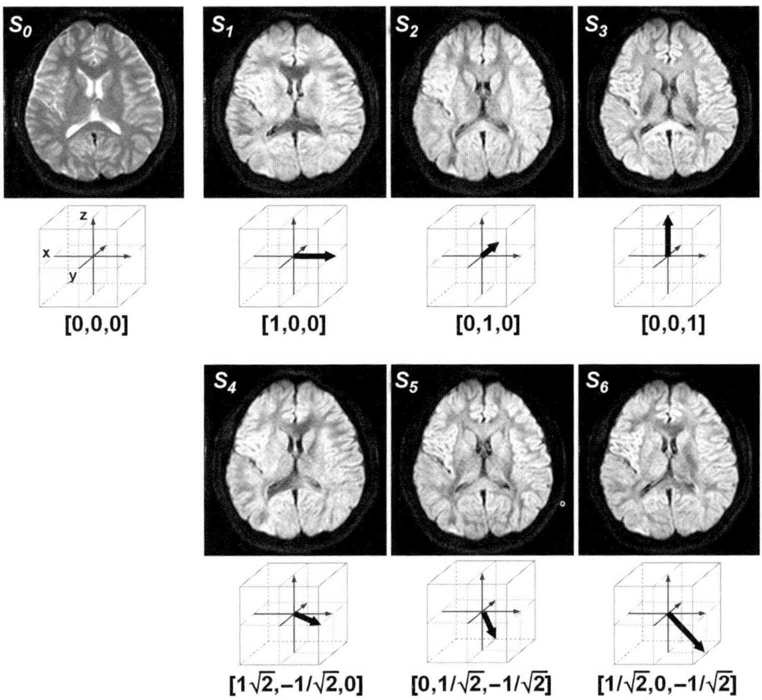

FIG. 7. Diffusion-weighted images measured along six magnetic gradients ($S_1,\ldots,S_6$) and non-diffusion-weighted image (S_0). The black arrow indicates the direction of the magnetic-field gradient used to derive each diffusion-weighted image. For high diffusion, the random phase shifts suppress the transverse magnetization signal; these are represented with dark pixels in the diffusion-weighted image. In contrast, the low diffusion leads to a small suppression of signal, causing the diffusion-weighted image's pixels to appear brighter. In the S_1 image measured along the left–right direction, the regions that diffuse highly along this direction, such as the corpus callosum, have high signal attenuation and thus low intensity. The internal capsule, which appears dark at the S_3 image, is measured along the inferior–superior direction. By combining these seven images, we can derive a diffusion-tensor image.

$$\ln\left[\frac{S_1}{S_0}\right] = -\overline{\mathbf{b}}_1 \cdot \overline{\mathbf{D}}, \ln\left[\frac{S_2}{S_0}\right] = -\overline{\mathbf{b}}_2 \cdot \overline{\mathbf{D}}, \ln\left[\frac{S_3}{S_0}\right] = -\overline{\mathbf{b}}_3 \cdot \overline{\mathbf{D}},$$

$$\ln\left[\frac{S_4}{S_0}\right] = -\overline{\mathbf{b}}_4 \cdot \overline{\mathbf{D}}, \ln\left[\frac{S_5}{S_0}\right] = -\overline{\mathbf{b}}_5 \cdot \overline{\mathbf{D}}, \ln\left[\frac{S_6}{S_0}\right] = -\overline{\mathbf{b}}_6 \cdot \overline{\mathbf{D}} \tag{10}$$

Since S_0, S_1, …, and S_6 are measured, and $\overline{\mathbf{b}}_1, \overline{\mathbf{b}}_2, \ldots, \overline{\mathbf{b}}_6$ are previously assigned, the diffusion coefficients D_{xx}, D_{xy}, D_{xz}, D_{yy}, D_{yz}, and D_{zz} can be derived from the solutions of the linear equations (Eq. 10). Note that the diffusion tensors are calculated voxel by voxel in the entire brain. The resultant diffusion-tensor components of the DWIs in Fig. 7 are displayed in Fig. 8.

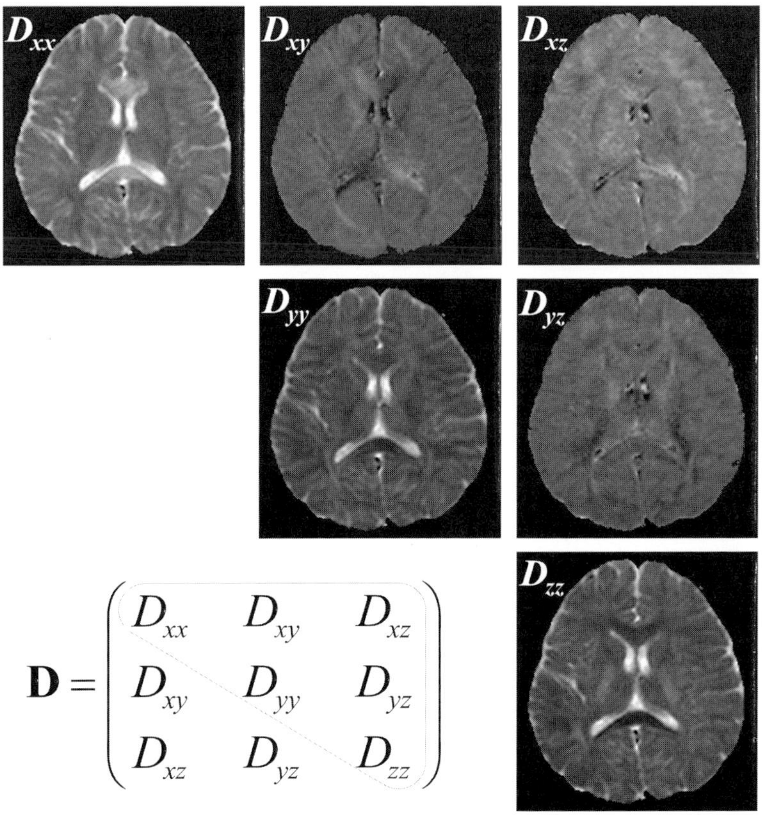

FIG. 8. Diffusion-tensor image. The six components of a diffusion-tensor image derived from the six diffusion-weighted images and a non-diffusion-weighted image (from Fig. 7) is displayed. High signal attenuation along the left–right direction in the corpus callosum is presented with low intensity in S_1 in Fig. 7 but with high intensity in the diffusion-tensor image, D_{xx}.

IV. Quantification of Diffusion Tensors

A. Scalars, Vectors, and Tensors

In physical applications, certain quantities (such as temperature and mass) possess only magnitude. A quantity that can be represented by a single real number is called a "scalar." Quantities such as force and velocity, which possess both magnitude and direction, can be represented by arrows that have length and direction from a given reference point, and these are called "vectors." More complex geometric or physical quantities, such as strain and diffusion, are multidirectional entities that can be represented by what are known as "tensors."

A tensor, which contains multidirectional information, can be resolved into three orthogonal vectors and their magnitudes by the mathematical operation known as diagonalization. Orthogonal unit vectors, also called eigenvectors, represent the principal directions of a tensor. The magnitudes are represented by scalars, or eigenvalues.

$$\begin{aligned}\mathbf{D} &= \begin{pmatrix} D_{xx} & D_{xy} & D_{xz} \\ D_{xy} & D_{yy} & D_{yz} \\ D_{xz} & D_{yz} & D_{zz} \end{pmatrix} \\ &= \begin{bmatrix} \mathbf{e}_1 & \mathbf{e}_2 & \mathbf{e}_3 \end{bmatrix}^T \begin{pmatrix} \lambda_1 & 0 & 0 \\ 0 & \lambda_2 & 0 \\ 0 & 0 & \lambda_3 \end{pmatrix} \begin{bmatrix} \mathbf{e}_1 & \mathbf{e}_2 & \mathbf{e}_3 \end{bmatrix} \\ &= \lambda_1 \mathbf{e}_1 \mathbf{e}_1^T + \lambda_2 \mathbf{e}_2 \mathbf{e}_2^T + \lambda_3 \mathbf{e}_3 \mathbf{e}_3^T = \sum_{k=1}^{3} \lambda_k \mathbf{e}_k \mathbf{e}_k^T \end{aligned} \tag{11}$$

where $\mathbf{e}_1$, $\mathbf{e}_2$, and $\mathbf{e}_3$ are eigenvectors and λ_1, λ_2, and λ_3 are eigenvalues obtained by diagonalization of the tensor matrix. For diffusion tensors, λ_1, λ_2, and λ_3 indicate diffusivity along the directions $\mathbf{e}_1$, $\mathbf{e}_2$, and $\mathbf{e}_3$, respectively. The eigenvalues are often denoted in a sorted order ($\lambda_1 \geq \lambda_2 \geq \lambda_3$), and the eigenvector $\mathbf{e}_1$ that is associated with the maximal eigenvalue $\lambda_1 > \lambda_2, \lambda_3$, (which will be called the major eigenvector throughout this paper) implies the principal direction of the diffusion.

As shown in Fig. 9, a diffusion tensor derived from more than directional measurements can be conceived in terms of the ellipsoidal model, where each eigenvector corresponds to the principal axis of the ellipsoid and the eigenvalues indicate the extent of the ellipsoid along the corresponding axis.

B. Quantification of Diffusion Tensors

Because the diffusion process of the tissue underlying a voxel can be modeled as an ellipsoid, the scalar indices for the size and shape (i.e., eccentricity) of the ellipsoidal diffusion can be represented by the combination of eigenvalues that

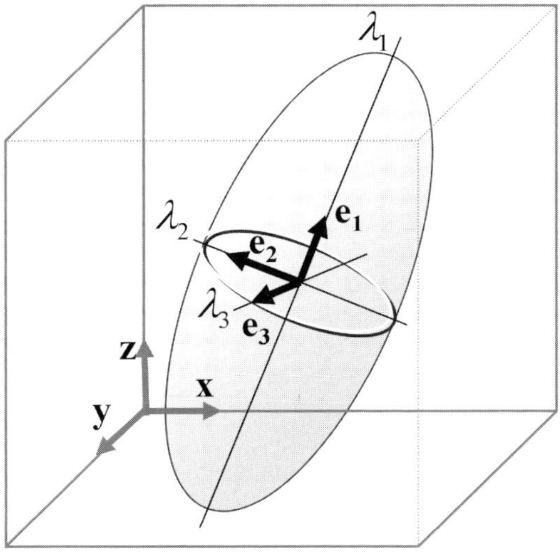

FIG. 9. Eigenvectors and eigenvalues of a diffusion tensor: Diffusion tensors can be better described by a new coordinate system of eigenvectors, $\mathbf{e_1}$, $\mathbf{e_2}$, and $\mathbf{e_3}$, and their eigenvalues λ_1, λ_2, and λ_3, derived by diagonalization of the diffusion tensor. $\mathbf{e_1}$, $\mathbf{e_2}$, and $\mathbf{e_3}$ are the principal axes of the ellipsoidal model of the diffusion. The eigenvalues are denoted in a sorted order (λ_1, λ_2, $\lambda_{3,}$), and an eigenvector corresponding to any of the largest eigenvalues (λ_1) is often denoted $\mathbf{e_1}$ and called a major eigenvector. The diffusion distance along an eigenvector for a unit of time is proportional to the square root of the corresponding eigen value, i.e., $\sqrt{i}$

are insensitive to the orientation of the diffusion, the direction of the gradient magnetic fields, and the laboratory coordinates. The most prevalent scalar indices used to quantify the diffusion tensor are the mean ADC and anisotropy.

1. *The Mean ADC*

The mean ADC, also called mean diffusivity, can be defined by a third of the trace, which is the sum of diagonal elements of the matrix, which also equals the sum of all the eigenvalues.

$$\begin{aligned}\langle ADC\rangle &= Trace(\mathbf{D})/3 = (D_{xx} + D_{yy} + D_{zz})/3 \\ &= (\lambda_1 + \lambda_2 + \lambda_3)/3 = \langle\lambda\rangle \end{aligned} \quad (12)$$

In isotropic diffusion, the mean ADC can be measured by an ADC measured along one of the x, the y, or the z direction. The mean ADC of anisotropic diffusion represents the ADC of the equivalent isotropic diffusion.

2. *The Anisotropy Index*

An anisotropic index is defined to characterize the eccentricity of the ellipsoidal diffusion in a scale-independent and coordinate-independent way. Relative anisotropy (RA) and fractional anisotropy (FA) (Basser and Pierpaoli, 1996) are the most representative examples of anisotropic indices.

$$RA = \frac{\sqrt{(\lambda_1 - \langle\lambda\rangle)^2 + (\lambda_2 - \langle\lambda\rangle)^2 + (\lambda_1 - \langle\lambda\rangle)^2}}{\sqrt{3\langle\lambda\rangle}}$$

$$FA = \frac{\sqrt{(\lambda_1 - \lambda_2)^2 + (\lambda_2 - \lambda_3)^2 + (\lambda_1 - \lambda_3)^2}}{\sqrt{2}\sqrt{(\lambda_1^2 + \lambda_2^2 + \lambda_3^2)}}$$

$$= \frac{\sqrt{3\left[(\lambda_1 - \langle\lambda\rangle)^2 + (\lambda_2 - \langle\lambda\rangle)^2 + (\lambda_1 - \langle\lambda\rangle)^2\right]}}{\sqrt{2(\lambda_1^2 + \lambda_2^2 + \lambda_3^2)}} \tag{13}$$

where $\langle\lambda\rangle = (\lambda_1 + \lambda_2 + \lambda_3)/3$.

The RA is the normalized standard deviation that represents the ratio between an isotropic and an anisotropic component. The FA is the fraction of the magnitude of diffusion due to anisotropic diffusion. Both RA and FA are 0 for isotropic diffusion ($\lambda_1 = \lambda_2 = \lambda_3$) and are 1 for ideal anisotropic diffusion (i.e., $\lambda_2 = \lambda_3 = 0$). RA changes linearly according to the changes of anisotropy, which is a good characteristic for the quantification of diffusion. However, RA is vulnerable to noise. FA is used prevalently, in spite of its nonlinearity, because of its robustness relative to noise.

Figure 10 shows examples of diffusion tensors with mean ADC and FA at the corpus callosum, prefrontal white matter, posterior limb of the internal capsule (PLIC), optic radiation, and cerebrospinal fluid. The FA at the posterior limb is close to 1, and the mean ADC is lower than that of any other tensor. The cerebrospinal fluid shows the highest mean ADC, with FA close to 0.

3. *The Geometric Index*

Diffusion can be decomposed into three geometric components: line, plane, and sphere (Westin *et al.*, 2002):

$$\mathbf{D} = c_l\mathbf{D}_l + c_p\mathbf{D}_p + c_s\mathbf{D}_s$$

where

$$\mathbf{D}_l = \lambda_1\mathbf{e}_1\mathbf{e}_1^T, \mathbf{D}_p = \lambda_1(\mathbf{e}_1\mathbf{e}_1^T + \mathbf{e}_2\mathbf{e}_2^T),$$

$$\mathbf{D}_s = \lambda_1(\mathbf{e}_1\mathbf{e}_1^T + \mathbf{e}_2\mathbf{e}_2^T + \mathbf{e}_3\mathbf{e}_3^T)$$

$$c_l = \frac{\lambda_1 - \lambda_2}{\lambda_1}, c_p = \frac{\lambda_2 - \lambda_3}{\lambda_1}, c_s = \frac{\lambda_3}{\lambda_1} \tag{14}$$

$\mathbf{D}_l$, $\mathbf{D}_p$, and $\mathbf{D}_s$ represent the geometric tensor matrix, composed solely of a line, a plane, and a sphere, respectively, and c_l, c_p, and c_s are fractions of these geometric tensor matrices, the sum of which is equal to 1.

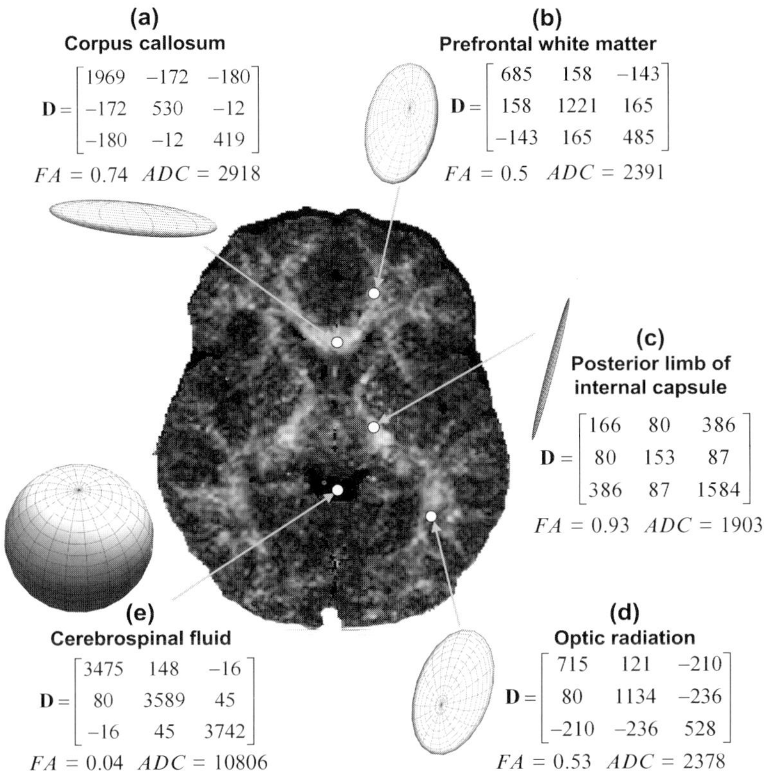

FIG. 10. Diffusion tensors in the brain. Diffusion tensors at (1) the corpus callosum, (2) the prefrontal white matter, (3) the posterior limb of the internal capsule, (4) the optic radiation, and (5) the cerebrospinal fluid are displayed with a fractional anisotropy image as a background image.

$$c_l + c_p + c_s = 1 \tag{15}$$

These geometric indices provide more detailed information about the shape of the diffusion at a specific region. Where diffusion occurs mostly in the direction associated with the largest eigenvector, i.e., $\lambda_1 \gg \lambda_2, \lambda_3$, c_l is 1 and the diffusion tensor $\mathbf{D}$ consists mostly of the $\mathbf{D}_l$ component, i.e., $\mathbf{D} \approx \mathbf{D}_l$. Diffusion is restricted to a plane spanned by two eigenvectors associated with the first and second largest eigenvalues ($\lambda_1 \approx \lambda_2 \gg \lambda_3$), in which case c_p is 1 and diffusion $\mathbf{D}$ can be reduced to $\mathbf{D}_p$. In the case of complete isotropic diffusion ($\lambda_1 = \lambda_2 = \lambda_3$), $\mathbf{D}$ is equal to $\mathbf{D}_s$.

4. *The Lattice Anisotropy Index*

In comparison with previous indices, which describe a diffusion quantity defined at a voxel, the scalar index for the pattern or distribution of diffusion within a local neighborhood is defined in terms of macrostructural diffusion coherency, which is known as the "lattice index" (LI). An example of the lattice index for a tensor D at a voxel v is defined as in Eq. 16 (Pierpaoli and Basser, 1996; Westin *et al.*, 2002):

$$LI(v) = \sum_{k \in \mathbf{N}_v} a_k \left(\sqrt{\frac{3}{8}} \frac{\sqrt{\langle \mathbf{T}, \mathbf{T}_k \rangle}}{\sqrt{\langle \mathbf{D}, \mathbf{D}_k \rangle}} + \frac{3}{4} \frac{\langle \mathbf{T}, \mathbf{T}_k \rangle}{\sqrt{\langle \mathbf{D}, \mathbf{D} \rangle} \sqrt{\langle \mathbf{D}, \mathbf{D}_k \rangle}} \right) \tag{16}$$

where k is an element of set $\mathbf{N}_v$ neighboring voxel v, a_k is a weighted spatial mask, $< >$ is the tensor dot product, and $\mathbf{T}$ is the anisotropic component of diffusion tensor $\mathbf{D}$, which can be written as

$$\mathbf{T} = \mathbf{D} - \frac{D_{xx} + D_{yy} + D_{zz}}{3} \tag{17}$$

The LI of a voxel is an intervoxel scalar quantity that depends on the directional coherence of the diffusion in the reference voxel with that in the neighboring voxels. The LI is very robust to random noise at a low signal-to-noise ratio (SNR). However, this robustness depends on acquisition parameters such as voxel size.

C. Maps of the Quantitative Diffusion Index

Scalar indices of each voxel listed above are used to contrast tissue types in the entire brain according to the characteristic of the index. Figure 11 shows exemplary contrast images according to the information that can be obtained from a diffusion-tensor image. A plain T2-weighted image provides anatomical references for gray matter, white matter, and cerebrospinal fluid spaces, as does a T1-weighted image. From three eigenvalues (i.e., λ_1, λ_2 and, λ_3), mean ADC, FA, c_l, c_p, and c_s images can be derived. Mean ADC implies the amount of diffusion, whereas FA provides the shape of diffusion, which is related to information about density and/or organization of the fiber tracts. An FA image conspicuously enhances white matter. A c_l component image shows similarity to an FA image. An LI image provides information regarding the coherence of the macroscopic fiber structures.

D. Physiology of Anisotropic Diffusion

1. *Sources of Anisotropy in the Brain*

The source of anisotropy in brain tissue is one of the most important issues in diffusion-tensor research, and it is not yet clearly understood. Among the several possibilities, the myelin sheath–encasing axon is the preferred hypothetical

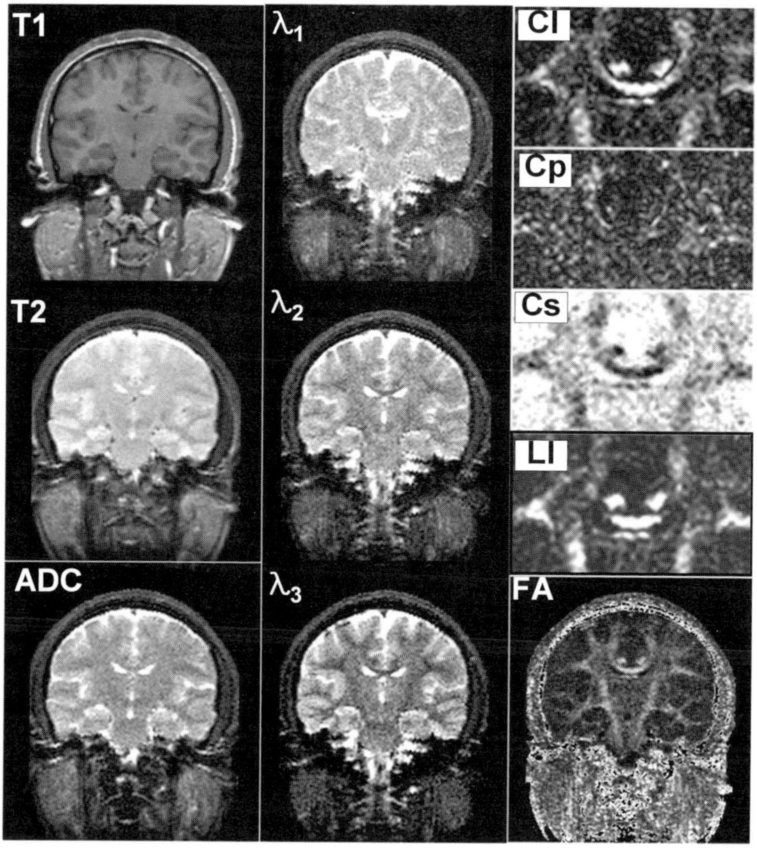

FIG. 11. Scalar indices of diffusion tensors. Coronal slices of several contrast images are displayed using scalar indices of the diffusion tensors. The T1 image is displayed for reference purposes. From three eigenvalues (i.e., λ_1, λ_2, and λ_3), images of the mean apparent diffusion coefficient (ADC), fractional anisotropy (FA), and line, place, sphere component (c_b, c_p, and c_s) images are derived. LI: lattice index.

explanation of diffusion anisotropy; myelin is composed of numerous lipid bilayers, which limit the permeability of water and hinder water diffusion perpendicular to the fibers. However, fibers without myelin sheaths, such as nonmyelinated garfish olfactory nerves (Beaulieu and Allen, 1994) and neonate brains in which myelin has not yet appeared (Huppi *et al.*, 1998; Wimberger *et al.*, 1995), also exhibit diffusion anisotropy. Therefore, we know that myelin is not an essential component for anisotropic diffusion. Cylindrical neurofibrils and fast axonal transportation linked to the microtubules have not been found to be dominantly related to diffusion anisotropy (see Beaulieu [2002] for a detailed review). As of this writing, it is believed that axonal membranes play a primary role, and that myelination, although it is not necessary for significant anisotropy, can modulate the degree of anisotropy.

2. *Macroscopic Anisotropy*

Although there are several factors that influence self-diffusion—for example, molecular weight, intermolecular interaction (viscosity), and temperature—the overall mobility of diffusion molecules in the brain depends mostly on their underlying microstructures. The degree of anisotropy is related to factors such as the density and diameter of axons, the degree of axon myelination, and the coherence of axon direction. It should be noted that structures that exhibit anisotropic diffusion at the molecular level can be isotropic at the macroscopic level when multiple fiber bundles cross or merge at a voxel. Therefore, it is not easy to directly relate a single factor to macroscopic anisotropy. Special care must be taken in the interpretation of any images obtained.

V. DTI: Acquisition and Artifacts

A. *b*-Values and Gradient Vectors

The *b*-value, seen in Eq. 5, is used in the order of 1000 s/mm^2 in most diffusion imaging sequences. This value can be obtained by increasing the amplitude ($|\mathbf{G}|$) and/or the duration (δ) of the diffusion-weighting gradients' lobes. The increased duration (δ) of the gradient, however, suffers from a long echo time, which results in longer acquisition time and unwanted contrast change due to the T2-weighted signal. In order to minimize the echo time, maximal amplitude of the magnet gradient ($|\mathbf{G}|$) with faster slew rate is used to achieve the desired *b*-values. However, the rapid change of high-amplitude gradients may induce undesirable stimulation of the peripheral nerve and may induce geometry-distorting eddy currents, which will be discussed in the following section.

Although six orientation DWIs can be used to construct a diffusion-tensor image, in general, DWIs with a higher quantity of gradient orientations are employed to reduce the orientation bias in the parameters derived from the diffusion-tensor image. Thus, proper application of the number of gradient orientations and their distributions in three-dimensional space are both crucial for the precise measurement of anisotropic diffusion within the acceptable imaging time. Jones (2004) showed that more than 30 diffusion-weighted gradients evenly distributed in three-dimensional space are required to make robust tensor estimates. These optimal gradients, obtained by the concepts of the electrostatic repulsion algorithm, lead to a higher SNR and an improved contrast in the diffusion parameter of interest (e.g., FA or mean ADC).

B. Imaging Artifacts

Many of the DTI techniques were developed in the process of reducing artifacts that often disturb the interpretation of data. Therefore, it is worthwhile to understand the artifacts in DTI.

1. *Motion Artifacts*

Since diffusion is a microscopic molecular movement, the outer macroscopic movement can disturb the detection of micro-scale diffusion. Macroscopic motion includes large-scale bulk motions and brain motions caused by pulsations in cerebrospinal fluid. These motions can cause artifacts such as ghosting, blurring, and misalignment of DWIs. Bulk motion is often removed by the application of rigid-body transformations in postprocessing. However, brain motion from pulsation of cerebrospinal fluid caused by the heartbeat is too complex to be removed with a rigid-body transformation. Although the pulsation effect is negligible for normal anatomical images, an extremely motion-sensitized DWI is substantially affected, especially in the period of 100–300 ms after systole (Norris, 2001). This pulsation of the brain can be partly reduced by the acquirement of images with cardiac triggering, but the cost is a longer acquisition time.

The most common approach to overcoming motion artifacts is to utilize a fast acquisition sequence, for example, a single-shot echo planar image EPI, which acquires a whole brain image in a single shot of about 100 ms. However, fast imaging sequences suffer from disadvantages such as low spatial resolution, geometric distortion (due to eddy currents and increased sensitivity to the susceptibility changes), low SNR, and image blurring. Multishot EPI is affected by less distortion than is single-shot EPI. However, it suffers from severe image artifacts due to subject motion, as a result of ultra-shot motion effects that are inconsistent between shots. To overcome this artifact, motion correction using navigator echoes is sometimes used for multishot DWI. Parallel imaging shows its usefulness in reducing acquisition time, reducing magnetic susceptibility artifacts, and increasing spatial resolution.

Another way to reduce motion artifacts is to use motion-insensitive sequences such as line scan diffusion-weighted imaging(LSDI) (Maier *et al.*, 1998). Acquirement of a two-dimensional image line by line using LSDI does not require phase-encoding steps and is insensitive to bulk motion and susceptibility artifacts but is limited by its time efficiency.

2. *Eddy Currents*

Eddy currents are small electrical currents that are found in electrically conductive structures of the MRI scanner. These eddy currents are induced by the high magnitude caused by rapidly switching on and off the magnetic gradient field during the DWI sequence. Eddy currents produce additional undesirable magnetic fields, which result in the geometric distortion of diffusion-weighted images and bring about a difference between the actual and prescribed local **b**-matrix values. When eddy currents are present, individual DWIs become misaligned by shifting, mostly along the phase-encoded direction. The mean ADC image and the FA image derived from these misaligned images have artifacts such as increased anisotropy and blurring, especially in the periphery of the brain. These artifacts can be reduced through the use of well-designed pulse sequences, such as the bipolar diffusion-encoding gradients (Alexander *et al.*, 1997), and the pulse-sequence-specific preemphasis calibration (Papadakis *et al.*, 2000). Eddy-current artifacts can also be reduced by postprocessing schemes that are based on image registration. Image registration of the whole DWIs to a template (in general the B_0 image) has been successfully applied as an eddy-current correction (Poupon *et al.*, 2000).

3. *Effects of Variations in Magnetic Susceptibility*

Adding an object with a locally heterogeneous magnetic susceptibility causes magnetic-field inhomogeneity and produces local magnetic-field gradients within the object. The magnetic-susceptibility variation in the brain adversely affects the DWIs because of the additional local gradients that behave like diffusion gradients. These may cause a spatially varying b-matrix and may thus cause distortion in the DWIs. Furthermore, susceptibility in homogeneity may result in a complete loss of signal. Susceptibility artifacts can easily be found at the regions adjacent to the sphenoid sinus or mastoid air, where a large difference between the air and tissues exists.

4. *Image Noise*

As is common in all MRI acquisitions, background noise degrades the true diffusion tensor. The noise level limits the maximal *b*-value. As the *b*-value increases to enhance contrast, the logarithm of the signal attenuation linearly decreases to the level of background noise, and the SNR is thus decreased. Moreover, background noise in DWIs biases tensor reconstruction and affects the degree of anisotropy by making isotropic structures appear anisotropic and by making anisotropic structures appear more anisotropic.

C. Practical Considerations in the Acquisition of Diffusion-Tensor Images

Diffusion-tensor-image acquisition is a balance between several factors, including SNR, spatial resolution, image distortion, and acquisition time. For example, a fast acquisition used to remove motion artifacts suffers from low spatial resolution and other eddy-current effects. An increase in the spatial resolution deteriorates the SNR and thus requires repetitive scanning in order to increase the image quality. This, in turn, causes a longer acquisition time and increases the possibility of motion. Another method by which to increase the spatial resolution is to use a high-Tesla MRI, which allows us to acquire thinner slices with the same duration and similar SNRs. Unfortunately, MRIs with higher field strengths are more vulnerable to artifacts.

At present, single-shot spin-echo EPI is the most prevalent sequence for DTI, due to its fast acquisition speed and its insensitivity to motion artifacts. However, eddy currents, innately low spatial resolutions, and susceptibility artifacts continue to be limitations in the use of fast imaging sequences in DTI.

In conclusion, there is no method that yields the "best" diffusion-tensor images. Therefore, imaging sequences should be chosen according to research purposes on a case-by-case basis.

VI. Visualization of Diffusion Tensors

In the previous sections, we discussed several scalar indices that quantify diffusion within a voxel. In this section, we will discuss various contrasting schemes of directional information about diffusion at each voxel. The directional information obtained from the diffusion tensor can be visualized by any of the following methods: (1) a color-coded vector map; (2) an arrow vector map; and (3) a three-dimensional tensor map.

A. Color-Coded Vector Maps

All natural colors can be represented by a mixture of three color components: red, green, and blue (RGB). Coincidentally, the principal direction of diffusion can be represented by the major eigenvector with three x, y, and z components. Therefore, if we assign each color component to the component of the major eigenvector, we can easily visualize the direction of the eigenvector according to its color tone. For example, the component in the direction from the left to the right brain is assigned to red; the posterior-to-anterior direction is assigned to green; and the inferior-to-superior direction is assigned to blue (Fig. 12b). The color purely indicates the major direction of tissue structure. In cases where the major direction is parallel to the axis from the left ear to the right ear, as it is in the corpus callosum, the color will appear red. The cingulum bundle, located in the anterior–posterior direction, will have a green color. In the case of the major direction of 45 degrees with lateral superior orientation in the axial slice of the brain, as in the corona radiata, the color at the voxel will be an equal

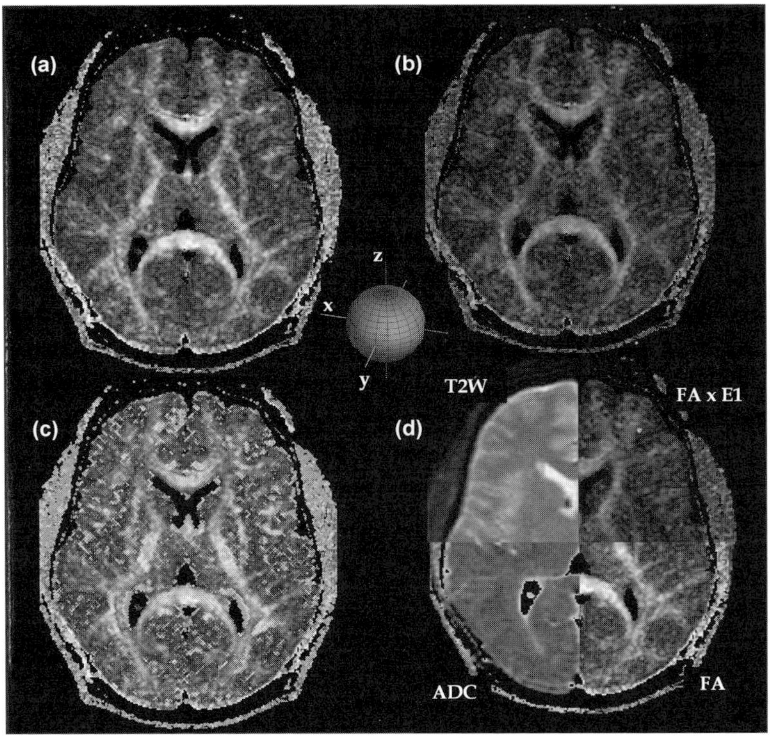

FIG. 12. Voxel-based visualization methods for the diffusion tensor. An axial slice of a diffusion-tensor image is displayed with the fractional anisotropy (FA) map (a), the color-coded vector map (b), the arrow vector map (c), and a combinatory display with T2, apparent diffusion coefficient (ADC), FA, and color-coded vector map (FA × E1) (d). (See Color Insert.)

combination of red and blue, i.e., purple. It is possible to incorporate the anisotropy by multiplying RGB color values by FA and representing anisotropy by color brightness. Dark indicates lower anisotropy and bright indicates higher anisotropy at the voxel.

B. Arrow-Vector Map

Although the color-coding scheme provides the viewer with a good intuitive feel of the fiber directions at each voxel, it does not provide directional information that is precise enough to allow him or her to follow the fiber tracts visually. Another way of representing principal direction is the arrow-vector map. The major eigenvector can be resolved into in-plane components and out-of-plane components. Arrows that have both orientation and length can represent in-plane components, which are 2D vectors. Out-of-plane components can be represented by color-scaled dots. For axial-slice images, we might see the following: the corpus callosum, where the in-plane component is dominant, can be mainly represented with an arrow, and the corona radiata in the PLIC, which has a dominant out-of-plane component, can be represented by hot color dots (Fig. 12c).

C. Three-Dimensional Tensor Maps

The two schemes of display introduced in Sections VI.A and VI.B are designed for use on a single vector field reduced from a tensor field. They do not provide shape information based on the second and third eigenvalues, and they may show changes in colors and arrows if the head position is not straight in the MRI coordinate. Because the diffusion tensor is modeled with an ellipse, the diffusion-tensor image can be better visualized by the display of an ellipse at each voxel. The three-dimensional rendering of an ellipse still captures the shape of the ellipse only ambiguously, however. We can partly overcome this ambiguity by displaying the tensor with a cube or a morphological mixture of line, plane, and sphere. Figure 13 shows an example of a tensor map that is displayed with ellipsoids and morphological objects.

VII. Basics of Fiber Tractography

Although we can understand local fiber structures using voxel-based visualization, it is often difficult to follow the fiber tracts across the slices visually. Therefore, computerized fiber-tracking methods are needed to help one understand the fiber structures in three-dimensional space. The fiber-tracking algorithm virtually creates a trace of a particle by following the local fiber orientation as defined by the diffusion-tensor field. It is based on the assumption that local fiber orientation (for each voxel) is parallel to the major eigenvector of the diffusion tensor. A number of fiber-tracking algorithms have been developed (Basser *et al.*, 2000; Conturo *et al.*, 1999; Jones *et al.*, 1999; Mori *et al.*, 1999; Poupon *et al.*, 2000) since the introduction of DTI. In this section, we shall discuss the basics of the most prevalent fiber-tracking methods.

A. Continuous Tensor Fields

Fiber tracking can be obtained by our connecting each discrete voxel (currently on the order of $2 \times 2 \times 2$ mm^3) to the adjacent one in the principal

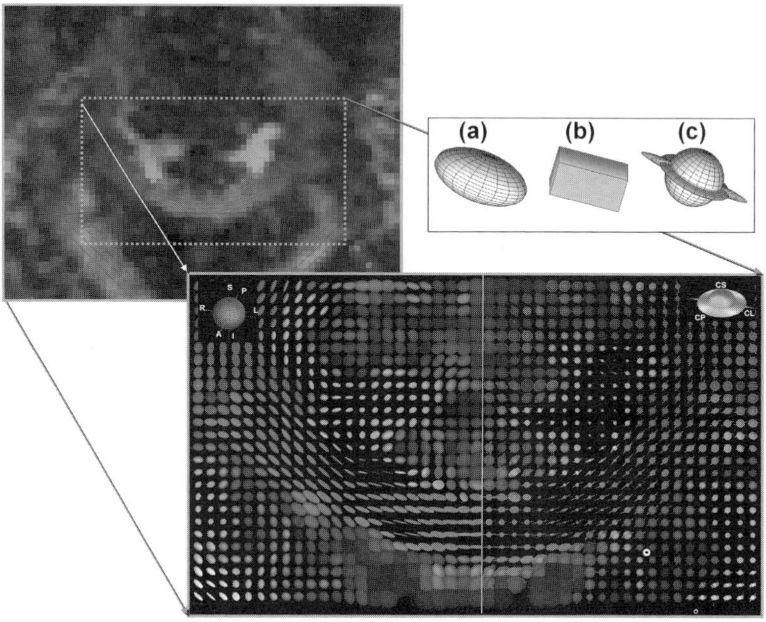

FIG. 13. Three-dimensional tensor map. A coronal section of the diffusion-tensor image can be displayed with a three-dimensional ellipse (a), a cubic object (b), and a geometric object (c). The zoomed figure shows the ellipsoidal display (left plane) and geometric-object display (right plane). The geometric objects are composed of lines, planes, and spheres according to the portion in the tensor of the linear, planar, and spherical components (c_l, c_p, and c_s). (See Color Insert.)

direction given by the local diffusion tensor. Jumping from one voxel to the neighboring voxel, however, may result in nonsmooth and erroneous trajectories. To prevent the errors caused by discrete quantization of the tensor field, fiber tracking is generally conducted on the continuous representation of the tensor field. In the continuous tensor field, the trajectory and its tangent vector (i.e., the major eigenvector) can be calculated not only restrictedly at the regular discrete voxel but also anywhere in the three-dimensional continuous field, as shown in Fig. 14a. This can be accomplished through the interpolation of the tensor field using linear interpolation, Lagrange polynomials, or B-spline functions. The explanations in this section are all based on this continuous tensor field.

B. Basic Streamline Propagation

Fiber tracts can be computed by propagation of an anterograde and a retrograde streamline from an initial seed point in the direction of the major

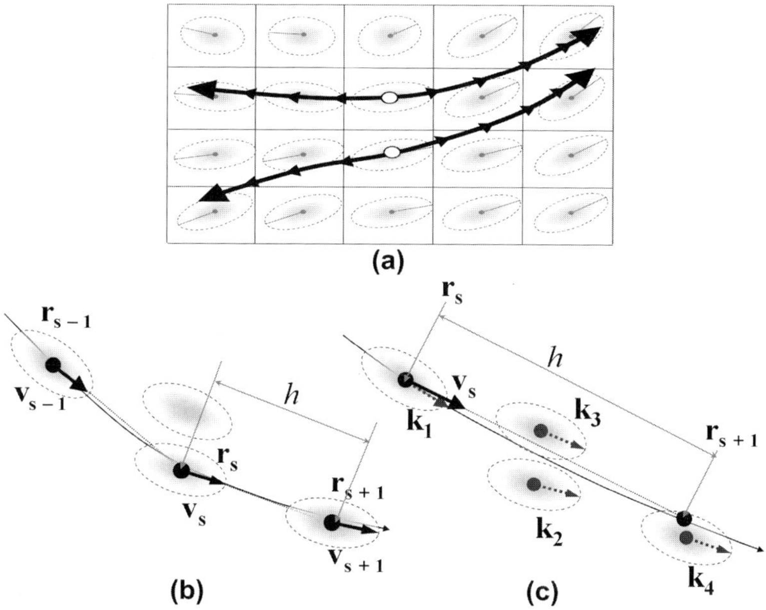

FIG. 14. The basic streamline algorithms in the continuous tensor field. From the seed point (the circle) in (a), fiber tracts are constructed in the anterograde and retrograde directions by following the principal direction of the diffusion vector. The propagation directions derived from major eigenvectors are calculated at any point in the continuous field (arrows in (a)), not restricted to a discrete voxel location such as the center of the voxel. The Euler algorithm (b) and the Runge–Kutta fourth-order algorithm (c) step for h distance in the propagation direction, $\mathbf{v}_s$. To decide the propagation direction $\mathbf{v}_s$ at the point $\mathbf{r}_s$, the Euler algorithm utilizes the major eigenvector at $\mathbf{r}_s$, whereas the Runge–Kutta fourth-order method utilizes four major eigenvectors from four trial points ($\mathbf{k}_1$, $\mathbf{k}_2$, $\mathbf{k}_3$, and $\mathbf{k}_4$). This reduces the accumulation errors that appear in Euler's method when the step size is big.

eigenvector for a small spatial step. The points on the trajectory $\mathbf{r}(s)$ can be derived by

$$\frac{d\mathbf{r}(s)}{ds} = \mathbf{v}(s) \tag{18}$$

where $\mathbf{v}(s)$ is a unit tangent vector to $\mathbf{r}(s)$ at the time s. With the initial condition, $\mathbf{r}(s = 0) = \mathbf{r}_0$, called the seed point, the solution for this differential equation can be derived by an iterative process:

$$\mathbf{r}_{s+1} = \mathbf{r}_s + h \cdot \mathbf{v}_s \tag{19}$$

where we calculate the next propagation location $\mathbf{r}_{s+1}$ by adding to the current location $\mathbf{r}_s$ the displacement with step size h along the propagation direction $\mathbf{v}_s$. A complete trajectory path from the initial seed point $\mathbf{r}_0$ can be defined with two sets of points in the anterograde propagation (i.e., where $\mathbf{e_1}$ is the major eigenvector of the tensor at $\mathbf{r}_0$) and the retrograde propagation (i.e., $\mathbf{v}_0 = -\mathbf{e}_1(\mathbf{r}_0)$).

1. *The Euler Method and the Runge–Kutta Method*

The simplest algorithm for solving our differential equation (Eq. 19) is the Euler method, which utilizes the major eigenvector as the propagation direction (Fig. 14b):

$$\mathbf{r}_{s+1} = \mathbf{r}_s + h \cdot \mathbf{e}_1(\mathbf{r}_s) \tag{20}$$

where $\mathbf{e}_1(\mathbf{r}_s)$ is the major eigenvector at point $\mathbf{r}_s$. However, the Euler method can suffer from large accumulated errors unless the step h is small enough to prevent the error.

The most well-known integration solver is the fourth-order Runge–Kutta method. To obtain the propagation direction, $\mathbf{v}_s$, four tangent vectors at neighboring points are evaluated, as in Fig. 14c: once at the initial point, twice at the trial midpoints, and once at the trial endpoint (Press *et al.*, 1992; Tench *et al.*, 2002).

$$\mathbf{v}_s = \frac{1}{6}(\mathbf{k}_1 + 2\mathbf{k}_2 + 2\mathbf{k}_3 + \mathbf{k}_4) \tag{21}$$

where

$$\begin{aligned}
\mathbf{k}_1 &= sign(\mathbf{v}_{s-1} \cdot \mathbf{e}_1(\mathbf{r}_s))\mathbf{e}_1(\mathbf{r}_s),\\
\mathbf{k}_2 &= sign(\mathbf{v}_{s-1} \cdot \mathbf{e}_1(\mathbf{r}_s + h/2 \cdot \mathbf{k}_1))\mathbf{e}_1(\mathbf{r}_s + h/2 \cdot \mathbf{k}_1),\\
\mathbf{k}_3 &= sign(\mathbf{v}_{s-1} \cdot \mathbf{e}_1(\mathbf{r}_s + h/2 \cdot \mathbf{k}_2))\mathbf{e}_1(\mathbf{r}_s + h/2 \cdot \mathbf{k}_2),\\
\mathbf{k}_4 &= sign(\mathbf{v}_{s-1} \cdot \mathbf{e}_1(\mathbf{r}_s + h\mathbf{k}_3))\mathbf{e}_1(\mathbf{r}_s + h\mathbf{k}_3)\\
sign(x) &= \begin{cases} 1, x > 0 \\ -1, x < 0 \end{cases}
\end{aligned} \tag{22}$$

where $\mathbf{e}_1(\mathbf{r}_s)$ is the major eigenvector at the position of $\mathbf{r}_s$ and $\mathbf{v}_{s-1}$ is the propagation direction in the previous step. Sign function is used to align the vectors $\mathbf{k}_1$, $\mathbf{k}_2$, $\mathbf{k}_3$, and $\mathbf{k}_4$ at four trial points in the direction of the propagation, since the major eigenvector at the tensor calculation may arbitrarily have either a forward or a reverse direction.

2. *Termination Criteria of Propagation*

The streamline propagation iteratively continues until it reaches outside white matter defined by the termination criteria. For example, the propagation stops where the anisotropy is lower than the threshold, which is regarded to be the anisotropy at the boundary between white matter (high anisotropy) and gray matter (low anisotropy). The rapid change in the propagation direction is also often used as a termination criterion due to the limited amount of curvature of trajectory per spatial increment in the physiology. However, the criterion of low anisotropy may mistakenly lead the propagation algorithm to stop at a fiber-crossing region that shows a low anisotropy, or to stop near subcortical gray matters that have relatively well-organized fibers but show low anisotropy. To overcome this mis-stopping problem, a stopping mask drawn on the structural MRI or a termination criterion based on the uncertainty of the diffusion direction (Behrens *et al.*, 2003) can be used.

C. Problems in Fiber Tractography

Fiber tractography, especially that based on streamline propagation, suffers from two major types of problems: partial volume effects and noise effects. We shall discuss these problems and several methods used to overcome them.

1. *Partial Volume Effects in Fiber Tracking*

The basic assumption of tractography is that the principal direction of diffusion aligns with a single fiber orientation. This may not be especially true for low spatial resolution (about $2 \times 2 \times 2\ \mathrm{mm}^3$) in the current system. Several tissue structures, such as two different pathway fibers, may lie within a specific voxel and contribute to form a single tensor at the voxel. Since the axonal diameter is on the order of 10 μm, the partial volume effect may not be easily overcome, despite progress in MRI techniques. We will discuss in detail the partial volume effect seen in fiber tracking.

a. Ambiguity of the White-Matter Boundary. The voxels located at the boundaries between white-matter fiber tracts and neighboring structures are typical examples of areas affected by the partial volume effect (Fig. 15a). The anisotropy of a voxel at this boundary is generally reduced in proportion to the amount of nonfiber structure. The orientation of the diffusion in this region is very susceptible to noise and to surrounding structures.

b. Fiber Crossing. The fiber crossings are easily found in the human brain; they include the regions between the corona radiata and radiation of the corpus callosum, between the corticospinal tracts and the pontocerebellar fibers, etc. (Wiegell *et al.*, 2000). The diffusion tensor at these crossing fibers tends to show increased planar or spherical components of diffusion and reduced anisotropy (Fig. 15b). In this region, the principal direction of diffusion is very sensitive to noise, and the ambiguity of the propagation direction is augmented. To overcome the ambiguity due to fiber crossing in tracking, several schemes for regularizing propagation have been developed by incorporation of the inertia term of the previous propagation with the current tensor. This method will be discussed in Section VII.D.

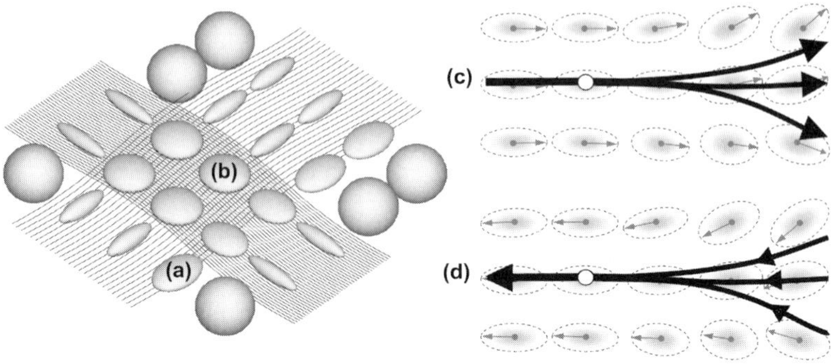

FIG. 15. Partial volume effects on the diffusion tensor. A tensor at the boundary of the fibers and nonfiber structures (a) and a tensor where two fiber bundles cross have highly augmented ambiguity. At the diverging fibers, the streamline approach cannot explain fiber branching from a single point to several fibers (c). Instead, fiber branching can be explained by the merging of multiple fibers starting from multiple seed points (d). (See Color Insert.)

c. Fiber Branching. For a streamline approach, it is difficult to represent fibers diverging from a single streamline starting at a location in the region of interest (ROI). This fiber-branching problem can be partly solved if we merge two separate fibers instead of splitting one fiber into two separate fibers, as in Fig. 15d. Practically, this can be implemented by reconstructing all the brain fibers with seed points at the voxels in the entire white matter and choosing fibers that penetrate a specific ROI.

2. *Noise and Fiber Tracking*

Fiber tracking is very sensitive to noise in the tensor field, which might be generated from acquisition artifacts. A slight deviation of the major eigenvector due to noise and its accumulation will cause severe changes in the trajectory. Therefore, reducing noise and regularizing the tensor field are critical for reliable fiber tracking.

One basic approach is to use general image-filtering methods. Gaussian filtering can be used to remove random Gaussian noises in DWIs. However, Gaussian filtering degrades spatial resolution and causes image blurring. Nonlinear anisotropic smoothing (Parker *et al.*, 2000) is designed to remove high-frequency noise while preserving the boundaries of structures of interest.

Another approach is to utilize *a priori* knowledge of the fiber tracts in regularizing the tensor field on which the fiber tracking is performed. The regularization of the tensor field is based on the global properties of the fibers, such as the limited curvature of the fiber trajectory. Spline-based approximation of the tensor field (Basser *et al.*, 2000) is one of the simple methods of incorporating the *a priori* hypothesis of limited curvature. A global restoration technique using the Bayesian framework (Poupon *et al.*, 2001) is another example that utilizes the low-curvature information in the regularization. The optimal direction map is a tradeoff between the bending energy of trajectories and their faithful following of the measured diffusion-tensor field. Instead of using eigenvectors directly derived from the original tensor field, fiber tracking is performed on regularized direction maps.

D. Regularized Streamline Algorithms

As discussed before, a straightforward implementation of eigenvector-tracking streamline algorithms has an ambiguity problem at locations where propagation faces fiber crossing. To overcome this problem, advanced streamline algorithms have regularized this propagation. The propagation vector, $\mathbf{v}_s$ in Eq. (19), is an outgoing vector from the current tensor at location $\mathbf{r}_s$, while $\mathbf{v}_{s-1}$, the propagation direction at previous step $\mathbf{r}_{s-1}$, is an incoming vector to the current tensor, denoted as $\mathbf{D}$. The regularization scheme is to derive the outgoing vector $\mathbf{v}_s$ by modulating the incoming vector $\mathbf{v}_{s-1}$ with the current tensor matrix $\mathbf{D}$.

Westin and colleagues (2002) suggested a projection method in which the outgoing vector is derived by projection of the incoming vector onto a tensor matrix operated with a tensor mapping function *F*, as demonstrated below:

$$\mathbf{v}_s = F(\mathbf{D}).\mathbf{v}_{s-1} \tag{23}$$

One example of the tensor projection operation is to use the geometric tensor matrices, i.e., $\mathbf{D}_l$, $\mathbf{D}_p$, and $\mathbf{D}_s$, and the geometric index, i.e., c_l, c_p, and c_s of the tensor $\mathbf{D}$ from Eq. 14, as below:

$$F(\mathbf{D}) = \begin{cases} c_l\mathbf{D}_l, & \text{if } c_l > c_p, c_s, \\ c_p\mathbf{D}_p, & \text{if } c_p > c_l, c_s, \\ c_s\mathbf{D}_s, & \text{if } c_s > c_l, c_p, \end{cases} \tag{24}$$

The tensor line method, also called the tensor deflection method (TEND) (suggested by Weinstein *et al.* [1999]), combines the incoming vector $\mathbf{v}_{s-1}$, the projected vector, and the major eigenvector of the current tensor $\mathbf{D}$ to determine the outgoing vector, $\mathbf{v}_s$.

$$\begin{aligned} \mathbf{v}_s &= f\mathbf{e}_1 + (1-f)\{((1-g)\mathbf{I} + g\mathbf{D}).\mathbf{v}_{s-1}\} \\ &= f\mathbf{e}_1 + (1-f)\{F(\mathbf{D}).\mathbf{v}_{s-1}\} \end{aligned} \tag{25}$$

where the line components c_l are used as a controlling variable *f*.

Both the tensor projection method and the tensor line method have the tensor mapping function F, which deflects the incoming vector towards the major eigenvector direction of **D** but limits the curvature of the deflection, as illustrated in Fig. 16a. When the diffusion is isotropic, c_l becomes zero, $F(\mathbf{D})$ equals a scaled identity matrix, and the outgoing vector is the continuation of the incoming vector, i.e., $\mathbf{v}_s = \mathbf{v}_{s-1}$. The continuation of the incoming vector also occurs when the incoming vector is parallel to the plane spanned by two eigenvectors $\mathbf{e}_1$ and $\mathbf{e}_2$, which correspond to the two largest eigenvalues.

The streamline algorithm with a regularization scheme is very efficient at penetrating ambiguous tensors, as shown in Fig. 16b. However, the regularized streamline algorithm may not correctly follow the fibers that have high curvature, due to the effect of the inertia component in tracking. Neither does it have a method to differentiate between fiber crossing and fiber kissing (i.e., the pattern of fibers diverging after drawing close to each other) in a voxel. In this situation, we should decide whether to stop, to rapidly turn the propagation direction, or to continue using regularization.

Figure 17 summarizes the procedure of fiber tracking discussed in this section. The extraction of fibers of interest from constructed fibers will be discussed in Section IX.C. Figure 18 shows a result of fiber tractography

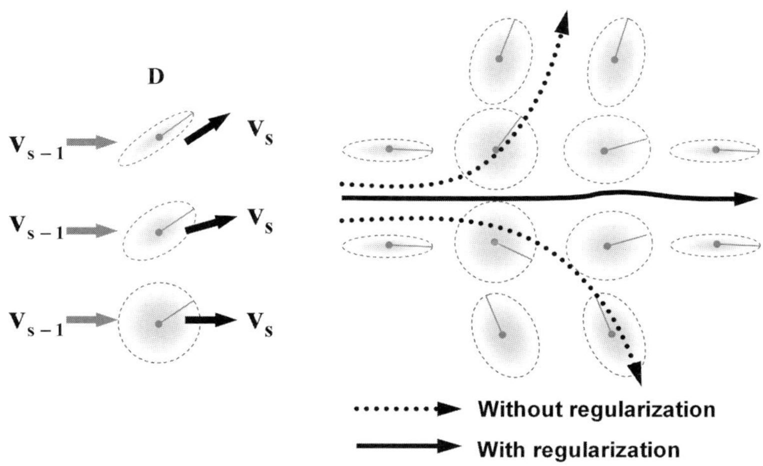

FIG. 16. The regularized streamline approach. When multiplied with the anisotropic diffusion tensor **D,** the incoming vector $\mathbf{v}_{s-1}$ is deflected towards the major eigenvector direction, **D.** When the tensor **D** is isotropic, **D** is equal to the identity matrix and the propagation continues toward the incoming vector, i.e., $\mathbf{v}_s = \mathbf{v}_{s-1}$ (a). When a streamline faces a tensor close to isotropic, the ambiguity is increased, and it may propagate in the wrong direction. Under regularization, the streamline can penetrate the ambiguous tensors in the direction of the previous propagation (b).

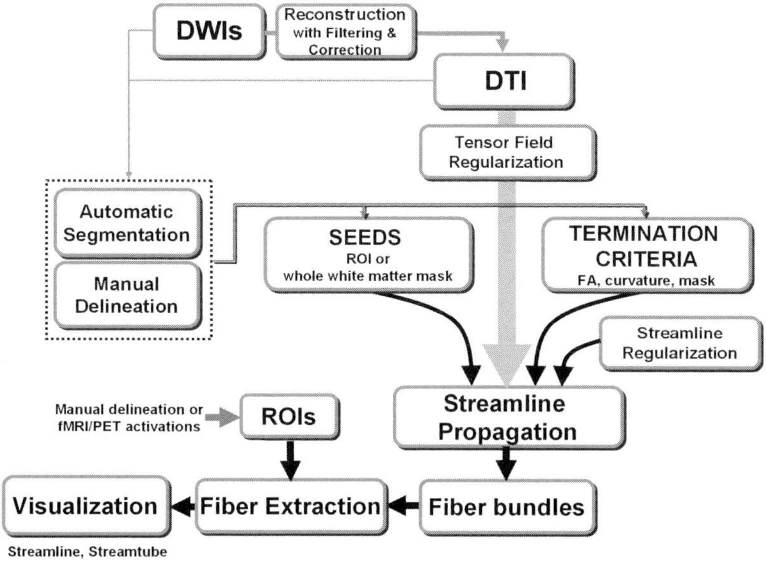

FIG. 17. A diagram of general fiber tractography. General fiber tractography is based on a streamline technique that requires seed points, termination criteria, and an integration solver, i.e., a streamline propagation mechanism. Seed points can be assigned by manual delineation of regions of interest or by generation of a white-matter mask from a segmentation algorithm. Fractional anisotropy and curvature changes are often used as the termination criteria. From the reconstructed fibers, fibers of interest can be extracted; this is explained in detail in Section IX.C.

reconstructed using the Runge–Kutta fourth-order algorithm with seed points of the entire white matter.

VIII. Advances in Fiber Tractography

A. Connectivity Mapping

The streamline techniques are deterministic in the sense that they utilize the eigenvector field reduced from the tensor field and do not allow for uncertainty of fiber direction. In the streamline technique, therefore, there is no mechanism to evaluate how reliably the path represents the true pathway or to evaluate how much more probable the connectivity from a seed point to a certain point is, compared to any other certain point. More recent approaches are statistical in nature and take

FIG. 18. Whole-brain fiber bundles in lateral view. The whole-brain fiber bundles are reconstructed using the fourth-order Runge–Kutta algorithm with seed points in the whole white matter. The diffusion-tensor image is acquired from 3-Tesla GE Echospeed systems using a line-scan diffusion-imaging sequence (LSDI). The fiber tractography was conducted using the DoDTI (*http://neuroimage.yonsei.ac.kr/dodti*). (See Color Insert.)

into consideration all of the diffusion information with the assumption that the probability of a fiber's propagating in a given direction is proportional to the corresponding diffusion coefficient. In contrast to the streamline approach, which is a one-to-one mapping, all connectivity-mapping approaches have in common that a single initial seed can have multiple end-points or a probability map, which makes it possible to evaluate the relative connectivity between the regions.

1. *Fiber Tractography by Solution of the Diffusion Equation*

One way to understand diffusion data is to spread a virtual concentration peak of heat by the law of diffusion, which can be formulated with partial differential equations (PDEs). We can achieve fiber tractography by solving PDEs where the diffusion coefficients of the heat equation equal the diffusion coefficients of the diffusion tensor. The solution can be obtained by iterative simulation of a peak concentration as it diffuses inside the brain based on the heat equation alone (Gembris *et al.*, 2001) or based on the heat equation with a convection term, to enhance the anisotropy of the diffusion equation (Batchelor *et al.*, 2002).

2. *Fast-Marching Tractography*

The fast-marching tube (Parker *et al.*, 2002) is a front propagation method wherein the interface or front propagates in the direction of high driving speed. The variable rate of the propagation is governed by the directionality of the tensor. In the direction of the major eigenvector, $\mathbf{e}_1$, the front propagates faster than the direction of the other two eigenvectors, $\mathbf{e}_2$ and $\mathbf{e}_3$. The front, started at a seed point, will cross regions in the brain at different times, and this generates a map of the arrival time at all points in the brain from the seed point. From this arrival-time map, we can calculate the path of connection from a seed point to any point in the volume by finding the minimum cost path back to the seed point from that point. Although this method is performed on the eigenvector field as the streamline approach is, its one-to-many mapping is the element that provides the evaluation of relative connectivity. Figure 19 shows an example of this fast-marching tractography (FMT).

3. *The Monte Carlo Random-Walk Simulation*

During the diffusion process, a particle performs a random walk through the medium. A Monte Carlo random-walk simulation (Hagmann *et al.*, 2003; Koch *et al.*, 2002) is based on this particle phenomenon with a local transition probability. The probability of transition in a given direction is chosen depending on the local diffusion coefficient along the propagation direction. The particle moves

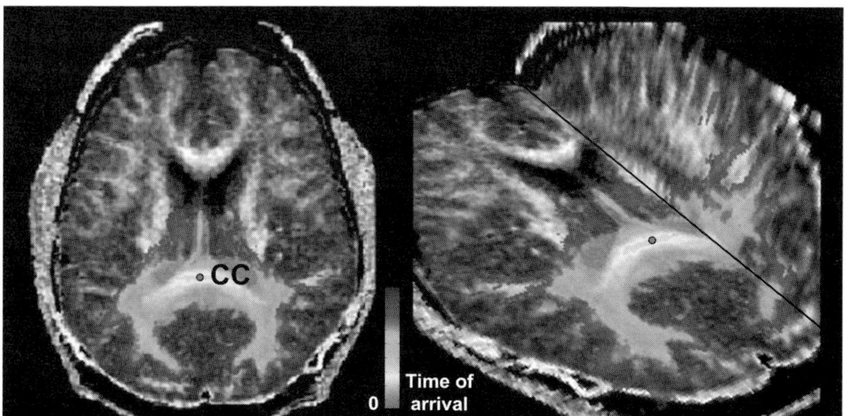

FIG. 19. Connectivity mapping using fast-marching tractography. A time-of-arrival map is derived using fast-marching tractography from a seed point located in the middle of the corpus callosum. In contrast to streamline approaches, fast-marching tractography has one-to-many mapping that makes it possible to evaluate the relative connectivity between the pairs of points in the brain. (See Color Insert.)

for a short time by stepping along a random direction, which is determined according to the transition probability at the initial location. In each iterative step, a new direction is randomly sampled and the new transition-probability distribution is determined by the diffusion tensor at each location. Toward the new direction, the virtual particle propagates with a diffusivity corresponding to the direction. Particles will move with a higher probability when traveling along the fiber direction than when they travel perpendicular to it. In a number of such random walks starting from a region, the frequency that some other region has been reached can be used as a relative measure of the anatomical connectivity between the two regions.

In order to constrain the random-walk process with the global properties of the fiber trajectory, hybrid approaches have been designed in order to find the optimal path between a pair of regions by combining the *a priori* information on the path curvature with the information from the diffusion-tensor data (Hagmann *et al.*, 2003).

4. *The Probabilistic Approach Using Uncertainty*

In general, analysis of the diffusion tensor has involved the fitting of a local diffusion model to the diffusion-weighted data at each voxel. It is based on the assumptions that the local diffusion profile is Gaussian and that the parameters of Gaussian diffusion are completely known. However, uncertainty exists in the relationship between the diffusion measurements and the underlying fiber structures. Noise in the NMR signal (both physical and physiological) can be one source of this uncertainty. Behrens and colleagues (2003) presented a method for the treatment of these uncertainties. They formulated the local uncertainty in the parameters of the generative diffusion model at a voxel level, in the form of posterior probability density functions on these parameters. Global connectivity is based on the probabilistic tractography, which incorporates every possible fiber orientation at every voxel with the probability given by the measured diffusion-tensor data.

B. High-Angular-Resolution Diffusion Imaging

In the previous sections, we have discussed the methods for enhancing the reliability of fiber tracking, especially in the case of the partial volume effect in terms of the postprocessing methodology. These techniques are mostly based on the assumption of a simple Gaussian diffusion on a voxel. However, with a Gaussian tensor model, it can be difficult to describe the fiber architecture in voxels where multiple directional fibers coexist. As discussed in the previous section, several regularization techniques on the traditional diffusion-tensor data have been designed to solve mixtures of multiple fiber directions. However, they may fail in cases of high curvature, as in the arcuate fascicles near the cortex. The problem of fiber crossing can be solved using high-angular-resolution diffusion imaging (HARD), in which over a hundred directional diffusivities evenly spaced out in a three-dimensional space are measured to generate a diffusion displacement profile. This diffusion displacement profile does not require any *a priori* knowledge of the diffusion characteristics of the structure. Several methods have been proposed for measuring HARD techniques. These include multitensor modeling (Tuch *et al.*, 2002), diffusion-spectrum imaging (Tuch *et al.*, 2002), and Q-ball imaging (Tuch, 2004; Tuch *et al.*, 2003). HARD is used to analyze multiple fiber bundles within a voxel (Frank, 2002). These imaging techniques have been receiving growing amounts of attention. However, the disadvantage of HARD techniques is that they are computationally intensive, and require a much greater acquisition time than is necessary for general tensor acquisition. Figure 20 shows an example of the Q-ball imaging method about the optic radiation.

IX. Quantification of White Matter using DTIs

Quantitative evaluation of the white matter benefits from the fact that the diffusion of water molecules in white-matter tissues is affected by the properties of neuronal fibers and their surrounding structures. The quantification of diffusion-tensor data for investigations of white matter has generally been approached using the following methods: (1) ROI-based methods; (2) voxel-based methods; and (3) fiber tractography–based methods.

A. ROI-Based Analysis of DTIs

ROI methods begin by identifying anatomical brain regions and comparing the anisotropy or the extent of the region. Due to its relatively easy accessibility, this method has been a basic tool in many researches using DTI (Kubicki *et al.*, 2003; Peled *et al.*, 1998). The ROI-based approach, however, requires *a priori* hypotheses to predefine the expected ROI. Therefore, ROIs are often chosen at the major fiber bundles that are relatively easy to separate from other structures. Since the boundary of a specific fiber bundle is not clear and a single voxel may contain multiple types of fibers due to the low spatial resolution of the diffusion-tensor image, care should be taken to ensure the consistent measurement of the ROI.

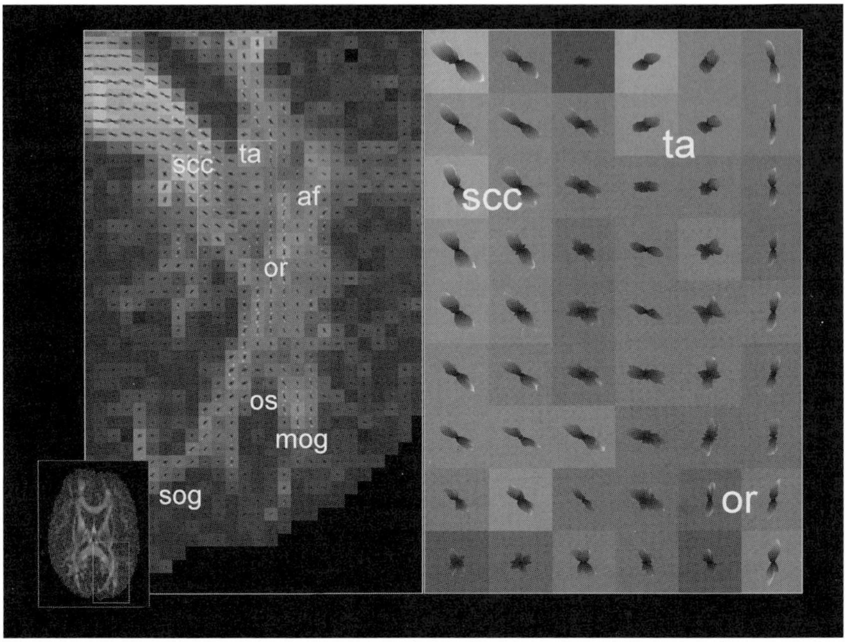

FIG. 20. Q-ball diffusion image (QBI) of three-way fiber crossing between the splenium of the corpus callosum, the tapetum, and the optic radiation. The QBI data were acquired on a 3-Tesla Siemens Trio with $b = 4000$ s/mm^2, 252 directions, and 2.2-mm isotropic resolution. af, arcuate fasciculus; mog, middle occipital gyrus; or, optic radiation; os, occipital sulcus; scc, splenium of corpus callosum; sog, superior occipital gyrus; ta, tapetum. Courtesy of David S. Tuch, Harvard Medical School. Adapted from Tuch (2004) with permission of Wiley-Liss, Inc., a subsidiary of John Wiley & Sons, Inc. © 2004 Wiley-Liss, Inc. (See Color Insert.)

B. Voxel-Based Statistical Parametric Mapping of DTIs

A voxel-based strategy is more exploratory and is suitable for identifying unanticipated or nonhypothesized areas of abnormal white-matter morphology. Voxel-based methods have also been used for studies where the ROI of potential abnormality is difficult to define precisely. The process for voxel-based quantification of diffusion tensors is basically similar to that for functional images such as fMRI or positron emission tomography (PET). The basic and critical requirement for voxel-based analysis is to perform spatial normalization with high reliability in order to minimize anatomical factors in the evaluation of the features of interest, such as FA and mean ADC. This spatial normalization is usually performed by spatial transformation of individual images to a template space, a process often called image registration. A template, which plays an important role inasmuch as it provides a standard space for the individual brains, is generally chosen to represent the group being tested; for example, a template could be a group-averaged image.

In general, a T2-weighted image acquired on the same scan with the diffusion-tensor image or non-diffusion-weighted reference image is used to estimate the nonlinear transformation function to a template. With this function, diffusion-parameter images (e.g., FA and mean ADC) are transformed to the template space. However, T2-weighted images do not include information essential for finding anatomical correspondences to the underlying white-matter structures (namely, the fiber tract's orientation and organization), and thus may not be appropriate for finding precise anatomical correspondences in DTI. Park and colleagues (2003) showed that registration using the whole tensor information yields better registration performance than just using an FA image or T2-weighted image in the registration of diffusion tensor images.

C. Fiber Tractography-Based Analysis of DTIs

White matter is a connectional architecture that cannot be fully described by a single voxel, and the most meaningful representations of fiber organization are obtained in terms of fiber bundles. Therefore, it is appropriate to evaluate the fiber bundles to increase the specificity of DTI.

It is of great importance to define and extract fibers of interest (FOIs) with high reliability in the analysis of fibers. The simplest way of generating FOIs is to perform fiber tracking starting at the seed points in the ROI drawn on the slices of the T2-weighted image or FA image. Another way of extracting FOIs is based on the whole-brain fiber bundles reconstructed from the seed points at the whole white matter. Among the reconstructed fiber of the entire brain, fibers that penetrate the ROI are selected as fiber bundles of interest. This method is particularly advantageous when branching occurs.

The group of fibers reconstructed from an ROI may contain unwanted fibers, including various directional fibers, due to the initial value sensitivity of fiber tracking. For example, by drawing an ROI at one or more than one slice of the PLIC, the corticospinal tract cannot be separated well enough from the fibers connecting the other cortical regions through the PLIC. This is because the subtypes of fibers in the PLIC that connect various cortical regions cannot easily be identified on the image slices. To emphasize the fibers that are of concern, researchers often use two or more ROIs to exclude fibers that started at the same ROI but ended at a different brain region. The example of FOI derived using multiple ROIs are displayed in Fig. 21. The parcellation of cortical gray matter using high-resolution structural MRI may be useful for the identification of fiber bundles, since the fiber bundles are interconnecting cortical regions in addition to interconnecting the cortex with the subcortical structures. Figure 22 shows an exemplary usage of cortical parcellation in identifying the fibers that originated from the right precentral gyrus (Park *et al.*, 2004).

A fiber bundle allows us to measure the anisotropy index, such as the mean FA of the particular fiber bundle as well as its size and shape. We can think of several ways of quantifying the extracted fiber bundles. The mean values of fiber characteristics such as anisotropy and cross-sectional area along the fiber bundles can be used for the statistical evaluation. The profile of anisotropy or cross-sectional area along the fibers can also be clinically important. However, the absolute number of axons connecting two cortical regions can hardly be calculated from tractography, because each streamline of the tractography represents a gross fiber bundle containing thousands or more of the axonal fibers, depending on the voxel size and anatomical location. Therefore, the number of fiber bundles reconstructed by current tractography may not be a direct index of the anatomy of the brain.

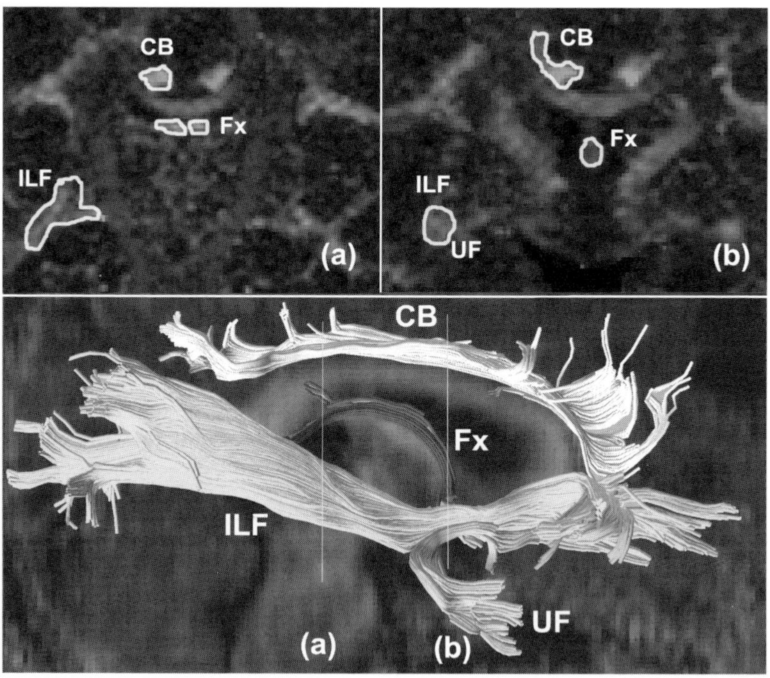

FIG. 21. Fibers of interest. The cingulum bundle (CB), inferior longitudinal fascicle (ILF), uncinate fascicle (UF), and fornix (Fx) are derived from multiple regions of interest delineated at two separate coronal slices. (See Color Insert.)

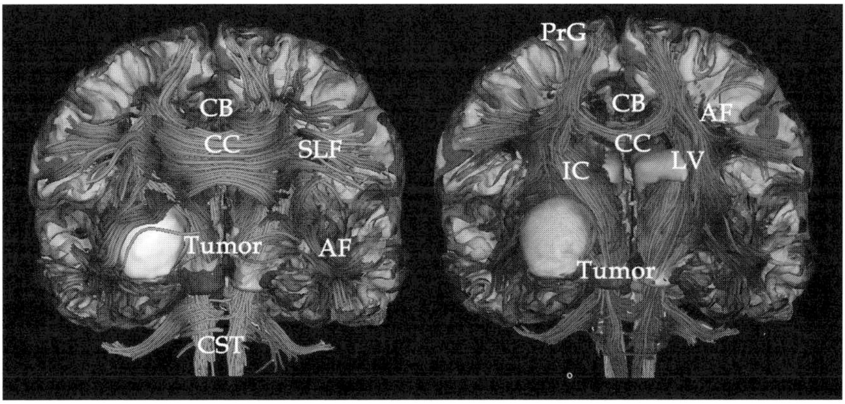

FIG. 22. Fiber tractography in tumor surgery. Fiber tractography can be used in the surgical planning for a tumor surgery. The left panel shows the tractography of the entire brain displayed with cross-sections of gray- and white-matter surface. The right panels show the fibers (in blue) that originated from the right precentral gyrus (blue color in cortex) and are located closest to the tumor. CC, corpus callosum; SLF, superior longitudinal fasciculus; CB, cingulum bundle; AF, arcuate fibers; IC, internal capsule; CST, cortico-spinal tract; PrG, precentral gyrus; LV, lateral ventricle. Adapted from Park *et al.* (2004) with permission from the American Society of Neuroradiology. (See Color Insert.)

D. Practical Considerations in the Analysis of DTIs

One of the most important issues in the analysis of diffusion-tensor images is the interpretation of findings, especially for studies that use the anisotropy index. Since the meaning of macroscopic anisotropy is affected by the many factors discussed earlier, it may not be correct to simply state that the difference of FA indicates the difference or abnormality in the connectivity. To reduce the complexity in the interpretation of the anisotropic difference, it will be helpful to reference other imaging techniques. For example, the magnetization transfer ratio, which is sensitive to the myelination, can be used to investigate the portion of anisotropy that is due mainly to either the cell type of the fiber or the structural coherence of fibers.

Exploration of fiber properties by fiber tractography requires great care in interpretation as well, since there are many considerations, not the least of which is the problem of how to evaluate fibers that cross. The fiber trajectories generated by following the largest and the most homogeneous fiber bundles cannot represent a diameter that is much smaller than the size of a voxel. If we can achieve a more reliable white-matter tractography, this will be an important step towards the quantification of fiber bundles of concern, which would lead to a further understanding of "neural connectivity."

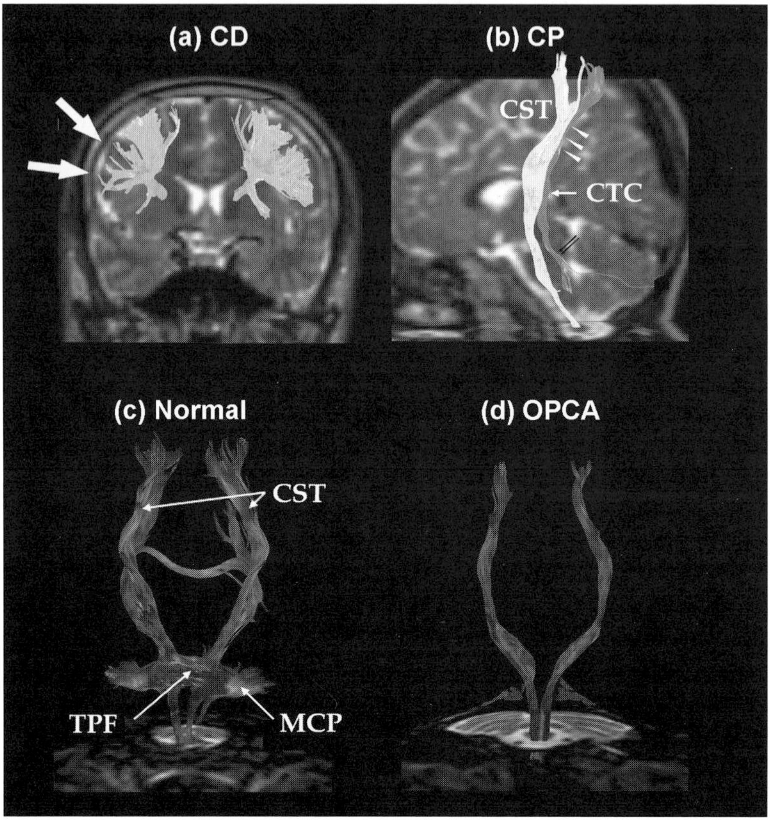

FIG. 23. Clinical applications of fiber tractography. A patient with cortical dysplasia (CD) shows reduced subcortico–cortical white-matter connectivity (a). In a cerebral palsy (CP) patient, the sensory fibers are reduced but motor fibers are normal (b). In a patient with olivopontocerebellar atrophy (OPCA), transverse pontine fibers (TPF) and the middle cerebellar peduncle (MCP) are reduced or absent (d); compare with the normal subject (c). CST, corticospinal tract; TPF, transverse pontine fibers; CTC, cortico–thalamic connection. Courtesy of Seung-Koo Lee, Department of Diagnostic Radiology, Yonsei University. (See Color Insert.)

X. Applications of DTI

Alteration of molecular mobility due to changes in structures, properties, or tissue compartments can affect the diffusion tensor; some examples include inflammation, demyelination, axonal loss and reorganization, cytotoxic edema, and infiltration of white-matter fibers by tumor. The clinical applications of DTI based on these white-matter pathologies include brain ischemia, trauma, tumor, focal epilepsy, multiple sclerosis, Alzheimer's disease, and schizophrenia. DTI has also been used for normal-development studies and aging studies. Most of these research projects have been conducted by exploring the geometry and the degree of anisotropy of the diffusion. There are many recent reviews that deal with specific cases (Dong *et al.*, 2004; Jellison *et al.*, 2004; Kubicki *et al.*, 2002; Neil *et al.*, 2002; Sullivan and Pfefferbaum, 2003; Sundgren *et al.*, 2004; Taylor *et al.*, 2004). As the anisotropic diffusion in white matter renders estimation of fiber tracts, diffusion tensors provide information on the connectivity between the regions using fiber tractography. Application areas of fiber tractography have been expanded gradually, and the usefulness of fiber tractography has been validated in some areas. As shown in Fig. 22, fiber tractography can be applied to explore the effect of tumors or space-occupying lesions on the white-matter architecture, and discoveries about these effects can be used in surgical planning (Clark *et al.*, 2003; Mori *et al.*, 2002; Park *et al.*, 2004). This method can be useful in many clinical applications, including central-nervous-system neoplasm (Melhem *et al.*, 2002), white-matter diseases such as cerebellar degeneration (Lee *et al.*, 2003) and Wallerian degeneration (Pierpaoli *et al.*, 2001), and pediatric brain abnormalities such as congenital anomalies (Lee *et al.*, 2004) and cerebral palsy with periventricular leukomalacia (Hoon *et al.*, 2002). The exemplary applications of fiber tractography in clinical research are displayed in Fig. 23.

XI. Conclusion

In this chapter, I introduced the basics, acquisition, processing, and applications of DTI. This method has provided a unique *in vivo* tool for the exploration of white matter, which lays the groundwork for the understanding of anatomical connectivity in the human brain. DTI has also been regarded as an important tool in the functional neuroimaging area using PET and functional MRI, and thus a growing amount of research has been conducted based on the combination of both imaging modalities. Although we cannot disregard current limitations of DTI, a number of recent techniques have proved its validity and usefulness in clinical applications. The advances in acquisition and postprocessing of DTI will certainly lead to a better understanding of white-matter pathology and contribute to more effective clinical application.

References

Alexander, A. L., Tsuruda, J. S., and Parker, D. L. (1997). Elimination of eddy current artifacts in diffusion-weighted echo-planar images: The use of bipolar gradients. *Magn. Reson. Med.* **38,** 1016–1021.

Basser, P. J., Mattiello, J., and Le Bihan, D. (1994). MR diffusion tensor spectroscopy and imaging. *Biophys. J.* **66,** 259–267.

Basser, P. J., Pajevic, S., Pierpaoli, C., Duda, J., and Aldroubi, A. (2000). *In vivo* fiber tractography using DT-MRI data. *Magn. Reson. Med.* **44,** 625–632.

Basser, P. J., and Pierpaoli, C. (1996). Microstructural and physiological features of tissues elucidated by quantitative-diffusion-tensor MRI. *J. Magn. Reson. B* **111,** 209–219.

Batchelor, P. G., Hill, D. L. G., Atkinson, D., Calamante, F., and Conelly, A. (2002). Fiber tracking by solving the diffusion-convection equation. *Proc. International Soc. Magn. Reson. Med.*, p. 1135. Honolulu, HI.

Beaulieu, C. (2002). The basis of anisotropic water diffusion in the nervous system - a technical review. *NMR Biomed.* **15,** 435–455.

Beaulieu, C., and Allen, P. S. (1994). Determinants of anisotropic water diffusion in nerves. *Magn. Reson. Med.* **31,** 394–400.

Behrens, T. E., Johansen-Berg, H., Woolrich, M. W., Smith, S. M., Wheeler-Kingshott, C. A., Boulby, P. A., Barker, G. J., Sillery, E. L., Sheehan, K., Ciccarelli, O., Thompson, A. J., Brady, J. M., and Matthews, P. M. (2003). Non-invasive mapping of connections between human thalamus and cortex using diffusion imaging. *Nat. Neurosci.* **6,** 750–757.

Clark, C. A., Barrick, T. R., Murphy, M. M., and Bell, B. A. (2003). White matter fiber tracking in patients with space-occupying lesions of the brain: A new technique for neurosurgical planning? *Neuroimage* **20,** 1601–1608.

Conturo, T. E., Lori, N. F., Cull, T. S., Akbudak, E., Snyder, A. Z., Shimony, J. S., McIntyre, R. C., Burton, H., and Rachel, M. E. (1999). Tracking neuronal fiber pathways in the living human brain. *Proc. Natl. Acad. Sci. USA* **96,** 10422–10427.

Dong, Q., Welsh, R. C., Chenevert, T. L., Carlos, R. C., Maly-Sundgren, P., Gomez-Hassan, D. M., and Mukherji, S. K. (2004). Clinical applications of diffusion tensor imaging. *J. Magn. Reson. Imaging* **19,** 6–18.

Frank, L. R. (2002). Characterization of anisotropy in high angular resolution diffusion-weighted MRI. *Magn. Reson. Med.* **47,** 1083–1099.

Gembris, D., Schumacher, H., and Suter, D. (2001). Solving the diffusion equation for fiber tracking in the living human brain. Proceedings of the International Society for Magnetic Resonance in Medicine (ISMRM), Glasgow, Scotland, 1529.

Hagmann, P., Thiran, J. P., Jonasson, L., Vandergheynst, P., Clarke, S., Maeder, P., and Meuli, R. (2003). DTI mapping of human brain connectivity: Statistical fibre tracking and virtual dissection. *Neuroimage* **19,** 545–554.

Hoon, A. H., Jr., Lawrie, W. T., Jr., Melhem, E. R., Reinhardt, E. M., Van Zijl, P. C., Solaiyappan, M., Jiang, H., Johnston, M. V., and Mori, S. (2002). Diffusion tensor imaging of periventricular leukomalacia shows affected sensory cortex white matter pathways. *Neurology* **59,** 752–756.

Huppi, P. S., Maier, S. E., Peled, S., Zientara, G. P., Barnes, P. D., Jolesz, F. A., and Volpe, J. J. (1998). Microstructural development of human newborn cerebral white matter assessed *in vivo* by diffusion tensor magnetic resonance imaging. *Pediatr. Res.* **44,** 584–590.

Jellison, B. J., Field, A. S., Medow, J., Lazar, M., Salamat, M. S., and Alexander, A. L. (2004). Diffusion tensor imaging of cerebral white matter: A pictorial review of physics, fiber tract anatomy, and tumor imaging patterns. *AJNR Am. J. Neuroradiol.* **25,** 356–369.

Jones, D. K. (2004). The effect of gradient sampling schemes on measures derived from diffusion tensor MRI: A Monte Carlo study. *Magn. Reson. Med.* **51,** 807–815.

Jones, D. K., Simmons, A., Williams, S. C., and Horsfield, M. A. (1999). Non-invasive assessment of axonal fiber connectivity in the human brain via diffusion tensor MRI. *Magn. Reson. Med.* **42,** 37–41.

Koch, M. A., Norris, D. G., and Hund-Georgiadis, M. (2002). An investigation of functional and anatomical connectivity using magnetic resonance imaging. *Neuroimage* **16,** 241–250.

Kubicki, M., Westin, C. F., Maier, S. E., Mamata, H., Frumin, M., Ersner-Hershfield, H., Kikinis, R., Jolesz, F. A., McCarley, R., and Shenton, M. E. (2002). Diffusion tensor imaging and its application to neuropsychiatric disorders. *Harv. Rev. Psychiatry* **10,** 324–336.

Kubicki, M., Westin, C. F., Nestor, P. G., Wible, C. G., Frumin, M., Maier, S. E., Kikinis, R., Jolesz, F. A., McCarley, R. W., and Shenton, M. E. (2003). Cingulate fasciculus integrity disruption in schizophrenia: A magnetic resonance diffusion tensor imaging study. *Biol. Psychiatry* **54,** 1171–1180.

Le Bihan, D., Breton, E., Lallemand, D., Grenier, P., Cabanis, E., and Laval-Jeantet, M. (1986). MR imaging of intravoxel incoherent motions: Application to diffusion and perfusion in neurologic disorders. *Radiology* **161,** 401–407.

Lee, S. K., Mori, S., Kim, D. J., Kim, S. Y., Chu, M., Heo, K., Lee, B. I., and Kim, D. I. (2003). Diffusion tensor MRI and fiber tractography of cerebellar atrophy in phenytoin users. *Epilepsia* **44,** 1536–1540.

Lee, S. K., Mori, S., Kim, D. J., Kim, S. Y., and Kim, D. I. (2004). Diffusion tensor MR imaging visualizes the altered hemispheric fiber connection in callosal dysgenesis. *AJNR Am. J. Neuroradiol.* **25,** 25–28.

Maier, S. E., Gudbjartsson, H., Patz, S., Hsu, L., Lovblad, K. O., Edelman, R. R., Warach, S., and Jolesz, F. A. (1998). Line scan diffusion imaging: Characterization in healthy subjects and stroke patients. *AJR Am. J. Roentgenol.* **171,** 85–93.

Melhem, E. R., Mori, S., Mukundan, G., Kraut, M. A., Pomper, M. G., and van Zijl, P. C. (2002). Diffusion tensor MR imaging of the brain and white matter tractography. *AJR Am. J. Roentgenol.* **178,** 3–16.

Mori, S., Crain, B. J., Chacko, V. P., and van Zijl, P. C. (1999). Three-dimensional tracking of axonal projections in the brain by magnetic resonance imaging. *Ann. Neurol.* **45,** 265–269.

Mori, S., Frederiksen, K., van Zijl, P. C., Stieltjes, B., Kraut, M. A., Solaiyappan, M., and Pomper, M. G. (2002). Brain white matter anatomy of tumor patients evaluated with diffusion tensor imaging. *Ann. Neurol.* **51,** 377–380.

Neil, J., Miller, J., Mukherjee, P., and Huppi, P. S. (2002). Diffusion tensor imaging of normal and injured developing human brain - a technical review. *NMR Biomed.* **15,** 543–552.

Norris, D. G. (2001). Implications of bulk motion for diffusion-weighted imaging experiments: Effects, mechanisms, and solutions. *J. Magn. Reson. Imaging* **13,** 486–495.

Papadakis, N. G., Martin, K. M., Pickard, J. D., Hall, L. D., Carpenter, T. A., and Huang, C. L. (2000). Gradient preemphasis calibration in diffusion-weighted echo-planar imaging. *Magn. Reson. Med.* **44,** 616–624.

Park, H. J., Kubicki, M., Shenton, M. E., Guimond, A., McCarley, R. W., Maier, S. E., Kikinis, R., Jolesz, F. A., and Westin, C. F. (2003). Spatial normalization of diffusion tensor MRI using multiple channels. *Neuroimage* **20,** 1995–2009.

Park, H. J., Kubicki, M., Westin, C. F., Talos, I. F., Brun, A., Peiper, S., Kikinis, R., Jolesz, F. A., McCarley, R. W., and Shenton, M. E. (2004). Method for combining information from white matter fiber tracking and gray matter parcellation. *AJNR Am. J. Neuroradiol.* **25,** 1318–1324.

Parker, G. J., Schnabel, J. A., Symms, M. R., Werring, D. J., and Barker, G. J. (2000). Nonlinear smoothing for reduction of systematic and random errors in diffusion tensor imaging. *J. Magn. Reson. Imaging* **11,** 702–710.

Parker, G. J., Wheeler-Kingshott, C. A., and Barker, G. J. (2002). Estimating distributed anatomical connectivity using fast marching methods and diffusion tensor imaging. *IEEE Trans. Med. Imaging* **21,** 505–512.

Peled, S., Gudbjartsson, H., Westin, C. F., Kikinis, R., and Jolesz, F. A. (1998). Magnetic resonance imaging shows orientation and asymmetry of white matter fiber tracts. *Brain Res.* **780,** 27–33.

Pierpaoli, C., Barnett, A., Pajevic, S., Chen, R., Penix, L. R., Virta, A., and Basser, P. (2001). Water diffusion changes in Wallerian degeneration and their dependence on white matter architecture. *Neuroimage* **13,** 1174–1185.

Pierpaoli, C., and Basser, P. J. (1996). Toward a quantitative assessment of diffusion anisotropy. *Magn. Reson. Med.* **36,** 893–906.

Poupon, C., Clark, C. A., Frouin, V., Regis, J., Bloch, I., Le Bihan, D., and Mangin, J. (2000). Regularization of diffusion-based direction maps for the tracking of brain white matter fascicles. *Neuroimage* **12,** 184–195.

Poupon, C., Mangin, J., Clark, C. A., Frouin, V., Regis, J., Le Bihan, D., and Bloch, I. (2001). Towards inference of human brain connectivity from MR diffusion tensor data. *Med. Image. Anal.* **5,** 1–15.

Press, W. H., Teukolsky, S. A., Vetterling, W. T., and Flannery, B. P. (1992). Numerical recipes in C. Cambridge, Cambridge University Press.

Stejskal, E. O., and Tanner, J. E. (1965). Spin diffusion measurements: Spin echoes in the presence of a time-dependent field gradient. *J. Chem. Phys.* **42,** 288–292.

Sullivan, E. V., and Pfefferbaum, A. (2003). Diffusion tensor imaging in normal aging and neuropsychiatric disorders. *Eur. J. Radiol.* **45,** 244–255.

Sundgren, P. C., Dong, Q., Gomez-Hassan, D., Mukherji, S. K., Maly, P., and Welsh, R. (2004). Diffusion tensor imaging of the brain: Review of clinical applications. *Neuroradiology* **46,** 339–350.

Taylor, W. D., Hsu, E., Krishnan, K. R., and Mac Fall, J. R. (2004). Diffusion tensor imaging: Background, potential, and utility in psychiatric research. *Biol. Psychiatry* **55,** 201–207.

Tench, C. R., Morgan, P. S., Wilson, M., and Blumhardt, L. D. (2002). White matter mapping using diffusion tensor MRI. *Magn. Reson. Med.* **47,** 967–972.

Tuch, D. S. (2004). Q-ball imaging. *Magn. Reson. Med.* **52,** 1358–1372.

Tuch, D. S., Reese, T. G., Wiegell, M. R., Makris, N., Belliveau, J. W., and Wedeen, V. J. (2002). High angular resolution diffusion imaging reveals intravoxel white matter fiber heterogeneity. *Magn. Reson. Med.* **48,** 577–582.

Tuch, D. S., Reese, T. G., Wiegell, M. R., and Wedeen, V. J. (2003). Diffusion MRI of complex neural architecture. *Neuron* **40,** 885–895.

Westin, C.-F., Maier, S. E., Mamata, H., Nabavi, A., Jolesz, F. A., and Kikinis, R. (2002). Processing and Visualization for Diffusion Tensor MRI. *Medical Image Analysis* **6,** 93–108.

Wiegell, M. R., Larsson, H. B., and Wedeen, V. J. (2000). Fiber crossing in human brain depicted with diffusion tensor MR imaging. *Radiology* **217,** 897–903.

Wimberger, D. M., Roberts, T. P., Barkovich, A. J., Prayer, L. M., Moseley, M. E., and Kucharczyk, J. (1995). Identification of "premyelination" by diffusion-weighted MRI. *J. Comput. Assist. Tomogr.* **19,** 28–33.

PERFUSION fMRI FOR FUNCTIONAL NEUROIMAGING

Geoffrey K. Aguirre,* John A. Detre,* and Jiongjiong Wang[†]

*Department of Neurology and
[†]Department of Radiology, University of Pennsylvania
Philadelphia, Pennsylvania 19104

Arterial spin labeling perfusion functional magnetic resonance imaging (fMRI) is a noninvasive method of obtaining time-series measurements of regional cerebral perfusion. Magnetically tagged protons in arterial water are labeled and allowed to perfuse the cerebral tissue. The perfusion-fMRI data are then generated from measurements of the difference between sequentially obtained label images and control images. This differencing step bestows signal properties upon perfusion data that are quite different from those of blood oxygen level–dependent (BOLD) fMRI. Unlike with BOLD, the resulting time-series data are quantifiable and stable over long time scales. As a consequence, perfusion fMRI is well suited to the study of mental operations and neural changes that evolve slowly, over a period of hours to days. The technique is currently limited by a low signal-to-noise ratio (SNR) and slow image acquisition. Ways to improve both of these weaknesses are available, including scanning at higher field strength, using surface coils, and using optimized three-dimensional image acquisition.

INTERNATIONAL REVIEW OF
NEUROBIOLOGY, VOL. 66
DOI: 10.1016/S0074-7742(05)66007-2

I. Introduction

The imaging revolution in cognitive neuroscience began with a quantifiable, stable measure of cerebral physiology. Positron emission tomography (PET) was applied to the measure of cerebral blood flow (CBF) in the 1980s (Raichle *et al.*, 1983). With the use of injected, radiolabeled (^{15}O) water and tracer kinetic theory, it was possible to create images that measured the perfusion of brain tissue in physiologic units (cc of blood per 100 g of tissue per minute). Exploiting the neurovascular coupling that links local changes in neural activity to local changes in blood flow, PET was subsequently used to study the relationship between cognition and neural activity (Petersen *et al.*, 1988). PET functional imaging had a far-reaching impact, although a consistent limitation of the method was its relatively poor temporal resolution (on the order of a minute), which often conflicted with the rapidity of the mental operations under study.

In the 1990s, blood oxygen level–dependent (BOLD) functional magnetic resonance imaging (fMRI) was introduced. Unlike PET images, the BOLD fMRI signal has no stable, absolute interpretation and tends to slowly drift up and down over time. Despite this limitation, BOLD fMRI quickly supplanted PET for cognitive-neuroscience applications. The success of BOLD fMRI lay, in part, in the ubiquity of the necessary hardware and its noninvasive nature. Also critical, however, was the temporal resolution the method could provide, which was greatly improved over that of PET—on the order of seconds. Technical refinements of BOLD fMRI have brought further improvements in temporal resolution. Through careful experimental design, BOLD fMRI is able to distinguish randomly ordered neural events that occur several times a second (Dale and Buckner, 1997) and to detect neural-onset asynchronies on the order of 100 ms (Menon *et al.*, 1998). Efforts to combine simultaneous electroencephalography (EEG) and fMRI measurements evidence a continuing need for more precise measurements of neural events in time (see Chapter 10 of this volume).

There are some cognitive processes, however, that evolve over much longer time scales. Many forms of perceptual or motor learning, for example, are manifested as enduring changes in performance that accumulate over minutes to hours. Emotional behavior, as another example, is marked by a gradual change from one mood state to another. These sorts of mental operations are only imperfectly studied using BOLD fMRI. The limitation is the aforementioned instability of the BOLD signal at long time scales, which makes it difficult to distinguish interesting changes in neural activity from noisy fluctuations of the BOLD signal. The ideal imaging method would provide the stable signal of PET imaging and the high temporal resolution and noninvasiveness of BOLD.

A relatively new imaging method, perfusion fMRI, provides this methodological synthesis. Arterial spin labeling (ASL) perfusion fMRI permits the noninvasive quantification of regional brain-tissue perfusion using labeled, inflowing arterial protons as endogenous tracers (Detre and Alsop, 1999). To date, ASL imaging has been used primarily as a clinical tool to measure resting CBF in pathological states (e.g., Alsop *et al.*, 2000). Now, after a series of technical improvements and methodological developments, perfusion fMRI is poised for widespread application to cognitive-neuroscience hypotheses. In this paper, we consider perfusion fMRI for functional neuroimaging. We will first discuss the basic physics and properties of the method. Next, we will compare different techniques for generating ASL images and consider the consequences of those techniques for the signal and noise properties of ASL. These properties of ASL will be explicitly contrasted with those of BOLD fMRI. We will consider the implications of the stability of the ASL signal for data analysis and experimental design. Finally, we will highlight some limitations of ASL methods and describe current efforts to overcome these restrictions.

II. Acquisition of Perfusion-fMRI Data

ASL imaging operates in a manner rather analogous to ^{15}O PET. The basic scheme is to label water in arterial blood as it flows through the neck into the brain. The label is generally provided by radiofrequency (RF) pulses, which alter the magnetization of protons in the arterial water (a process referred to as inversion). The labeled protons collectively act as a diffusible tracer, and once they reach the capillaries they pass into the brain tissue. There, they affect the total magnetization present, and thus alter a subsequently acquired MRI image of the brain (Fig. 1). Indeed, virtually any pulse sequence (gradient echo, spin echo, or spiral) can be used to "read out" the magnetic effects of the label. The "magnetic" tracer has a decay rate of T1 (about 1–2 seconds), which is sufficiently long to allow detection of any perfusion of the microvasculature and tissue, but short enough to allow for monitoring of dynamic changes.

There are several possible ways to perform the magnetic labeling, and these can be grouped into two categories: continuous techniques and pulsed techniques (Wong *et al.*, 1998) (Fig. 2). In continuous ASL (CASL), flow-driven adiabatic inversion is used to label arterial blood water flowing through the inversion plane during a time window of about a few seconds (Williams *et al.*, 1992). Pulsed ASL (PASL) uses nearly instantaneous inversion of spins in a slab proximal to the imaged tissue (Kim and Tsekos, 1997). In general, CASL techniques provide greater perfusion contrast than do PASL methods, but they are technically more

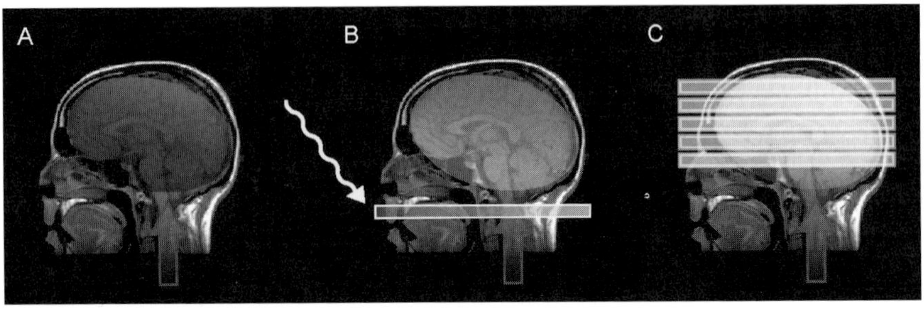

FIG. 1. Arterial spin labeling. (A) Arterial blood is carried to the brain via the carotid and vertebral arteries, which pass through the neck. (B) A slice-selective radiofrequency pulse is directed at an imaging plane in the neck, altering the magnetization of arterial water protons as they pass through the plane. The labeled protons continue through the vascular tree until they arrive at the cerebral capillary beds and pass into the brain tissue. (C) The effects of the alteration of the magnetization of the tissue are present in the MRI image slices that are then acquired.

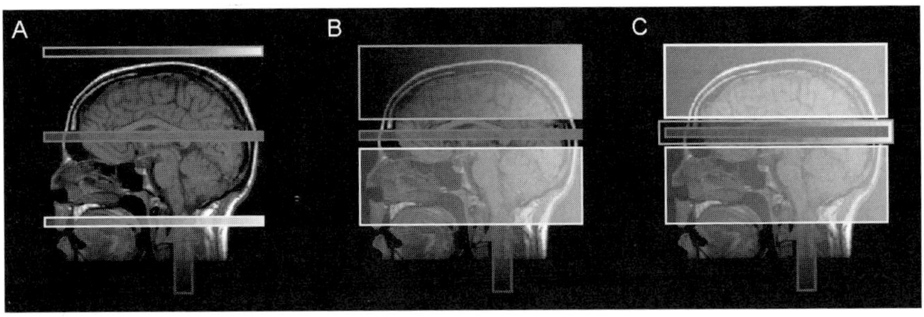

FIG. 2. Comparison of arterial spin labeling (ASL) techniques. Yellow indicates the labeling plane and gray the site of control labeling. (A) In continuous adiabatic labeling, a long-duration, spatially narrow radiofrequency pulse is used to tag protons as they flow through the imaging plane. The control image is produced either by application of the label distal to the imaged slice or (in multislice continuous ASL) by the performance of a "double inversion" proximal to the imaged slice. Quantification is based upon the known label efficiency. (B, C) Pulsed ASL using either (B) selective slab or (C) selective or nonselective inversion. A larger volume of blood is labeled using a very brief pulse. Pulsed ASL images can be acquired more rapidly, and their quantification is based upon the label duration. The control inversion is applied distal to the imaging plane in selective inversion, or to the imaging slice itself in selective or nonselective inversion.

difficult to implement. Continuous irradiation can also deposit significant RF power into the subject, which may be limiting at higher magnetic fields. It has been argued that the efficiencies of the two labeling schemes are comparable, as PASL is generally sampled at a higher rate despite the fact that its SNR is lower than CASL's (Wong, 1998). We observed that CASL perfusion image series are more stable in time than PASL series, probably because long labeling pulses in CASL average the effects of flow variability from the systolic to the diastolic cycle. Finally, the image coverage is generally larger with CASL than with PASL, which has to make room for a relatively thick inversion slab for labeling.

A further refinement, a postlabeling delay, can be introduced prior to the slice acquisitions. Generally, longer delays optimize accurate quantification of regional CBF, as the delay reduces the effects of different arterial transit times to different brain regions (Alsop and Detre, 1996). This improvement is balanced by a reduction in the magnitude of perfusion functional activation, as longer delays allow more of an opportunity for the spins to relax (Gonzalez-At *et al.*, 2000). In practice, once the labeling period, postlabeling delay, and image-acquisition times are considered, each labeled image requires 2–3 seconds to acquire.

The effects of the label are quantified by a pairwise comparison of the label images with separate images acquired with control labeling. The control condition attempts to replicate the frequency-dependent and off-resonance effects of the labeling without producing significant ASL. There are several methods by which to accomplish this, such as positioning the labeling pulses outside the brain volume or performing a "double inversion" of the spins (Detre and Alsop, 1999). For our purposes here, it is sufficient to note that most fMRI applications of ASL involve the interleaved acquisition of label and control images. The comparison of these temporally adjacent label and control images reveals the perfusion effect, and it is also the source of the distinct noise properties of ASL as compared to BOLD fMRI, which we will discuss later. Each perfusion image represents a 4–6 second interval of time, as the image is the difference between adjacent label and control images, which themselves took 2–3 seconds to acquire.

Several methods are then available for the conversion of the raw, differenced image into a quantitative measure of CBF (e.g., Wang *et al.*, 2002). CBF measurements using CASL perfusion MRI have been compared to ^{15}O-PET CBF in the same subjects, and an excellent correlation was observed between these modalities both at rest (Ye *et al.*, 2000a) and with functional activation (Feng *et al.*, 2004). Although it is possible to analyze perfusion-fMRI data without performing the conversion to CBF values, we will argue in Section IV.D that quantification of the perfusion effect improves cross-subject and cross-session statistical power. Figure 3 shows an example of the resulting quantified perfusion image.

As mentioned, the effects of ASL are independent of the pulse sequence used to obtain images and sample the resulting changes in magnetization. This gives researchers the potential to acquire images with little susceptibility weighting, such as spin-echo or spiral images, while maintaining the same sensitivity to the perfusion effect. This is an advantage in the examination of brain regions with high susceptibility gradients, such as the inferior frontal and temporal lobes (Fig. 4) (Wang *et al.*, 2004). However, to maximize slice coverage and minimize acquisition times, many ASL implementations use gradient-echo echo-planar

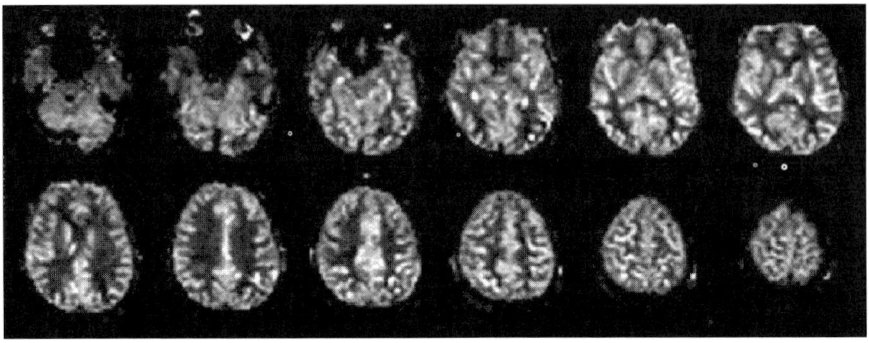

FIG. 3. Twelve axial slices, arranged inferior to superior, through a volume of ASL data. Higher CBF is evident within the cortical gray matter than within the white matter.

images for measuring magnetization. Importantly, the raw image data in this case also contain BOLD contrast. This allows BOLD and perfusion effects to be compared within the same data set (Wong *et al.*, 1997).

III. Noise and Signal Properties of Perfusion fMRI

The most important properties of perfusion fMRI for functional neuroimaging result from the altered temporal noise of perfusion data as compared to BOLD fMRI. These noise properties, in turn, are a consequence of the manner in which the perfusion signal is generated as the difference between adjacent images. In this section, we will discuss (1) techniques for obtaining the perfusion signal from the raw, label, and control images; (2) the noise properties of perfusion-fMRI data; (3) temporal resolution; and (4) spatial resolution.

A. Differencing Methods

In most cases, perfusion-fMRI data are acquired as interleaved label and control images, although the label acquisitions can be used independently to detect perfusion changes (Duyn *et al.*, 2001). The simplest way to obtain the perfusion signal is to perform a pair-wise subtraction of adjacent label and control images (Fig. 5). In the ideal case, label and control images would be acquired at precisely the same time, so that the only difference between the two images would be attributable to the effect of the label. However, because of the practical limitations of the method, such as the transit time of blood from the neck to the distal cortex, at least two seconds elapse between the acquisition of a labeled image and the acquisition of the next control. During this time, fluctuations in the

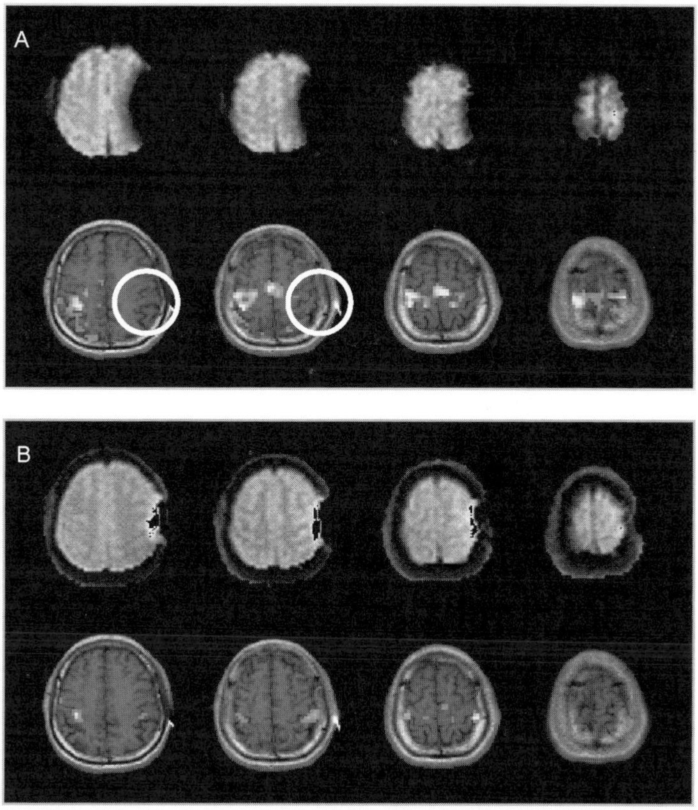

FIG. 4. Arterial spin labeling (ASL) data may be acquired by a variety of imaging sequences (Wang *et al.*, 2004). (A) BOLD fMRI data collected while a subject performed a simple motor movement (as opposed to a resting condition). During data acquisition, a metal bar was placed near the subject's head, producing a susceptibility artifact and loss of signal in the echo-planar images and corresponding activation maps. (B) Under the same circumstances, perfusion fMRI data were collected with a spin-echo sequence. The spin-echo sequence is less vulnerable to susceptibility artifacts, and a difference in CBF between the motor movement condition and the resting condition can be seen despite the presence of the metal bar. Note that the activation as measured using ASL in (B) is smaller than that seen with BOLD in (A). This difference is discussed in Section III.D.

signal value can create changes between the label and the control that are not the result of the effect of the label. For example, when gradient-echo images are used to measure tissue magnetization, and the subject is engaged in a task that has a periodic structure, then BOLD signal changes can contaminate the difference between the label and control images. As a result, other methods have been used to derive the perfusion signal, including "surround subtraction," in which the label is subtracted from the average of the two adjacent control images (Wong

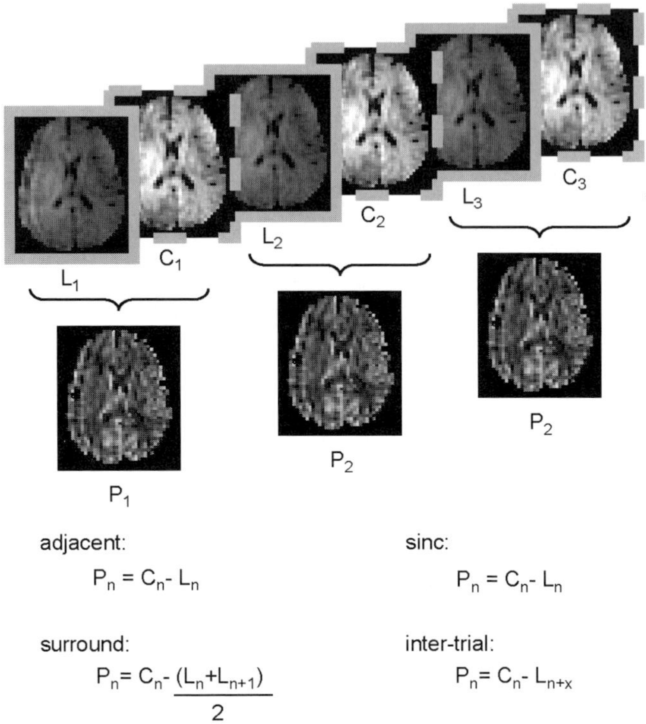

FIG. 5. Differencing methods available to generate perfusion data from label and control images. The source arterial spin labeling data are composed of interleaved label (L) and control (C) images. The perfusion (P) images are generated by taking the difference between label and control. Adjacent subtraction is the simple difference between adjacent image pairs. Surround subtraction averages pairs of label images. Sinc subtraction uses sinc interpolation to obtain an estimate of what the label time series would have been had it been acquired one repetition time (TR) earlier. Intertrial subtraction pairs label and control images from identical points in time with respect to the onset of experimental trials, which occur every x TRs.

et al., 1997), and "sinc subtraction," in which sinc-interpolated label and control images are subtracted (Aguirre *et al.*, 2002). Either approach is suitable for analysis of blocked patterns or other slowly changing patterns of neural activity (neural activity is likely to be the most common type of experimental design studied with perfusion fMRI).

A final option, "intertrial subtraction" (Yang *et al.*, 2000), forgoes the attempt to place label and control images in an absolute temporal register and instead pairs label and control images by their timing with relation to experimental events. A simple example of an experiment that is appropriately analyzed with intertrial subtraction is a sparse event-related design in which an identical stimulus is presented briefly every 18 seconds, while images are acquired at a repetition time (TR) of 2 seconds. The derivation of the perfusion signal then involves identifying pairs of label and control images that are acquired at equal points in time in relation to the onset of stimuli. By design, these label and control images will be drawn from separate trials. An advantage of the intertrial subtraction technique is that it effectively removes any artifact in the measured perfusion signal introduced by BOLD effect and therefore most accurately recovers the true shape of the perfusion hemodynamic response. Recently, Liu and Wong (2005) provided a unified model within which the properties of these differencing methods can be considered as specific cases of modulator functions.

B. Temporal Noise Properties of Perfusion fMRI

The character of perfusion-fMRI data is best understood first in contrast with the properties of BOLD fMRI. BOLD fMRI data collected from human subjects in the absence of any experimental task or time-varying stimuli (i.e., under the null hypothesis) demonstrate greater power at some frequencies than at others. Specifically, there is increasing power at low frequencies, a distribution of power that is well characterized by a I/frequency (1/f) function (Zarahn *et al.*, 1997) and other models (e.g., the autoregressive model [Purdon and Weisskoff, 1998]). There are two important consequences of this low-frequency noise. First, the uneven distribution of noise means that BOLD data are not temporally independent under the null hypothesis. As a consequence, standard parametric (Aguirre *et al.*, 1997; Zarahn *et al.*, 1997) and nonparametric (Aguirre *et al.*, 1998a) statistical tests are invalid for the analysis of BOLD data (they require instead the use of modified general linear models, such as are implemented in statistical parametric modeling [SPM]). Second, the greater noise at low frequencies will cause a relative reduction in sensitivity for some experimental designs. Specifically, experiments with fundamental frequencies in the lower range (e.g., a boxcar design with 60-second epochs) will have reduced sensitivity due to the presence of greater noise at these lower frequencies.

It is to be expected, however, that the temporal autocorrelation of ASL perfusion data will differ from that of BOLD. Specifically, the subtraction methods that derive the perfusion time series from adjacent, temporally interleaved images will dampen long-time-scale autocorrelation present in the source noise. In both simulations and empirical data (Aguirre *et al.*, 2002; Wang *et al.*, 2003a), we have demonstrated that the power spectrum of perfusion data is essentially flat (Fig. 6). The uniformity of the noise is best ensured by sinc subtraction, although the difference between the differencing methods is slight (Aguirre *et al.*, 2002; Liu and Wong, 2005; Wang *et al.*, 2003a). A further verification of the independence of perfusion data is provided by the Durbin–Watson statistic, which tests for autocorrelation against a first-order

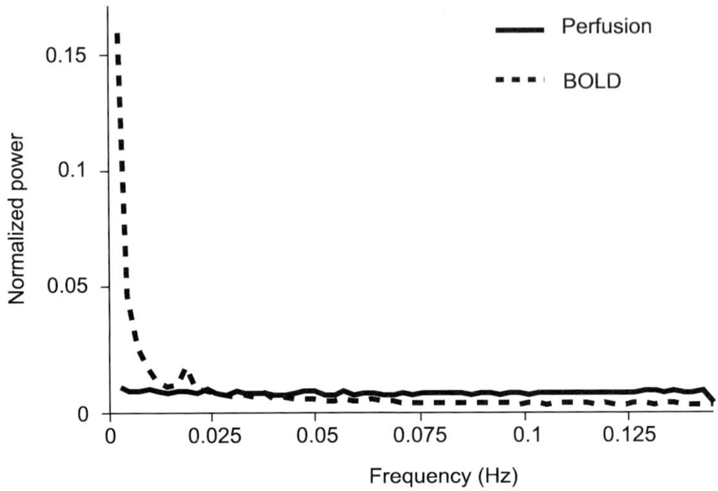

FIG. 6. Perfusion fMRI data are independent under the null hypothesis. The power spectrum typically observed in BOLD fMRI data (dashed line) has ever-increasing power at lower frequencies. In contrast, the power spectrum of perfusion data (solid line) is flat, indicating that the observations are independent in time under the null hypothesis.

autoregressive model (Fig. 7). Thus, perfusion data are nearly independent under the null hypothesis. The temporal independence of perfusion data has implications on data analysis and experimental design, both of which are discussed in the following sections.

C. Hemodynamic Response and Temporal Resolution

Because each perfusion image is derived from two time points, the temporal sampling of perfusion data is twice as coarse as in the source images. In practice, each perfusion image reflects events over a 4- to 6-second period. This relatively poor temporal resolution would seem to preclude the use of perfusion-fMRI methods to accurately measure the shape of the perfusion hemodynamic response that is evoked by changes in neural activity. In fact, however, it is possible to capture it by systematically adjusting the timing of the stimulus presentation with respect to the image acquisition (Josephs *et al.*, 1997) and thereby sampling the shape of the evoked response at an arbitrarily high temporal resolution (although at a corresponding cost to the accuracy of the estimate of the response at each time point). Using this approach with the intertrial subtraction method, Yang and colleagues measured the shape of the ASL hemodynamic response (Yang *et al.*, 2000). They found that the evoked perfusion response was quicker to

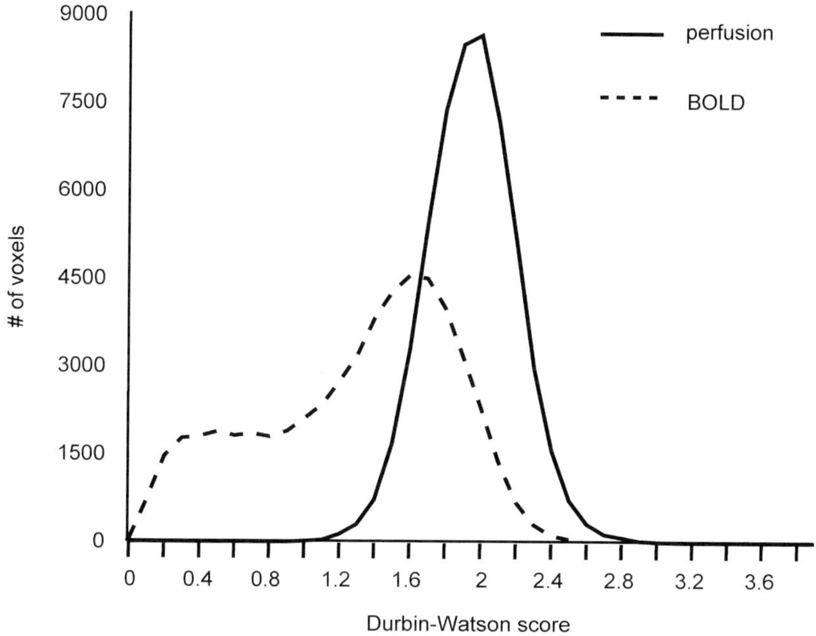

FIG. 7. Histogram of Durbin–Watson scores across subjects and across voxels. Null-hypothesis, pulsed–arterial spin labeling perfusion, and BOLD fMRI data were collected from a group of 10 subjects. The distribution of the perfusion scores (solid line) is symmetric about 2, providing no evidence for nonindependence of the time-series data. The distribution of the BOLD Durbin–Watson scores (dashed line) however, is strongly skewed, consistent with the known low-frequency autocorrelation in the noise (Aguirre *et al.*, 2003).

begin and narrower in its duration than the BOLD response. Others have obtained similar results (Fig. 8) (Aguirre *et al.*, 2002; Liu and Gao, 1999). This is to be expected, as ASL is primarily sensitive to signal changes at capillary sites and therefore does not reflect the more sluggish venous signal that dominates the BOLD response. High-field studies of perfusion activation using ASL contrast further indicate that the perfusion signal begins to increase as early as 600 ms after the onset of functional stimulation (Silva *et al.*, 2000), which is earlier than the 'initial dip' in BOLD contrast.

D. Spatial Resolution

Because BOLD contrast is due primarily to changes in intravascular deoxyhemoglobin concentration, functional activation may be observed overlying venous structures draining the activated region. In some studies, BOLD

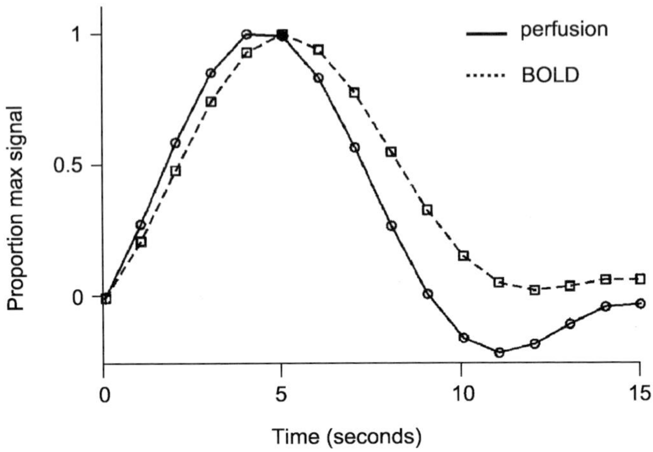

FIG. 8. Average shape of the BOLD (dashed) and perfusion (solid) hemodynamic response. Note that the perfusion response is narrower in time and earlier in onset.

activation has been seen in direct correspondence with macroscopic vessels, with susceptibility effects extending into the adjacent cortex (Hoogenraad *et al.*, 2001). In contrast to BOLD fMRI, perfusion fMRI uses a diffusible tracer (i.e., magnetically labeled arterial water), which can exchange with tissue water. Because the decay time of this tracer is quite brief (on the order of 1 second), there is minimal accumulation of the tracer in venous structures. As a result, signal changes in perfusion fMRI are not observed over veins, resulting in better localization of signal changes over activated cortex (Silva *et al.*, 1999). It has further been demonstrated that ASL perfusion fMRI provides a spatial resolution equivalent to the 'initial dip' in BOLD, which has been demonstrated to have superior spatial precision than the subsequent response (Duong *et al.*, 2000). The point-spread function of BOLD fMRI at full width at half maximum (FWHM) has been measured at 3.5 mm in the visual cortex (Engel *et al.*, 1997). It is unclear to what extent this resolution is dictated by the hemodynamic point-spread; lateral connections between adjacent neural areas may also produce spreading neural activity. Perfusion fMRI may be demonstrated to have a smaller point-spread function, although this is technically difficult to assess, given limitations in SNR that require rather large voxels to be acquired by multislice perfusion fMRI (discussed in Section VI). Finally, the limited responsiveness of perfusion methods to venous signals may explain in part why areas of activation consistently appear smaller in perfusion studies than with BOLD fMRI (e.g., in Fig. 4). A relatively weaker signal change may also contribute to this common observation.

IV. Preprocessing and Statistical Analysis with Perfusion fMRI

The primary characteristic of data preprocessing and statistical analysis for perfusion fMRI is that many of the idiosyncratic properties of BOLD fMRI that dominate neuroimaging data analysis are not of concern in the analysis of ASL data. We will discuss how perfusion fMRI data (1) may be analyzed with any parametric or nonparametric test, including permutation methods; (2) do not require, and are weakened by, the application of extrinsic temporal smoothing; (3) do not have temporally dependent spatial correlations, so they uniformly benefit from spatial smoothing; and (4) provide for improved population inferences.

A. Permutation Analysis for Perfusion fMRI

The absence of serial correlation between the error terms in perfusion fMRI renders unnecessary the use of the "modified" general linear model (Worsley and Friston, 1995) for the analysis of perfusion activation experiments. Indeed, perfusion fMRI data that have been appropriately differenced (as discussed above) can be analyzed using SPM software with the "PET" settings. Further, as with PET data, it is possible to assess the significance of statistical results from within-subject perfusion-fMRI studies using permutation methods (Nichols and Holmes, 2002), an approach that is not valid for BOLD fMRI data. Working with Tom Nichols, we have shown that permutation methods control the map-wise false-positive rate and provide appropriately uniform distributions of *p*-values in null-hypothesis perfusion data. Additionally, when used to analyze experimental data, permutation approaches provide for lower map-wise significance thresholds than do traditional Gaussian random field approaches (Fig. 9). The essential points are (1) that statistical thresholds for detecting activation are lower when significance is assessed using permutation than when it is assessed with traditional GRF methods and (2) that perfusion-fMRI data are eligible for this benefit, whereas within-subject BOLD fMRI data are not.

B. Temporal Smoothing

The presence of low-frequency temporal noise in BOLD fMRI data, and its variability from voxel-to-voxel, have led to the practice of high-pass filtering of BOLD fMRI to minimize low-frequency noise and low-pass smoothing to "condition" the autocorrelation structure and reduce bias (Friston *et al.*, 2000).

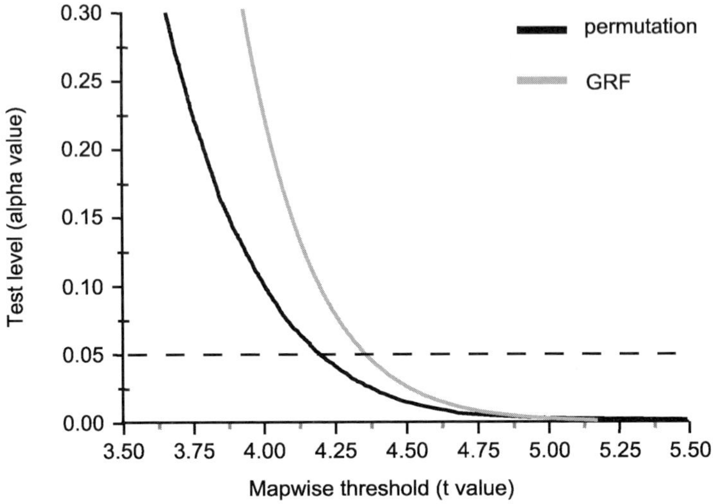

FIG. 9. Map-wise thresholds using permutation for perfusion-fMRI data. Shown are the probability density functions averaged across a group of 10 subjects performing a motor response task studied with pulsed arterial spin labeling. The plots show the map-wise *t*-threshold as a function of a level for different methods of assessing significance. Permutation thresholds in this method were lower than those obtained by Gaussian random field theory (or Bonferroni correction) at an *a* of 0.05 (Aguirre *et al.*, 2003).

Because perfusion-fMRI data do not possess temporal autocorrelation under the null hypothesis, these measures may not be necessary for the analysis of ASL data. Indeed, our evaluation of the receiver–operator characteristics of different analysis approaches (Wang *et al.*, 2005a) confirmed that such temporal manipulation degrades the performance of perfusion fMRI. Thus, considered together with our previous point, no temporal autocorrelation need be assumed for perfusion-fMRI data, and none need be imposed.

C. Spatial Coherence and Smoothing

An additional issue in fMRI analysis concerns the "spatial coherence" of the voxel time series in imaging data (Zarahn *et al.*, 1997). Spatial coherence is akin to "smoothness," but it also includes components of spatial correlation that cannot be captured by a continuously differentiable auto-covariance function (e.g., measurements of FWHM smoothness). Perhaps surprisingly, spatial coherence

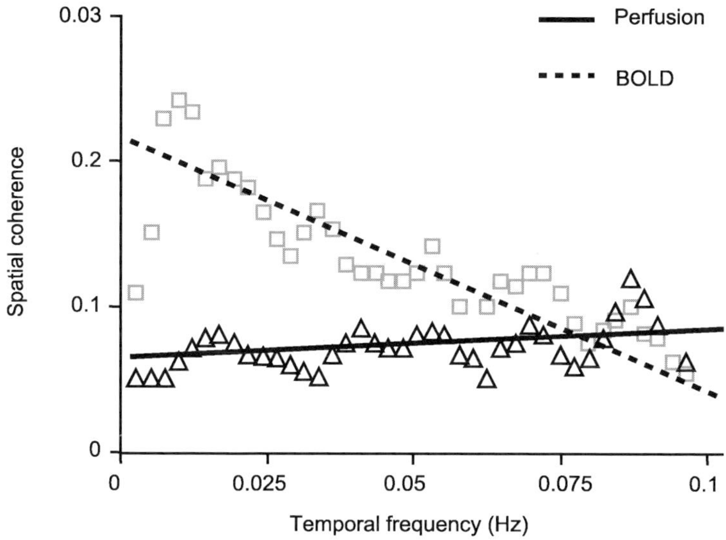

FIG. 10. Spatial coherence as a function of temporal frequency for perfusion-(solid line) and BOLD (dashed line) fMRI data. (See Zarahn *et al.*, 1997 for details of the spatial-coherence measure). The negative slope seen for the relationship of the BOLD data indicates that slow fluctuations in the BOLD noise tend to be correlated across space.

can be assessed (and can vary) at different temporal frequencies—it is, in effect, a measure of the degree to which power at a particular temporal frequency shares phase across space. In BOLD fMRI, spatial coherence has been found to vary systematically across temporal frequency, in that lower temporal frequencies tend to share phase to a greater extent across space than do high frequencies (Zarahn *et al.*, 1997). As a consequence, spatial smoothing of BOLD data acts to augment temporal noise in the low-frequency range and can deleteriously impact experimental power (Aguirre *et al.*, 1997).

Because perfusion-fMRI data do not possess temporal autocorrelation in time under the null hypothesis, we might predict that temporal frequency will have less effect on spatial coherence. Indeed, this is what we have found (Wang *et al.*, 2003a) (Fig. 10). Perfusion data derive a statistical power benefit from spatial smoothing that is greater than that seen for BOLD fMRI (Wang *et al.*, 2005a). The consequence of this is that perfusion-fMRI data will usually benefit from spatial smoothing during preprocessing, although the standard caveats regarding the optimal detection of activations of different scales still apply (Worsley *et al.*, 1996).

D. Population Inference

Neuroimaging questions are sometimes asked about groups of subjects, as opposed to about results from a particular subject. For example, one might ask whether the population from which a set of subjects is drawn possesses a hypothesized effect, or whether two different populations differ in the evocation of some effect. These types of hypotheses are appropriately tested within the context of a random-effects model (Friston *et al.*, 1999), in which a single effect measurement is obtained from each subject. A "second-level" statistical test is then performed upon this group of effect values. In practice, these effect measures are the scaling values calculated for parameters of interest that model each subject's data. Such designs appropriately account for variability in the magnitude of the effect across subjects.

In the ideal case, all variability present in cross-subject data would be explained by variability in the magnitude of evoked neural activity. In reality, there are several other sources of variability in the BOLD fMRI effect size. For example, between-subject differences in physiology likely produce variability in neurovascular coupling, which, even in the presence of identical magnitudes of neural activity, would lead to different BOLD-fMRI signal changes. These sources of between-subject variability in the BOLD effect act to reduce the statistical power of random-effects analyses.

We might expect that between-subject variability in task effect sizes would be different in measurements taken with perfusion fMRI from those taken with BOLD fMRI, as perfusion measures a single, normalized physiologic parameter (as opposed to BOLD, which measures blood oxygenation, volume, and flow). We have confirmed that between-subject variabilities in task-effect sizes are smaller when measured with perfusion fMRI than when measured with BOLD fMRI (Aguirre *et al.*, 2002; Wang *et al.*, 2003a) and that this beneficial effect is maximized when perfusion data are converted to absolute CBF values. Therefore, group analyses may be more powerful when conducted with perfusion imaging than with BOLD imaging, even if the magnitude of signal change (relative to within-subject noise) is lower for perfusion data than for BOLD.

V. Experimental Design with Perfusion fMRI

In the introduction to this chapter, we suggested that perfusion fMRI may be particularly well suited to the detection of slow changes in neural activity. We will now consider this important aspect of perfusion-fMRI experimental design. We will also describe the combined use of BOLD and perfusion contrast in a single experiment.

A. Detection of Slow Changes in Neural Activity

Consider an experiment that seeks to detect a slow, continuous change in neural activity over the course of several minutes. Such a neuronal profile might be produced by motor-sequence learning, in which a subject learns a pattern of finger movements in response to visual cues over 20 minutes of continuous training. While performing this kind of task, subjects evidence faster reaction times to the stimuli, even if they are unaware that there is a pattern to the responses that they are producing (Nissen and Bullemer, 1987). Ideally, we would want to follow the profile of neural change and relate that to cognitive states (such as awareness of the presence of a pattern) and behavioral measures (such as response time and error rate). Such a study cannot be performed with BOLD fMRI, however. This is because the power of the experimental paradigm would lie within the greatly elevated noise range of the data, and it would therefore not be possible to distinguish neuronal change from signal drift.

An alternative approach often employed in these circumstances is to interleave the learning task and the control task during the scan session. This "chops" the behavior to a higher temporal frequency, placing it within the range that BOLD fMRI can detect. Changes in neural activity during the learning condition can then be compared to the presumably stable signal obtained during the control condition, and the effects of signal drift will be removed. Although this has been a profitable approach in a number of studies (e.g., Shadmehr and Holcomb, 1997), there are limitations to this behavioral manipulation. First, it prevents continuous performance. Typically, the learning and control conditions must be alternated every 30 seconds or so, but the mental operations that support sequence learning may be altered by frequent interruption. Second, it requires the assumption that the control condition does not change over time, but subjects may engage in rehearsal or preparatory strategies during the "control" task, in which case the baseline itself would be altered over time, rendering the comparison with the learning condition suspect.

With perfusion fMRI, however, it is possible to directly study the time course of a change in neural activity, even if it evolves over hours or days. This is because the perfusion-fMRI signal has uniform noise across temporal frequencies and thus should maintain the power to detect a given change in neural activity whether it evolves over 30 seconds or 30 minutes. We have recently studied slow motor learning using perfusion imaging (Olson *et al.*, 2005). Ten subjects performed a serial-response-time task continuously for 20 minutes. For the first 15 minutes, there was a pattern in the sequence of stimuli to which they responded, although they were initially unaware that the stimuli had any particular order. During this training, reaction times fell linearly with time. After 15 minutes, the sequence changed without warning or comment to a new, random order of stimuli (termed the "transfer" sequence), which lasted for another 5 minutes.

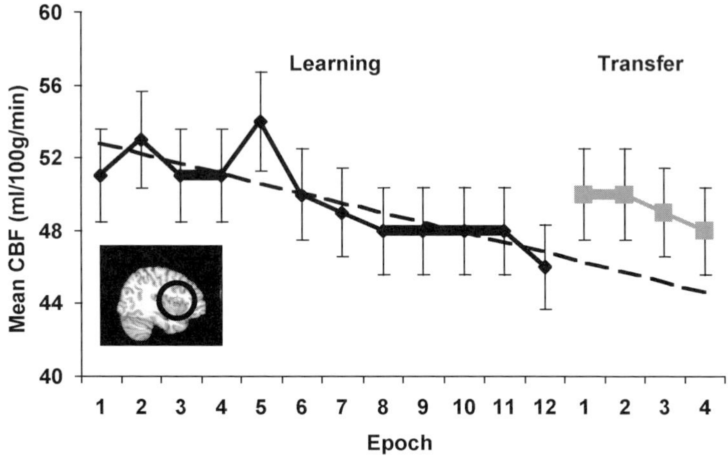

FIG. 11. Average cerebral blood flow across 10 subjects from the left-inferior premotor cortex during 15 minutes of continuous-sequence learning followed by a 5-minute "transfer" period with a new sequence. Each "epoch" corresponds to a 75-second period during which the 11-item sequence repeated seven times.

Figure 11 shows the time course of CBF measured from the left-ventral premotor cortex during this period. A steady decline in CBF is clearly visible during the training, followed by a discontinuous rise in CBF in response to the switch to the "transfer" sequence. These data demonstrate that perfusion fMRI can be used to reliably detect even very gradual changes in neural activity and can provide us with the ability to relate dynamic changes in neural processing with behavior over long time scales.

The stability of the perfusion signal is such that the notion of a "long time scale" can be pushed to the extreme. The perfusion signal is sufficiently stable that the CBF difference between rest and finger movements, for example, can be reliably measured even if these images are obtained on different days (Wang *et al.*, 2003b) (Fig. 12). This emphasizes that the quantified perfusion image has an absolute value—it is not necessary that the signal during one condition be compared to a signal obtained in a different condition during the same session. One might, for example, relate the absolute regional CBF during a particular behavioral state to a personality measure across subjects—for example, CBF within the amygdala during viewing of unpleasant pictures might be found to correlate with a clinical-depression rating scale.

Thus, perfusion fMRI provides researchers with an opportunity to study slowly evolving behaviors or perform longitudinal studies that would be awkward to conduct using BOLD fMRI. One can readily imagine other sorts of behaviors (e.g., emotional state) that would be greatly altered by rapid switching between

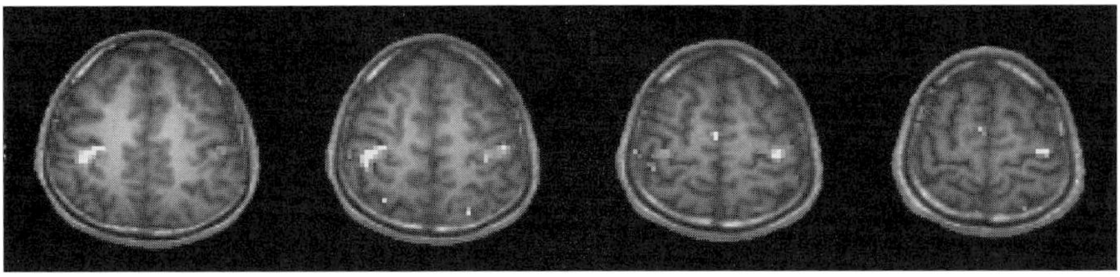

FIG. 12. Perfusion-fMRI activation for motor activity studied on two different days. This subject participated in a perfusion-fMRI scan on the first day during which he remained still. The next day, he performed a continuous motor task involving finger opposition. The quantified perfusion signal is sufficiently stable that the difference in CBF in the motor cortex is revealed even when studied on different days (Wang *et al.*, 2003b).

conditions or experimental effects with a time course that cannot be manipulated (e.g., pharmacokinetics of an administered drug). Studies of emotion, pharmacologic treatment, rehabilitation, learning, and sleep are a few of the domains of cognition and clinical practice for which perfusion may be a relevant tool.

Theoretical (Aguirre *et al.*, 2002) and empirical (Wang *et al.*, 2003a) analyses have shown that, for behaviors that evolve over time scales longer than 1 or 2 minutes, perfusion fMRI will provide superior within-subject sensitivity to changes in neural activity relative to BOLD fMRI. It is, of course, worth emphasizing that for changes in neural activity that proceed more rapidly, BOLD fMRI continues to provide markedly better within-subject power, in some cases by a factor of 2–3. This is particularly true for event-related designs. Because BOLD fMRI continues to be superior to perfusion for many classes of experimental design, it may be advantageous in some circumstances to use both methods in concert, as will be discussed next.

B. Combined BOLD and Perfusion Studies

As was noted earlier, the perfusion effects of ASL are independent of the pulse sequence used to obtain the images after the label period. Thus, if echo-planar images are used, the raw image data also contain BOLD contrast, allowing BOLD and perfusion effects to be compared within the same data set. It is the case, however, that when optimized for ASL acquisition, the echo-planar sequence is no longer ideal for the BOLD effect. In particular, a shorter echo time (TE) is desirable for ASL data collection, whereas a longer TE is preferable for BOLD SNR. Despite this fact, BOLD signal changes can readily be detected in the source label and control image data.

BOLD and ASL effects are best combined when they are used to seek different temporal structures of neural activity within the same dataset. In our sequence-learning example, our main focus was the slow changes in neural activity that accompany learning. It was the case, however, that a subject occasionally pressed the wrong button of his or her response pad while performing the task. Would it be possible for us to detect neural profiles associated with error responses? These errors were sporadic and brief; they had a temporal structure similar in many ways to a "sparse" event-related design. Such changes in neural activity would be difficult to detect using perfusion, but they are within the sensitivity range for BOLD fMRI. We were able to identify regional changes in BOLD signal in the unsubtracted, label, and control images associated with error trials (Olson *et al.*, 2005).

It appears that the voxel-wise BOLD and perfusion time-series data are uncorrelated under the null hypothesis (G. K. Aguirre, unpublished observations). As a result, with clever experimental design, one may use the BOLD signal effect to define regions of interest and may then, within those regions, probe for subtler, slower changes in the perfusion signal. These approaches can combine the best of these two different imaging modalities.

VI. Limitations and Developing Areas

The potential benefits that perfusion fMRI can provide are balanced by several significant limitations. Principal among these is the SNR for ASL imaging, which is smaller than that for BOLD. The change in raw image intensity between the control and labeled images is about 1%, as compared to 2–4% in BOLD. A practical consequence of this limited SNR is that ASL images are typically acquired with relatively large voxels or thick slices in an effort to improve sensitivity. A second limitation is the speed of image acquisition. Because perfusion volumes are calculated using pairs of images, the temporal sampling of the resulting data is half of the source data. Moreover, the time necessary to acquire a single image is lengthy: roughly a second each for label, delay, and image acquisition. A corollary problem is that the need to reduce image acquisition time prompts the use of two-dimensional echo-planar imaging for read-out. Echo-planar images can be acquired quickly, but they are distorted and severely degraded in the presence of static susceptibility gradients, such as are found near the frontal sinuses and auditory canals, causing loss of signal from the orbitofrontal and temporal cortex.

There are, however, several ways in which these limitations might be ameliorated. For example, we can obtain improved ASL data by scanning at higher field strengths. The stronger magnetic field has two beneficial effects. First, the raw SNR is improved, as it is proportional to the main field strength. Second, the T1 relaxation time of the label is prolonged at higher field strengths. As a result, the loss of spin labeling while blood is traveling from the labeling plane to the cerebral tissue is much less than that at standard field strength, producing a greater perfusion signal in brain tissue while reducing arterial transit–related artifacts and quantification errors. As was mentioned, application of CASL at high field strengths can be difficult, as the deposition of RF energy could potentially exceed acceptable levels. There are ways, however, to finesse these limitations and obtain CASL images with acceptable energy levels (Garraux *et al.*, 2005; Wang *et al.*, 2005b). More efficient labeling approaches are also being developed to improve the SNR of the perfusion signal while reducing RF power at high magnetic fields. For activation studies at high fields, the use of optimal image acquisition schemes, particularly shorter TE, has been proven crucial for improved functional sensitivity (Wang *et al.*, 2002). This is because the higher blood flow that accompanies neuronal activation may lead to direct outflow of portions of labeled arterial spins into veins, which have very low signal or may even be invisible due to strong susceptibility effects at high fields.

Additional improvement can be provided by phased-array coils, which combine a set of smaller receiver coils to cover a larger image volume. When placed in the proximity of the head, an array coil with eight channels is able to yield an SNR gain of more than fourfold in the cortex at 1.5 Tesla (de Zwart *et al.*, 2002). While surface-coil methods can introduce undesirable inhomogeneity into the image (e.g., the center of the image will be "darker" than the periphery), these effects are minimized in quantitative ASL-perfusion images due to calibration using the control images. Phase-array coils can also be used to perform "parallel" imaging, to obtain data with shorter echo times and thus reduced susceptibility artifacts (Wang *et al.*, 2005c). Finally, perfusion SNR in time can be improved by background suppression, in which the magnetization of the static brain tissue is reduced and the dynamic range for the detection of the ASL approach is thereby increased (Ye *et al.*, 2000b). This method requires, however, the use of a three-dimensional method for image readout, as background suppression can be accomplished at only a single acquisition time (which is not provided, for example, by echo-planar imaging).

As mentioned, methods other than echo-planar imaging can be used to obtain the images that are sensitive to the perfusion effect. Spin-echo imaging is one example that has decreased susceptibility artifacts (Fig. 4), but it still suffers from field distortion. Furthermore, spin-echo imaging is a two-dimensional method, so it does not allow for background suppression. Ideally, perfusion images would be acquired using a three-dimensional sequence that is not sensitive to susceptibility and can be obtained rapidly. The rapidity is a challenge for three-dimensional methods, but it is essential if the data are to be used for fMRI in which multiple images must be acquired quickly to provide a time series.

Candidate fast imaging methods include 3D fast spin echo (FSE) and 3D GRASE (Feinberg *et al.*, 1995; Fernandez-Seara *et al.*, 2005). The combination of 3D algorithms with parallel imaging and array coils provides a promising approach to whole-brain perfusion imaging free of susceptibility effects.

It is possible that, because of the combination of these refinements, ASL-based perfusion fMRI will see an improvement in SNR of fourfold or more. With improved sensitivity, it will be possible to obtain data with greater spatial resolution, taking full advantage of the selectivity of ASL methods to capillary signals. The promise of ASL is therefore a fully quantifiable, stable physiological measure with improved spatial resolution and minimal anatomical distortion. It should prove useful now and in the years ahead as a tool for the study of cognition and neural function.

References

Aguirre, G. K., Detre, J. A., Zarahn, E., and Alsop, D. C. (2002). Experimental design and the relative sensitivity of BOLD and perfusion fMRI. *Neuroimage* **15,** 488–500.

Aguirre, G. K., Nichols, T. E., and Wang, J.-J. (2003). Permutation tests for perfusion fMRI. *In* "Human Brain Mapping." New York, New York.

Aguirre, G. K., Zarahn, E., and D'Esposito, M. (1998a). A critique of the use of the Kolmogorov-Smirnov (KS) statistic for the analysis of BOLD fMRI data. *Magn. Reson. Med.* **39,** 500–505.

Aguirre, G. K., Zarahn, E., and D'Esposito, M. (1997). Empirical analyses of BOLD fMRI statistics. II. Spatially smoothed data collected under null-hypothesis and experimental conditions. *Neuroimage* **5,** 199–212.

Alsop, D. C., and Detre, J. A. (1996). Reduced transit-time sensitivity in noninvasive magnetic resonance imaging of human cerebral blood flow. *J. Cereb. Blood Flow Metab.* **16,** 1236–1249.

Alsop, D. C., Detre, J. A., and Grossman, M. (2000). Assessment of cerebral blood flow in Alzheimer's disease by spin-labeled magnetic resonance imaging. *Ann. Neurol.* **47,** 93–100.

Dale, A. M., and Buckner, R. L. (1997). Selective averaging of rapidly presented individual trials using fMRI. *Hum. Brain Mapp.* **5,** 329–340.

de Zwart, J. A., Ledden, P. J., Kellman, P., van Gelderen, P., and Duyn, J. H. (2002). Design of a SENSE-optimized high-sensitivity MRI receive coil for brain imaging. *Magn. Reson. Med.* **47,** 1218–1227.

Detre, J. A., and Alsop, D. C. (1999). Perfusion fMRI with arterial spin labeling. *In* "Functional MRI" (P. A. Bandettini and C. Moonen, Eds.), pp. 369–380. Springer Verlag, Berlin.

Duong, T. Q., Kim, D. S., Ugurbil, K., and Kim, S. G. (2000). Spatiotemporal dynamics of the BOLD fMRI signals: Toward mapping submillimeter cortical columns using the early negative response. *Magn. Reson. Med.* **44,** 231–242.

Duyn, J. H., Tan, C. X., van Gelderen, P., and Yongbi, M. N. (2001). High-sensitivity single-shot perfusion-weighted fMRI. *Magn. Reson. Med.* **46,** 88–94.

Engel, S. A., Glover, G. H., and Wandell, B. A. (1997). Retinotopic organization in human visual cortex and the spatial precision of functional MRI. *Cereb. Cortex* **7,** 181–192.

Feinberg, D. A., Kiefer, B., and Johnson, G. (1995). GRASE improves spatial resolution in single shot imaging. *Magn. Reson. Med.* **33,** 529–533.

Feng, C. M., Narayana, S., Lancaster, J. L., Jerabek, P. A., Arnow, T. L., Zhu, F., Tan, L. H., Fox, P. T., and Gao, J. H. (2004). CBF changes during brain activation: fMRI vs. PET. *Neuroimage* **22,** 443–446.

Fernandez-Seara, M. A., Wang, Z., Wang, J., Rao, H., Guenther, M., Feinberg, D. A., and Detre, J. A. (2005). Continuous arterial spin labelling perfusion measurements using single shot 3D GRASE at 3T. *Magn. Reson. Imaging,* submitted.

Friston, K. J., Holmes, A. P., Price, C. J., Buchel, C., and Worsley, K. J. (1999). Multisubject fMRI studies and conjunction analyses. *Neuroimage* **10,** 385–396.

Friston, K. J., Josephs, O., Zarahn, E., Holmes, A. P., Rouquette, S., and Poline, J. (2000). To smooth or not to smooth? *Neuroimage* **12,** 196–208.

Garraux, G., Hallett, M., and Talagala, S. L. (2005). CASL fMRI of subcortico-cortical perfusion changes during memory-guided finger sequences. *Neuroimage* **25,** 122–132.

Gonzalez-At, J. B., Alsop, D. C., and Detre, J. A. (2000). Cerebral perfusion and arterial transit time changes during task activation determined with continuous arterial spin labeling. *Magn. Reson. Med.* **43,** 739–746.

Hoogenraad, F. G., Pouwels, P. J., Hofman, M. B., Reichenbach, J. R., Sprenger, M., and Haacke, E. M. (2001). Quantitative differentiation between BOLD models in fMRI. *Magn. Reson. Med.* **45,** 233–246.

Josephs, O., Rees, G., Turner, R., and Friston, K. J. (1997). Event-related fMRI. *NeuroImage* **5,** S483.

Kim, S. G., and Tsekos, N. V. (1997). Perfusion imaging by a flow-sensitive alternating inversion recovery (FAIR) technique: Application to functional brain imaging. *Magn. Reson. Med.* **37,** 425–435.

Liu, H. L., and Gao, J. H. (1999). Perfusion-based event-related functional MRI. *Magn. Reson. Med.* **42,** 1011–1013.

Liu, T. T., and Wong, E. C. (2005). A signal processing model for arterial spin labeling functional MRI. *Neuroimage* **24,** in press.

Menon, R. S., Luknowsky, D. C., and Gati, J. C. (1998). Mental chronometry using latency-resolved functional MRI. *Proc. Natl. Acad. Sci. USA* **95,** 10902–10907.

Nichols, T. E., and Holmes, A. P. (2002). Nonparametric permutation tests for functional neuroimaging: A primer with examples. *Hum. Brain Mapp.* **15,** 1–25.

Nissen, M. J., and Bullemer, P. (1987). Attentional requirements of learning: Evidence from performance measures. *Cognit. Psychol.* **18,** 1–32.

Olson, H., Sledge, K., Rao, J., Wang, J., Detre, J. A., and Aguirre, G. K. (2005). Continuous sequence learning studied with perfusion fMRI. *Brain and Cognition.* Submitted.

Petersen, S. E., Fox, P. T., Posner, M. I., Mintun, M., and Raichle, M. E. (1988). Positron emission tomographic studies of the cortical anatomy of single-word processing. *Nature* **331,** 585–589.

Purdon, P. L., and Weisskoff, R. M. (1998). Effect of temporal autocorrelation due to physiological noise and stimulus paradigm on voxel-level false-positive rates in fMRI. *Hum. Brain Mapp.* **6,** 239–249.

Raichle, M., Martin, W., Herscovitch, P., Mintun, M., and Markham, J. (1983). Brain blood flow measured with intravenous H215O. *J. Nucl. Med.* **24,** 790–798.

Shadmehr, R., and Holcomb, H. H. (1997). Neural correlates of motor memory consolidation. *Science* **277,** 821–825.

Silva, A. C., Lee, S. P., Iadecola, C., and Kim, S. G. (2000). Early temporal characteristics of cerebral blood flow and deoxyhemoglobin changes during somatosensory stimulation. *J. Cereb. Blood Flow Metab.* **20,** 201–206.

Silva, A. C., Lee, S. P., Yang, G., Iadecola, C., and Kim, S. G. (1999). Simultaneous blood oxygenation level-dependent and cerebral blood flow functional magnetic resonance imaging during forepaw stimulation in the rat. *J. Cereb. Blood Flow Metab.* **19,** 871–879.

Wang, J., Aguirre, G. K., Kimberg, D. Y., and Detre, J. A. (2003a). Empirical analyses of null-hypothesis perfusion FMRI data at 1.5 and 4 T. *Neuroimage* **19,** 1449–1462.

Wang, J., Aguirre, G. K., Kimberg, D. Y., Roc, A. C., Li, L., and Detre, J. A. (2003b). Arterial spin labeling perfusion fMRI with very low task frequency. *Magn. Reson. Med.* **49,** 796–802.

Wang, J., Alsop, D. C., Li, L., Listerud, J., Gonzalez-At, J. B., Schnall, M. D., and Detre, J. A. (2002). Comparison of quantitative perfusion imaging using arterial spin labeling at 1.5 and 4.0 Tesla. *Magn. Reson. Med.* **48,** 242–254.

Wang, J., Li, L., Roc, A. C., Alsop, D. C., Tang, K., Butler, N. S., Schnall, M. D., and Detre, J. A. (2004). Reduced susceptibility effects in perfusion fMRI with single-shot spin-echo EPI acquisitions at 1.5 Tesla. *Magn. Reson. Imaging* **22,** 1–7.

Wang, J., Wang, Z., Aguirre, G. K., and Detre, J. A. (2005a). To smooth or not to smooth? ROC analysis of perfusion fMRI data. *Magn. Reson. Imaging* **23,** 75–81.

Wang, J., Zhang, Y., Wolf, R. L., Roc, A. C., Alsop, D. C., and Detre, J. A. (2005b). Amplitude-modulated continuous arterial spin-labeling 3.0-T perfusion MR imaging with a single coil: Feasibility study. *Radiology* **235,** 218–228.

Wang, Z., Wang, J., Connick, T. J., Wetmore, G. S., and Detre, J. A. (2005c). Continuous ASL perfusion MRI with an array coil and parallel imaging at 3T. *Magn. Reson. Med.* Submitted.

Williams, D. S., Detre, J. A., Leigh, J. S., and Koretsky, A. P. (1992). Magnetic resonance imaging of perfusion using spin inversion of arterial water. *Proc. Natl. Acad. Sci. USA* **89,** 212–216.

Wong, E. C., Buxton, R. B., and Frank, L. R. (1997). Implementation of quantitative perfusion imaging techniques for functional brain mapping using pulsed arterial spin labeling [see comments]. *NMR Biomed.* **10,** 237–249.

Wong, E. C., Buxton, R. B., and Frank, L. R. (1998). A theoretical and experimental comparison of continuous and pulsed arterial spin labeling techniques for quantitative perfusion imaging. *Magn. Reson. Med.* **40,** 348–355.

Worsley, K. J., and Friston, K. J. (1995). Analysis of fMRI time-series revisited-again. *Neuroimage* **2,** 173–182.

Worsley, K. J., Marrett, S., Neelin, P., and Evans, A. C. (1996). Searching scale space for activation in PET images. *Hum. Brain. Mapp.* **4,** 74–90.

Yang, Y., Engelien, W., Pan, H., Xu, S., Silbersweig, D. A., and Stern, E. (2000). A CBF-based event-related brain activation paradigm: Characterization of impulse-response function and comparison to BOLD. *Neuroimage* **12,** 287–297.

Ye, F. Q., Berman, K. F., Ellmore, T., Esposito, G., van Horn, J. D., Yang, Y., Duyn, J., Smith, A. M., Frank, J. A., Weinberger, D. R., and McLaughlin, A. C. (2000a). H(2)(15)O PET validation of steady-state arterial spin tagging cerebral blood flow measurements in humans. *Magn. Reson. Med.* **44,** 450–456.

Ye, F. Q., Frank, J. A., Weinberger, D. R., and McLaughlin, A. C. (2000b). Noise reduction in 3D perfusion imaging by attenuating the static signal in arterial spin tagging (ASSIST). *Magn. Reson. Med.* **44,** 92–100.

Zarahn, E., Aguirre, G. K., and D'Esposito, M. (1997). Empirical analyses of BOLD fMRI statistics. I. Spatially unsmoothed data collected under null-hypothesis conditions. *Neuroimage* **5,** 179–197.

FUNCTIONAL NEAR-INFRARED SPECTROSCOPY: POTENTIAL AND LIMITATIONS IN NEUROIMAGING STUDIES

Yoko Hoshi

Department of Integrated Neuroscience, Tokyo Institute of Psychiatry
Tokyo 156-8585, Japan

Near-infrared spectroscopy (NIRS), which was originally designed for clinical monitoring of tissue oxygenation, has been in development since the 1990s as a useful tool for neuroimaging studies, for the purposes of which it is called functional near-infrared spectroscopy (fNIRS). In this paper, I first outline the basic principles of NIRS, including types of measurements, types of instrumentation, and near-infrared light propagation in the human head. Then, I review fNIRS applications, in which the strengths and advantages of NIRS measurements over other neuroimaging methods are discussed. Following a brief review of optical imaging, I also clarify the problems and identify the limitations of NIRS measurements. Finally, I describe its future prospects in neuroimaging studies, with emphasis on the fact that although there are still many problems to be solved, the potential benefits of NIRS are considerable: NIRS not only helps to obtain

INTERNATIONAL REVIEW OF
NEUROBIOLOGY, VOL. 66
DOI: 10.1016/S0074-7742(05)66008-4

0074-7742/05 $35.00

further insights into brain functions but provides a new direction for cognitive neuroscience research.

I. Introduction

Near-infrared spectroscopy (NIRS), a noninvasive optical technique, can measure changes in hemoglobin (Hb) concentrations in brain tissue. Jöbsis first described the *in vivo* application of NIRS (Jöbsis, 1977), and this technique was originally designed for clinical monitoring of tissue oxygenation (Brazy *et al.*, 1985; Rea *et al.*, 1985; Wyatt *et al.*, 1986). Since the early 1990s, it has also been in development as a useful tool for neuroimaging studies called functional near-infrared spectroscopy (fNIRS) (Hoshi and Tamura, 1993a; Kato *et al.*, 1993; Villringer *et al.*, 1993). For about 25 years, the technology has been advancing, and a wide range of NIRS instruments have been developed. Among them, the instruments for continuous wave (CW) measurements based on the modified Beer–Lambert law (CW-type instruments), which include the earliest NIRS instruments, are the most readily available commercially. Instruments of this type allow us to observe dynamic changes in regional cerebral blood flow (rCBF) in real time by measuring concentration changes in cerebral Hb. Several types of brain activities, such as motor and cognitive activities, have been assessed using NIRS (Colier *et al.*, 1999; Heekeren *et al.*, 1997; Sato *et al*, 1999; Shimada *et al.*, 2004; Tanosaki *et al.*, 2001). Since NIRS is completely noninvasive, does not require strict motion restriction, and can be used in a daily-life environment, it enables neuroimaging studies on children, the elderly, and patients with psychoneurological problems who have not previously been fully examined because of the difficulty of measurement by other neuroimaging techniques, such as positron emission tomography (PET) and functional magnetic resonance imaging (fMRI). Currently, applications of NIRS are expanding; its usefulness is also being reported in psychiatry (Shinba *et al.*, 2004; Suto *et al.*, 2004) and developmental psychology (Baird *et al.*, 2002; Hoshi and Chen, 2002).

At the same time, the accuracy and reliability of NIRS have not yet been widely accepted. This is mainly attributable to incomplete knowledge of which region in the brain is sampled by near-infrared (NIR) light, the difficulty of selective detection of NIRS signals arising from the cerebral tissue, and the problem of quantification. Despite a number of theoretical and experimental investigations, NIR light propagation in the human head remains to be fully understood. When light is applied to the scalp, and diffusely reflected light is then collected at a position on the scalp a few centimeters away from the incident point, the detected light carries information about both the cerebral tissue and the extracerebral tissue. Because changes in extracerebral blood flow influence the determination of changes in cerebral Hb concentration, it is necessary to separate signals originating in the cerebral tissue from those coming from the extracerebral tissue. For this purpose, a multidetector system consisting of CW-type instruments has been developed (Germon *et al.*, 1999; McCormick *et al.*, 1991). However, separation of NIR signals was incomplete, and other methods are being explored.

The major problem with NIRS has been that concentration changes in Hb cannot be quantified with CW-type instruments, which has hindered the wide employment of NIRS in clinical medicine and research. Over the past 25 years, work in the field of NIRS has concentrated on solving this problem. Many different approaches to quantification have been tried, and the quantitative accuracy of time-resolved spectroscopy (TRS) and phase-resolved spectroscopy (PRS) has been established. However, the difficulty of quantification has not yet been completely overcome (see Section VII). Because of the problem of quantification, the amplitudes of NIRS signals cannot be compared between subjects and/or between regions within a given subject. However, such comparisons have actually been based on an invalid assumption in a large number of published studies.

Although there remain many problems to be solved, the potential benefits of NIRS are considerable. NIRS allows new neuroimaging studies that are different from those done with PET and fMRI. It is thus expected that NIRS will open new dimensions in brain research and clinical applications. In this paper, I will first outline the basic principles of NIRS, which will aid understanding of the potential and limitations of the technique. Then, focusing mainly on CW measurements, I will give specific examples of the strengths and advantages of NIRS measurements over other neuroimaging modalities, and also clarify the problems and identify the limitations of NIRS measurements. Finally, I will describe the future prospects of NIRS.

II. Basic Principles of NIRS

NIR light, especially that between 700 and 900 nm, easily passes through biological tissue because the light in this region is less scattered and is absorbed by only a few biological chromophores, such as Hb, myoglobin (Mb), and cytochrome oxidase (CytOx) in mitochondria. The spectra of Hb and Mb in the NIR region vary with their oxygenation states. The spectrum of CytOx also varies with its oxidation state. Thus, by measuring the transmitted light through the tissue, we can obtain information about the oxygenation–deoxygenation states of Hb and Mb and the oxidation-reduction state (the "redox" state) of CytOx. Many different types of NIRS measurements, each associated with one type of instrumentation, have been developed. In the following sections, the three main categories of NIRS measurements (CW, TRS, and PRS) will be described.

A. CW Measurement

In CW-type instruments, the light sources (laser diodes [LDs] or light-emitting diodes [LEDs]) emit light continuously at constant amplitude, and the light intensity at a position a few centimeters away from the incident point is measured. Instruments of this type calculate relative concentration changes in chromophores according to the modified Beer–Lambert law (Eq. 1) (Delpy *et al.*,1988):

$$A = -\log I/I_0 = \varepsilon CL + S \tag{1}$$

where I and I_0 are the intensities of detected and illuminated light, respectively, ε is the molar absorption coefficient, C is concentration, and L is the length over which light travels through the tissue (optical path length, also known as "total path length" [t-PL]), and S denotes optical attenuation mainly due to scattering.

When the scalp is illuminated by light of a certain wavelength (λ_1), absorbance ($A_{\lambda 1}$) is written as

$$A_{\lambda 1} = \varepsilon_{\lambda 1}[\text{oxy-Hb}]L_{\lambda 1} + \varepsilon'_{\lambda 1}[\text{deoxy-Hb}]L_{\lambda 1} + \varepsilon''_{\lambda 1}[\text{oxidized CytOx}]L_{\lambda 1} + S_{\lambda 1} \tag{2}$$

where [oxy-Hb], [deoxy-Hb], and [oxidized CytOx] are concentrations of oxygenated Hb, deoxygenated Hb, and oxidized CytOx, respectively. To minimize the effect of $S_{\lambda 1}$, a dual-wavelength method is often used, in which absorbance differences between measurement and reference wavelengths ($A_{\lambda 1} - A_{\lambda r} = A_{\lambda 1-r}$) are calculated. When the condition changes from one to another, $S_{\lambda 1-r}$ can be considered a constant. In the case that concentration changes in each chromophore are global within the head, the change in the absorbance difference ($\Delta A_{\lambda 1-r}$) is expressed as

$$\begin{aligned}\Delta A_{\lambda 1-r} = {} & \varepsilon_{\lambda 1-r}\Delta[\text{oxy-Hb}]L_{\lambda 1-r} + \varepsilon'_{\lambda 1-r}\Delta[\text{deoxy-Hb}]L_{\lambda 1-r} \\ & + \varepsilon''_{\lambda 1-r}\Delta[\text{oxidized CytOx}]L_{\lambda 1-r}\end{aligned} \tag{3}$$

For measurements of living tissue, such as brain tissue, multiple wavelengths are used, and Eq. 3 stands for each pair of wavelengths. By solving the simultaneous equations on the assumption that $L_{\lambda n-r}$ is the same for all pairs of measurement and reference wavelengths, and constant during the measurement period, we can calculate concentration changes in oxy-Hb, deoxy-Hb, and oxidized CytOx. A sum of changes in oxy-Hb and deoxy-Hb provides the changes in total Hb (t-Hb). The t-PL is longer than the distance between the

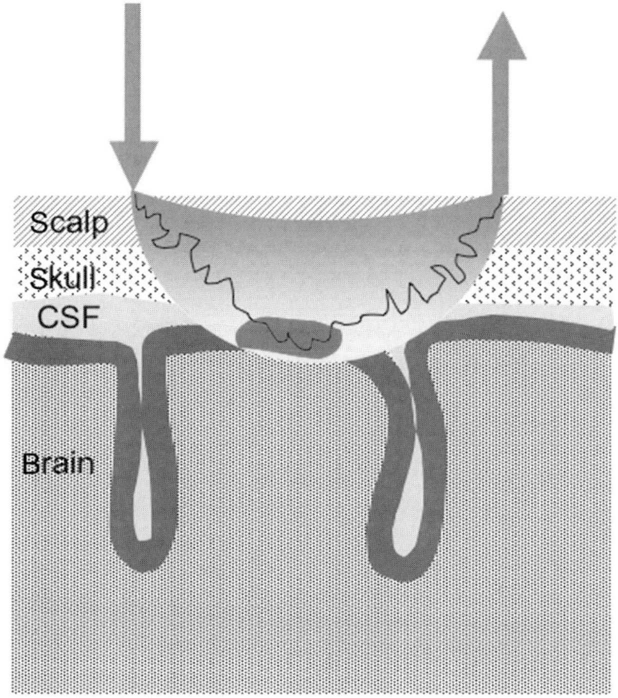

FIG. 1. Schematic representation of near-infrared light traveling through the head. The red area denotes a region of cortical activation. Determination of the path length through this region is required for quantification of changes in cerebral hemoglobin concentration. CSF, cerebrospinal fluid. (See Color Insert.)

source and detector because of the scattering effects. When the change in concentrations is not global, for example, a regional brain activation, the t-PL in Eq. 3 is replaced by a partial path length (p-PL) through a region of uniform absorption change (Fig. 1). Since CW-type instruments cannot measure either t-PL or p-PL, instruments of this type do not provide absolute values of concentration changes; they provide relative ones with an arbitrary unit (a.u.). However, the temporal resolution of CW-type instruments is high (in general, faster than 500 msec), and these instruments enable continuous measurement for a prolonged time.

The redox state of CytOx is very useful for the monitoring of tissue oxygenation. However, because changes in the redox state of CytOx occur only under severely hypoxic conditions (Hoshi *et al.*, 1997), CytOx is not measured in neuroimaging studies with NIRS.

B. Time-Domain Measurements (TRS)

In TRS, ultrashort (picosecond-order) laser pulses are applied to the tissue, and the emerging intensity is detected as a function of time (the temporal point spread function [TPSF]) with picosecond resolution (Chance *et al.*, 1988a; Deply *et al.*, 1988). Two detection schemes have been applied in TRS: the streak camera system and the time-correlated single-photon counting system. The former has the advantage of high temporal resolution, but it is large and expensive and has a limited dynamic range. The latter has the advantages of requiring cheaper components and having a wide dynamic range, though there are also several disadvantages, such as its low speed. Because time-resolved instruments have been developed as laboratory-based devices, they are difficult to use in the clinical environment. However, recent advances in technology have made possible the development of a compact 64-channel time-resolved optical imaging system (Eda *et al.*, 1999). A single-channel TRS instrument that will expand clinical applications of NIRS has also been developed and is commercially available (Oda *et al.*, 1999).

A temporal profile of the intensity of the detected light (Fig. 2) shows information about optical properties of the head, that is, grand-absorption (μ_a) and reduced-scattering (μ_s') coefficients of the scalp, the skull, the cerebrospinal

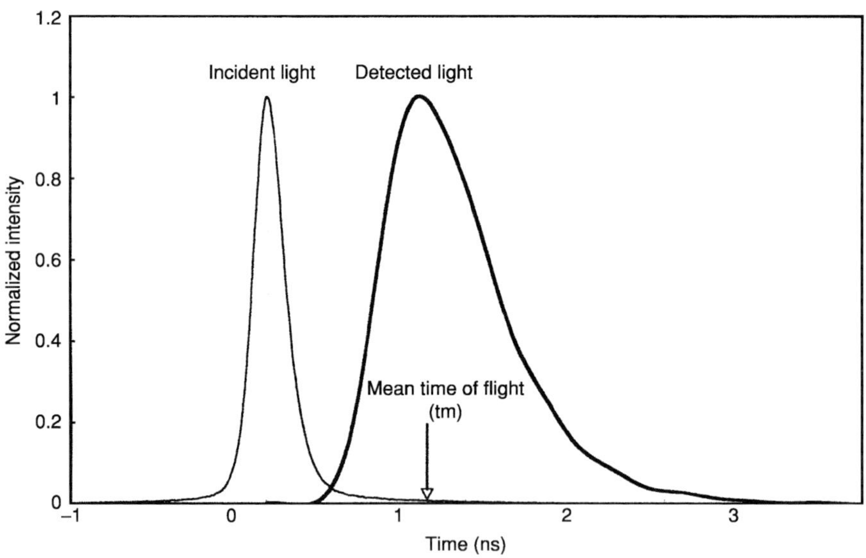

FIG. 2. Temporal profile of the intensity of transmitted light.

fluid (CSF), and the brain. Propagation of photons in living tissue, which is assumed to be a homogeneous medium, can be described by the diffusion approximation (Eq. 4) (Ishimaru, 1978),

$$\frac{1}{c}\frac{\partial \phi(\mathbf{r},t)}{\partial t} = \nabla[D(\mathbf{r})\nabla\phi(\mathbf{r},t)] - \mu_a(\boldsymbol{r})\phi(\boldsymbol{r},t) + S(r,t) \tag{4}$$

where $\phi(\mathbf{r}, t)$ is the diffuse photon fluence rate at position $\boldsymbol{r}$; the time t, c is the speed of light in the tissue; $D = 1/3\ \mu_s'(\boldsymbol{r})$ is the optical-diffusion coefficient; μ_a is the absorption coefficient; μ_s' is the reduced scattering coefficient; and $S(r, t)$ is the photon source. For a semi-infinite half-space geometry, the solution of Eq. 4 for the impulse input is expressed by Eq. 5 (Patterson *et al.*, 1989),

$$R(l,t) = \frac{z_0 \exp(-\mu_a ct)}{(4\pi Dct)^{3/2}t}\exp\left[-\frac{l^2 + z_0^2}{4Dct}\right] \tag{5}$$

where $R(l, t)$ is the intensity of the reflected light; l and t are the distance and time from the impulse input, respectively; and $z_0 = 1/\mu_s'$. By fitting Eq. 5 to the observed temporal profiles, we determine μ_a and μ_s'. As with CW measurements, through the use of multiple wavelengths and the determination of μ_a for each wavelength, absolute values of Hb concentrations in the head are estimated.

The mean transit time of scattered photons is calculated from the TPSF (t_m in Fig. 2). We determine the mean t-PL by multiplying the light speed in the media by the mean transit time of scattered photons. It should be noted, however, that the mean t-PL determined by TRS measurements of the head is a sum of p-PLs within the cerebral and extracerebral tissues.

C. Frequency-Domain Measurements (PRS)

The information contained in the temporal profile of the emerging light intensity through TRS can also be obtained by PRS, the relation between the time and frequency information being the Fourier transform (Duncan *et al.*, 1993; Sevic *et al.*, 1991). In frequency-domain instruments, the light source is intensity modulated at radio frequencies, and measurements are made both of the detected light intensity and of its phase shift and modulation depth with respect to the input light (Lakowicz and Berndt, 1990). It has been demonstrated that for typical tissues, and at frequencies below 200 MHz, the phase shift is linearly related to the mean t-PL (Arridge *et al.*,1992). Frequency-domain methods are low-cost alternatives to pulse-based methods for time-resolved methods. Several multichannel frequency-domain instruments are now in common use (Toronov *et al.*, 2001).

III. NIR Light Propagation in the Head

Knowledge about which regions in the brain are sampled by NIRS light is still incomplete, although a number of theoretical and experimental investigations on NIR light propagation in the human head have been performed (Boas *et al.*, 2002; Firbank *et al.*, 1998; Fukui *et al.*, 2003; Gratton *et al.*, 1994). Recent models for the theoretical analysis of light propagation in the head are based on the assumption of an inhomogeneous structure. These models consist of multiple layers, such as the scalp, skull, CSF, and brain; the brain can be divided into gray matter and white matter (Okada *et al.*, 1997). Theoretical analyses of models consisting of three- or four-layered slabs, the latter incorporating a clear CSF layer, have demonstrated that light propagation in the adult head is highly affected by the presence of a low-scattering CSF layer (Okada *et al.*, 1997). This indicated that light penetration in the adult brain might be limited to the outer cortical gray matter, which was confirmed by their later study, in which light propagation in head models generated from an MRI scan was predicted by Monte Carlo simulation (Firbank *et al.*, 1998). A more recent study has reported that a large source-detector spacing only broadens the sampling region on the brain surface and affects the penetration depth in the adult head to a lesser degree, whereas the intensely sensitive region in the neonatal head is confined to the gray matter; however, the deeper region of the white matter is sampled with a large source-detector spacing (Fukui *et al.*, 2003). In theoretical analysis, optical properties in each layer of the head are critical for prediction of the light propagation, though the μ_a and μ_s' in each layer used for analysis are different from study to study. This is attributable to the fact that *in situ* measurements of the optical properties are not feasible. Thus, further investigation is required for confirmation of the validity of the theoretically predicted light propagation.

IV. Interpretation of NIRS Signals

A. NIRS Signals in Activated Areas

Regional brain activation is accompanied by increases in rCBF and regional cerebral oxygen metabolic rate ($rCMRO_2$). It is widely accepted that the degree of the increase in rCBF exceeds that of the increase in $rCMRO_2$ (Fox and Raichle, 1986), which results in a decrease in deoxy-Hb in venous blood. Thus, increases in t-Hb and oxy-Hb and a decrease in deoxy-Hb are expected to be observed in activated areas in NIRS measurements (Fig. 3, top). However, deoxy-Hb and t-Hb do not necessarily show these changes: researchers observed (1) no

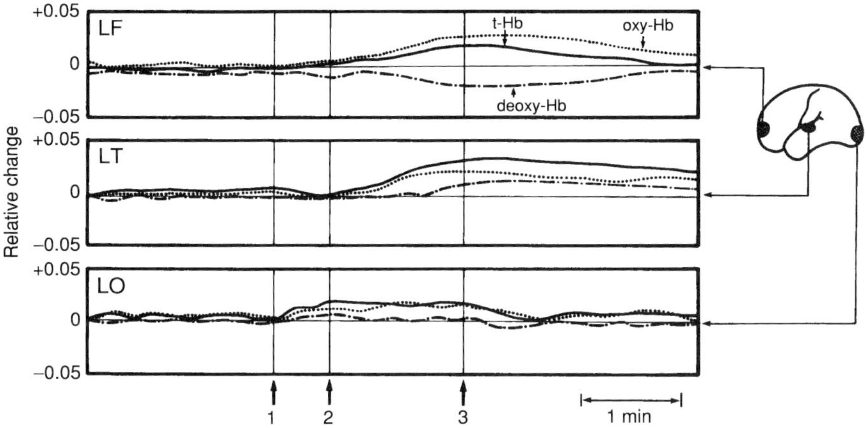

FIG. 3. Relative concentration changes in oxy-Hb, deoxy-Hb, and t-Hb in the left-frontal (LF), temporal (LT), and occipital (LO) regions while the subject is solving a mathematical problem. The subject read a text (between 1 and 2) and then solved a mathematical problem (between 2 and 3). Baselines were selected from the resting state, and these values are taken as zero for each signal. Changes from the baseline are represented as a relative amount with an arbitrary unit. Upward (plus) and downward (minus) trends show increases and decreases in values, respectively. From Hoshi and Tamura (1993b), with permission.

change in t-Hb alongside an increase in oxy-Hb and a reciprocal decrease in deoxy-Hb and (2) an increase or no change in deoxy-Hb accompanying increases in t-Hb and oxy-Hb (Fig. 3, middle and bottom) (Hoshi and Tamura, 1993b; Kato *et al.*, 1993; Kleinschmidt *et al.*, 1996). Comparing cerebral hemodynamic signals obtained simultaneously by NIRS and fMRI, some labs have reported that deoxy-Hb measured optically does not necessarily decrease in BOLD fMRI signal–increased areas (Hess *et al.*, 2000; Kleinschmidt *et al.*, 1996), while others have observed simultaneous reduction of deoxy-Hb in activation areas detected by fMRI (Tronov *et al.*, 2001). Hess *et al.* (2000) have hypothesized that the hemodynamic basis of BOLD fMRI signals is not necessarily a washout of deoxy-Hb as is generally assumed. Rather, a positive BOLD signal can also be caused by a local increase in blood volume, even if deoxy-Hb levels are persistently elevated. Using a newly developed perfused-rat-brain model (Fig. 4), we recently examined the direct effects of each change in CBF and $CMRO_2$ on cerebral Hb oxygenation to interpret NIRS signals (Hoshi *et al.*, 2001). We confirmed that directions of changes in oxy-Hb are always the same as those of rCBF, whereas the direction of changes in deoxy-Hb is determined by changes in venous blood oxygenation and volume. It has also been confirmed that small changes in CBF are not accompanied by such in t-Hb (Fig. 5A and B). Thus, oxy-Hb is the most sensitive indicator of changes in rCBF in NIRS measurements.

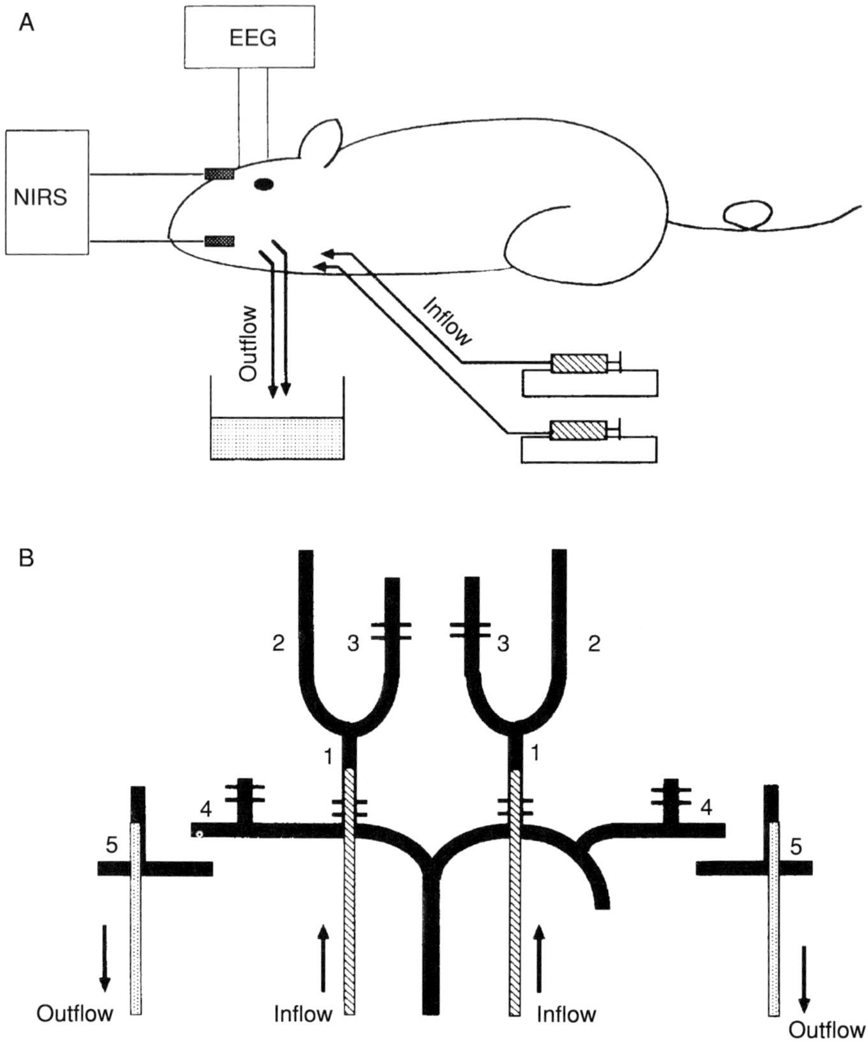

FIG. 4. (A) A perfused-rat-brain model maintaining the connection between the central and peripheral nervous systems. (B) Method of brain perfusion. 1, common carotid artery; 2, internal carotid artery; 3, external carotid artery; 4, vertebral artery; 5, external jugular vein. EEG, electroencephalography; NIRS, near-infrared spectroscopy. From Hoshi *et al.* (2001), with permission.

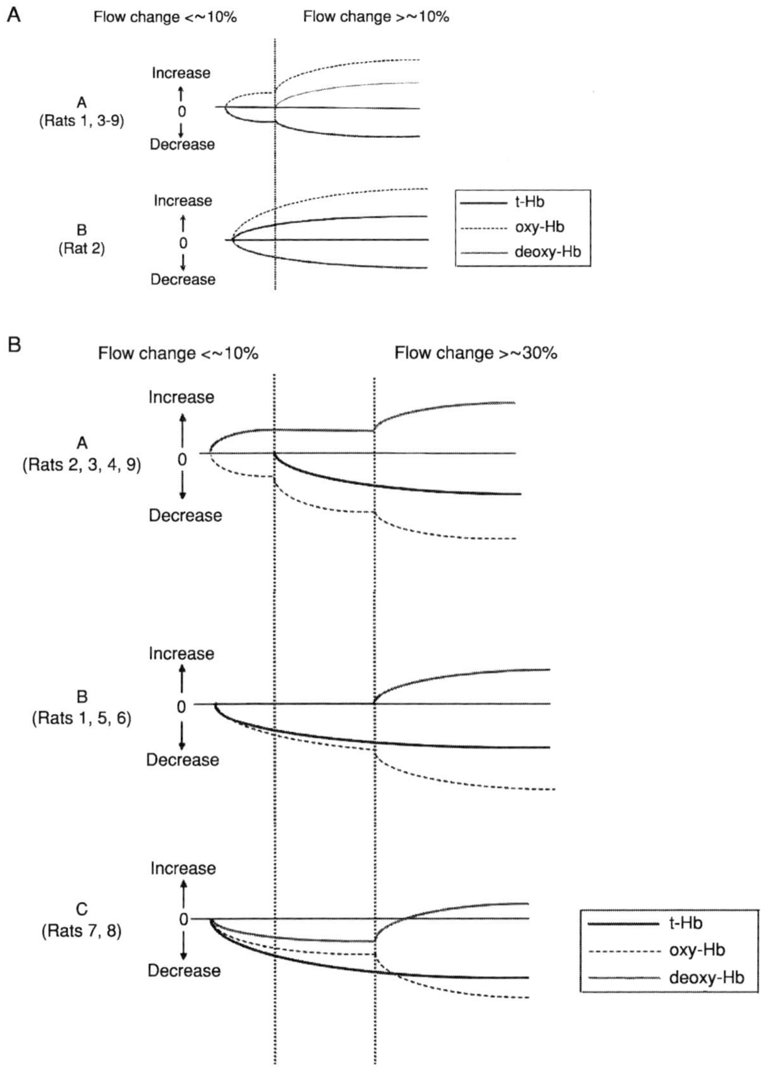

FIG. 5. (A) Pattern of changes in near-infrared spectroscopy (NIRS) signals while the flow rate was increasing. The pattern of changes in NIRS signals was divided into A and B according to the direction of changes in total hemoglobin (t-Hb) when the flow change was less than ~10%. (B) Pattern of changes in NIRS signals while the flow rate was decreasing. The pattern was divided into A, B, and C, according to the direction of changes in t-Hb and deoxy-Hb when the flow change was less than ~10%. From Hoshi *et al.* (2001), with permission.

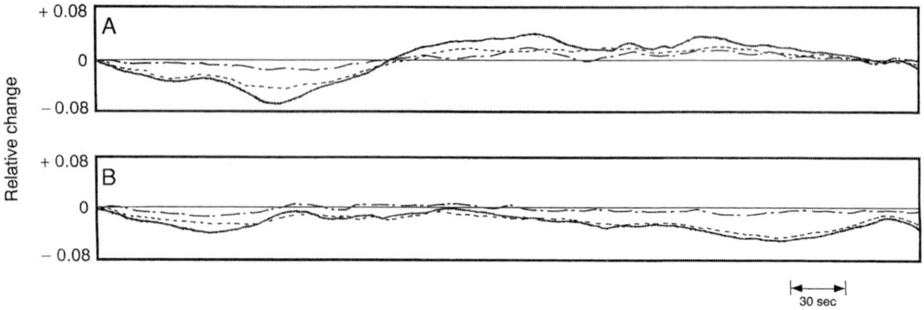

FIG. 6. Fluctuations in concentrations of oxygenated hemoglobin (dotted line), deoxygenated hemoglobin (broken line), and total hemoglobin (solid line) in the frontal (A) and occipital (B) regions during the resting state. The two regions were measured simultaneously.

B. Fluctuations in NIRS Signals at Rest

CW measurements have revealed that, even under resting conditions, the Hb oxygenation state fluctuates (Hoshi and Tamura, 1997; Toronov *et al.*, 2000). These fluctuations are divided into two types. The first is fluctuations whose general patterns are systemic and that are related to such physiological activities as the systemic arterial pulse oscillations (~1 Hz) and respiration (0.2–0.3 Hz). The second is slower Hb waves (frequency < 0.05 Hz), of which the temporal pattern varies with each brain region (Fig. 6). It is well known that the CBF velocity (CBFV) measured by transcranial Doppler ultrasound shows slow oscillation, which is thought to be attributable to small-pial-artery oscillations (Auer and Sayama, 1983; Lindegaard *et al.*, 1987). The frequency and characteristics of fluctuations in the Hb oxygenation state are similar to those of the CBFV oscillations. It is thus likely that the fluctuations in the Hb oxygenation state also originate from small-artery oscillations. Our simultaneous measurements with NIRS and electroencephalography (EEG) have suggested the possibility that these oscillations are a result of vasomotor responses to spontaneous neuronal activity (Hoshi *et al.*, 1998). Both the faster and the slower fluctuations of the Hb oxygenation state sometimes have amplitudes comparable to those of the signals evoked by functional activity. Thus, taking into account these fluctuations is critical for interpretation of NIRS signals.

V. fNIRS

NIRS has been used in a wide variety of studies on brain function (Colier *et al.*, 1999; Heekeren *et al.*, 1997; Horovitz and Gore, 2004; Sato *et al.*, 1999; Schroeter *et al.*, 2002; Shimada *et al.*, 2004; Tanosaki *et al.*, 2001). In earlier studies, which were performed with single- or two-channel instruments, the validity and accuracy of NIRS were confirmed by comparison with results obtained from other neuroimaging modalities, whereas recent studies, which include measurements using multichannel NIRS instruments, have revealed new information about brain functions. Examples of NIRS measurements that make the most of its strengths and advantages are described below.

A. Examination of Spatio-Temporal Characteristics of Cerebral Hemodynamic Changes Associated with Brain Activation

Working memory (WM) is a neural mechanism that provides and supports temporary storage and manipulation of information used in complex cognitive tasks like learning and reasoning (Baddeley, 1986). Baddeley and Hitch (1974) proposed a multicomponent model of WM, which is comprised of the central executive (CE) system and two slave systems as a theoretical framework. It is widely accepted that the prefrontal cortex (PFC) plays a critical role in the neural network that serves WM and executive control operations in humans. However, the fundamental question has not yet been definitively answered: Is the CE a unitary anatomical substrate with multiple functions or an agglomeration of independent but interacting control processes that are dependent on the ventrolateral PFC (VLPFC) and the dorsolateral PFC (DLPFC)? It is postulated that, if the CE could be localized as a unitary system, there would be common sustained regional brain activation within the PFC throughout WM tasks. It is expected that two working memory tasks—n-back and random-number generation (RNG) tasks—activate different brain regions, but that there can also be overlaps in brain activations, not only because both tasks require the same functions (such as sustained attention) but also because selected frontal regions are involved in broad functions (Duncan, 2001). Thus, the existence of common sustained regional brain activation across subjects and tasks does not necessarily mean that the CE is unitary. In contrast, if such brain activation does not exist, it can be concluded that the CE is not the unitary localized system. It was therefore expected that dynamic recording of changes in oxy-Hb as an indicator of changes in rCBF associated with multiple regions of the lateral PFC by NIRS would help clarify the nature of the CE.

We employed a 20-channel CW-type NIRS instrument to measure the bilateral lateral PFC during the performance of n-back and RNG tasks in 12 healthy adults (Fig. 7A) (Hoshi *et al.*, 2003). In most participants, increases in oxy-Hb were localized within one subdivison during the n-back task, whereas oxy-Hb increased more diffusely during the RNG task. Activation of VLPFC was prominent in the n-back task: both sustained and transient dynamics were observed (Fig. 7B and C). Transient dynamics are those in which oxy-Hb first

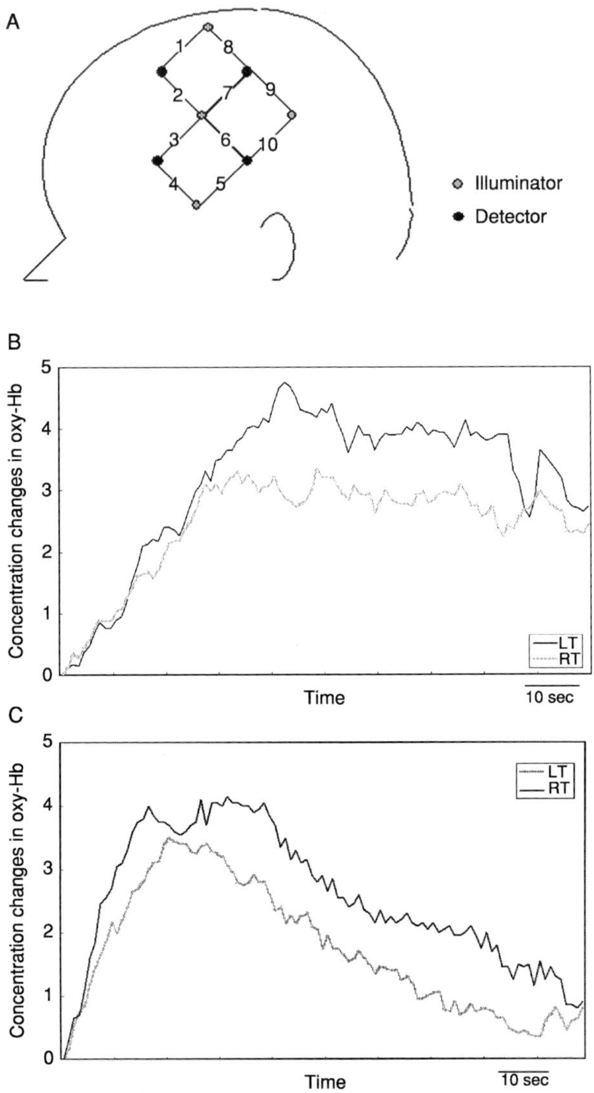

FIG. 7. (A) Positions of light guides on the left side. Numbers denote channels. (B) Sustained oxygenated hemoglobin (oxy-Hb) increases during the 3–0 back task in the bilateral ventrolateral prefrontal cortices (VLPFCs). (C) Nonsustained oxy-Hb increases during the 3–0 back task in the bilateral VLPFCs. From Hoshi *et al.* (2003), with permission. LT: left VLPFC; RT: right VLPFC.

increases but then decreases to less than 50% of the peak value or below the baseline level before the end of the task. For the RNG task, sustained activity was also observed in DLPFC, especially in the right hemisphere. However, details of patterns of activation varied across participants; the subdivisions commonly activated during the performance of the two tasks were the bilateral VLPFCs, each side of the VLPFC, and each side of the DLPFC in four, two, and four of the 12 participants, respectively. The remaining two of the 12 participants had no regions in common activated by these tasks. These results suggest that although PFC is implicated in the CE, there is no stereotyped anatomical PFC substrate for the CE.

Transient hemodynamic responses are also observed in the secondary somatosensory cortex during electric median nerve stimulation (Tanosaki *et al.*, 2001). This implies that the results obtained in PET and fMRI studies can vary between measurement periods. It is therefore necessary to examine the spatial and temporal characteristics of cerebral hemodynamic changes during brain activation.

B. Applications in Developmental Psychology

Normative development of the brain and mind has been investigated in a variety of fields of inquiry (e.g., cognition, development, education, pediatrics, and neurology); however, we are still relatively ignorant about a great many important issues concerning the development of the brain and mind. This is attributable mainly to the fact that there are only a few ways to investigate cognitive and socio-emotional development during infancy and preschool years. Traditionally, psychological tests and observations of behavior have been employed for this purpose. In neonates and younger infants, who cannot yet speak or perform tasks, changes in facial expression, as well as physiological parameters, such as heart rate, are often used as indicators of cognition. It is expected that neuroimaging techniques will provide further insight into the developmental brain.

Since NIRS has been oriented toward use for clinical monitoring of neonatal brain oxygenation, a number of investigations of cerebral hemodynamics and oxygen metabolism with NIRS in neonates have been reported (Brazy *et al.*, 1985; Pryds *et al.*, 1988; Wyatt *et al.*, 1986). However, investigations of brain functions themselves are relatively few in neonates and infants, and most have been investigations of evoked response to visual stimuli (Hoshi *et al.*, 2000; Meek *et al.*, 1998), olfactory stimuli (Bartocci *et al.*, 2000), and auditory stimuli (Sakatani *et al.*, 1999; Zaramella *et al.*, 2001), and of passive knee movement (Isobe *et al.*, 2001). Even more recently, higher-order functions such as response to language have been investigated in neonates (Peña *et al.*, 2003). These studies have shown that, in contrast to measurements of adult subjects, deoxy-Hb often increases accompanying increases in oxy-Hb and t-Hb, which corresponds to an inverse BOLD signal (Born *et al.*, 1998). This stimulus-related increase in deoxy-Hb was explained by a lower increase in rCBF than in $rCMRO_2$ (Meek *et al.*, 1998). However, our study has demonstrated that the direction of activation-related changes in deoxy-Hb in neonates varies with each measurement, even in the same subject during photic stimulation, which has provided another possible explanation for the increase in deoxy-Hb (Hoshi *et al.*, 2000). That is, it may be attributable to venous dilation.

Employing a single- or multi-channel NIRS instrument, we have recently succeeded in detecting hemodynamic changes associated with emotion and cognitive function in babies (1–7 months old) and infants (4–6 years old) (Hoshi and Chen, 2002). These studies demonstrate that NIRS enables a neuroscientific approach to developmental psychology. Using the wearable NIRS system (see Section V.D), we were able to monitor hemodynamic responses to certain movements, such as walking on a balance beam, in preschool and elementary-school children (Fig. 9B). Although only a few NIRS studies on cognitive and socio-emotional development have been so far reported (Baird *et al.*, 2002; Hoshi and Chen, 2002), the importance of NIRS in developmental psychology will soon increase.

C. Applications in Psychiatry

Okada *et al.* (1994) first applied NIRS to the evaluation of the frontal function in chronic schizophrenics. Using two single-channel CW-type instruments, they measured Hb oxygenation in the bilateral frontal regions during a mirror drawing task (MDT) in patients with schizophrenia. They also measured age- and sex-matched healthy subjects, who showed bilateral increases in oxy-Hb during the MDT (Fig. 8A). In contrast, half the patients showed unusual patterns of changes in Hb that never appeared in healthy subjects (Fig. 8B). The results suggested that certain schizophrenic symptoms might be related to defective interhemispheric integration. In the few years following this study, only a few psychiatric applications were reported (Fallgatter and Strik, 2000; Saito *et al.*, 1995). Lately, however, NIRS has become an increasingly popular method in psychiatry (Matsuo *et al.*, 2002; Shinba *et al.*, 2004; Suto *et al.*, 2004). Several research groups have examined task-related hemodynamic changes in psychiatric patients and found task-dependent abnormalities in frontal hemodynamics in schizophrenia (Fallgatter and Strik, 2000; Shinba *et al.*, 2004) and depression (Matsuo *et al.*, 2002; Suto *et al.*, 2004). Such task-dependent abnormalities were also found in patients with Alzheimer's disease (Fallgatter *et al.*, 1997; Hock *et al.*, 1997). These results underline the usefulness of NIRS in investigations of frontal-lobe dysfunction and evaluations of psychopathologic conditions in psychiatric patients.

D. Free-Motion Neuroimaging Studies

Measurements with less motion restriction in daily-life environments open new dimensions in neuroimaging studies. Using a 30-channel CW-type instrument, Miyai *et al.* (2001) succeeded in visualizing cortical activation patterns associated with human gait. Mere alternating foot movements activated regions similar to those associated with walking, but the regions were less broad, while imagery during an actual gait showed increased activity in different regions. From these findings, they proposed the use of NIRS for evaluating cerebral activation patterns during pathological gait and rehabilitative intervention.

The newly developed wearable NIRS system (Fig. 9A) allows subjects to move about freely, even during measurement. The details of the portable NIRS instrument used here (HEO 200, Omron Ltd. Inc., Japan) have been reported elsewhere (Shiga *et al.*, 1997). Briefly, it consists of a main unit and a probe unit. The main unit consists of a one-chip central processing unit (CPU), an LED driver, an amplifier, a memory-card interface, a liquid-crystal display, and a

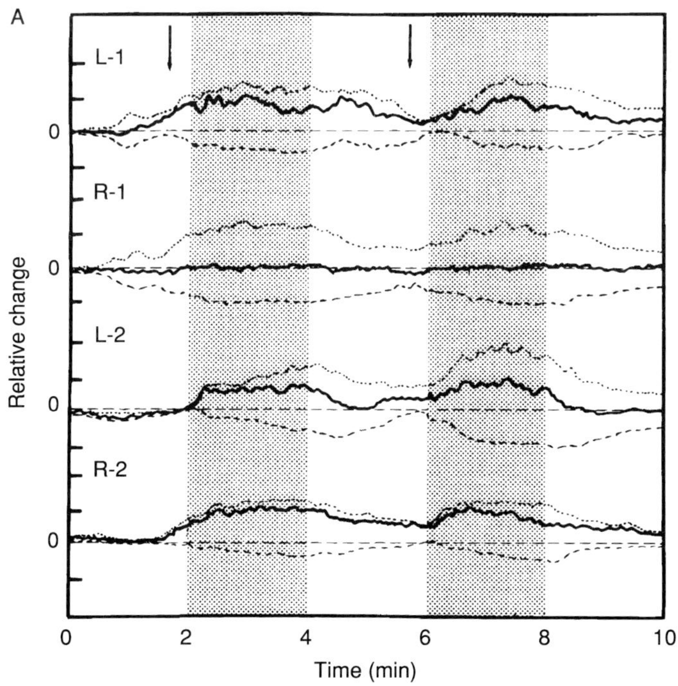

Fig. 8. (*Continued*)

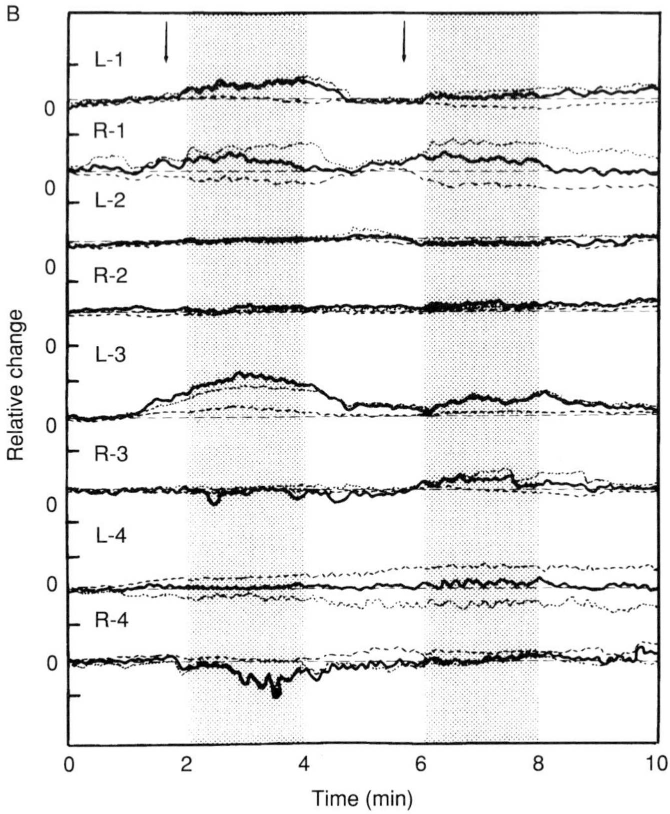

FIG. 8. (A) Relative concentration changes in oxygenated hemoglobin (oxy-Hb) (dotted line), deoxygenated hemoglobin (deoxy-Hb) (broken line), and total hemoglobin (t-Hb) (solid line) during the mirror drawing tasks (MDTs) in two right-handed healthy subjects. The MDTs were performed during the period shown by the shaded areas (the first column shows tasks performed with the right hand; the second shows those with the left hand). L-1 and R-1 show changes in the left and right frontal regions, respectively, in Subject 1. L-2 and R-2 show changes in Subject 2. (B) Relative concentration changes in oxy-Hb, deoxy-Hb, and t-Hb during the MDTs in four right-handed schizophrenics. L-1 and R-1, patient 1; L-2 and R-2, patient 2; L-3 and R-3, patient 3; L-4 and R-4, patient 4. From Okada *et al.* (1994), with permission.

power source. The probe unit, molded in elastic black silicon rubber, has a photodiode (the detector) in the center and a two-wavelength (760 and 840 nm) NIR LED on each side. This instrument is connected to the transmitter of a wireless system by an RS-232C cable. The NIRS instrument (8 × 16 × 4 cm, 340 g) is about the same size as the wireless transmitter. These are packed in a small bag that a subject carries (Fig. 9B). NIRS signals are sent by wireless to the

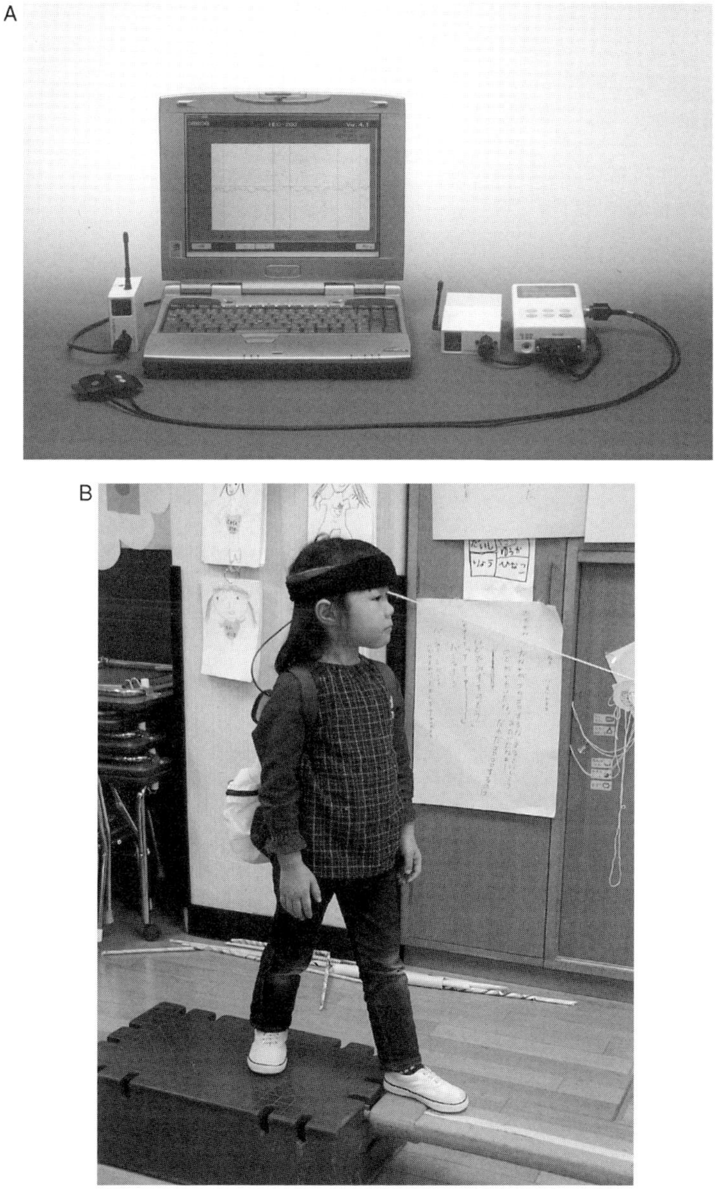

FIG. 9. (A) Wearable near-infrared spectroscopy (NIRS) system. (B) Measurement being performed using the wearable NIRS system. (See Color Insert.)

receiver, which is connected to a laptop computer on which data are displayed in real time. NIRS signals can be transferred to a place at a maximum distance of 30 m in an open field, but the maximum range inside a building is only about 10 m. This NIRS system makes it possible to monitor brain activity of freely moving subjects outside of laboratories. Since the current wearable NIRS system is, however, single-channel, the instrumentation and data analysis of the technique are being improved to enable a multichannel system.

VI. Optical Imaging

Work on improvements in quantification and localization has been extending to the development of NIR optical imaging of the brain. Optical imaging provides great promise for neuroimaging studies on subjects who are difficult to examine with PET and fMRI physiologically and/or physically, such as neonates and subjects who are doing exercise, as mentioned in Section V. There are two approaches to imaging: topography and tomography. The topography generates two-dimensional images of tissues parallel to the source-detector plane, and surface maps have no depth information. The earliest attempts at optical topography were performed using single-channel frequency-domain instruments, which measured the time of flight of photons during rest and during photic stimulation (Gratton *et al.*, 1995). More recently, multichannel frequency-domain instruments for optical topography have been developed. These instruments have mostly produced images of measured values such as phase shift and absorbance by back-projecting these values along the line between neighboring sources and detectors (Chance *et al.*, 1998b). However, qualitative images of absorption and scattering (Danen *et al.*, 1998) and images of relative concentration of oxy-Hb and deoxy-Hb (Franceschini *et al.*, 2000) have also been generated.

At the same time, we have employed TRS to generate quantitative images of Hb concentration changes (Eda *et al.*, 1999). The 64-channel time-resolved optical imaging system that we developed in the research and development project for optical tomographic imaging systems, in the New Energy and Industrial Technology Development Organization and the Ministry of International Trade and Industry of the Japanese government, has 3 100-ps-pulsed LDs (with wavelengths of 761, 797, and 834 nm) and 64 fiber bundles that are connected to 64 corresponding time-resolved detection systems (Fig. 10A). Each of the detecting systems consists of an optical attenuator; a high-speed, high-sensitivity photomultiplier tube; and a time-correlated single-photon counting circuit that contains a miniaturized constant fraction discriminator/time-to-amplitude converter module and a signal-acquisition unit with an analog to digital converter. To reduce the number of optical fibers, a special coaxial bundle was developed:

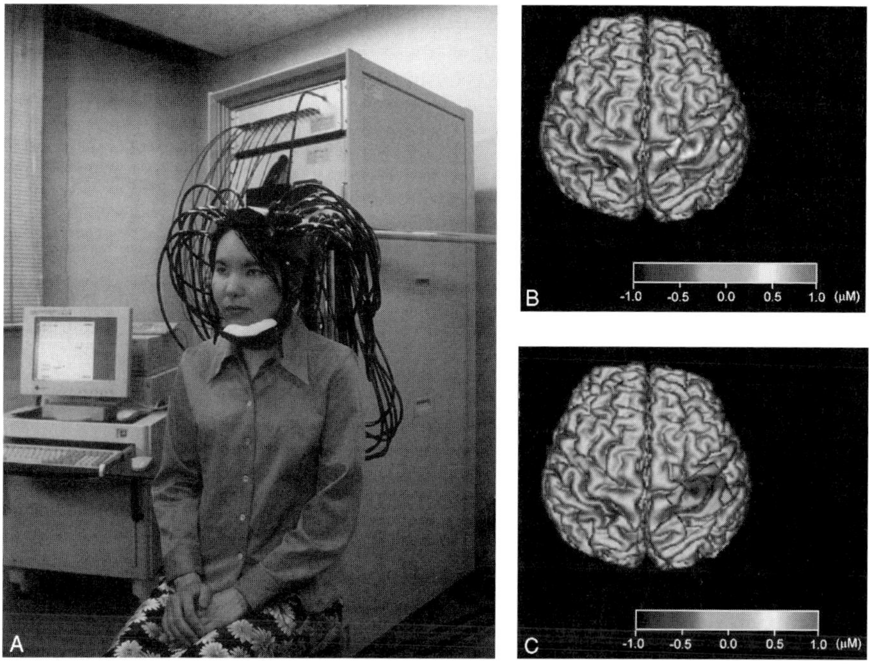

FIG. 10. Multichannel time-resolved optical imaging system (A) and topographical images of oxygenated hemoglobin (B) and deoxygenated hemoglobin (C) during left finger tapping. (See Color Insert.)

the center part is for illumination and the annular part for detection. Figure 10B and C show topographical images generated by this system of oxy-Hb and deoxy-Hb concentration changes, respectively, during left finger tapping.

The advent of multichannel CW-type instruments for optical topography has greatly increased the use of fNIRS in a variety of fields (Maki *et al.*, 1995). Current commercial instruments, which are compact and are the most sophisticated, are being developed as a real-time imaging modality that can display images at a rate of 1 per second or higher. However, the problem of quantification, which will be described in Section VI.A, has not been solved, even by multichannel systems.

Tomography is a more challenging approach; it generates a transverse slice or three-dimensional image. Tomography requires detection of photons penetrating deep into the brain tissues, which is quite difficult for adults in both the reflectance and the transmittance mode because of the overwhelming scattering. Thus, the optical tomography being developed is mainly aimed at neonates. Benaron *et al.* (2000; Hintz *et al.*, 1998) first demonstrated optical tomography using TRS. Three-dimensional imaging of the entire head is being pursued at University College London using a 32-channel time-resolved system (Schmidt *et al.*, 2000), and preliminary three-dimensional images have been obtained (Hebden *et al.*, 2002). Although recent technological and methodological advances have been encouraging, optical tomography is still far from clinical use, and further basic research is required.

VII. Problems with NIRS

A. Quantification of Cerebral Hb Concentration

Quantification of NIRS data has been a central issue in the NIRS field. When Hb concentration changes are global within the tissue, quantification is possible with TRS and PRS. In the case that Hb concentration changes are localized, however—as with functional brain activation—those changes cannot be quantified accurately. The optical path length determined by TRS and PRS is the mean t-PL but not the mean p-PL in the cerebral tissue. Since the t-PL is much longer than the p-PL (Firbank *et al.*, 1998; Fukui *et al.*, 2003), Hb concentration changes are underestimated when t-PL is substituted in Eq. 3 (the partial volume effect). However, measurement of p-PL is not feasible. To quantify NIRS data obtained from CW measurements without measuring t-PL, the assumption that the ratio of the source-detector separation to the t-PL (called the differential path length factor [DPF]) is constant has often been made (Delpy *et al.*, 1988). Arranging the source-detector separation for each pair at the same distance, in which the t-PL can be considered a constant if the assumption is correct, multichannel CW-type instruments generate topographical images of relative concentration changes in Hb. However, this assumption is not correct. As is shown in Fig. 11, the DPF (Deply *et al.*, 1988) varies with each position. Furthermore, the ratio of the p-PL to the t-PL varies with each wavelength and each measurement position (Hoshi *et al.*, 2005). Therefore, comparing the amplitudes of data obtained by CW measurements is not valid for comparison of changes in cerebral Hb concentration. This should be considered in the interpretation of topographical images produced by CW instruments.

The μ_a of the head estimated by TRS represents the μ_a of the cerebral tissue rather than that of the extracerebral tissue, but it is not equal to the former (Patterson *et al.*, 1989; Steinbrink *et al.*, 2001). Theoretical analyses with two-layered head models have demonstrated that when the μ_a of the upper layer is smaller than that of the lower layer, the estimated μ_a of the head is smaller than that of the lower layer (Kienle *et al.*, 1998). Reported optical properties indicate that the μ_a of the extracerebral tissue is smaller than that of the cerebral tissue

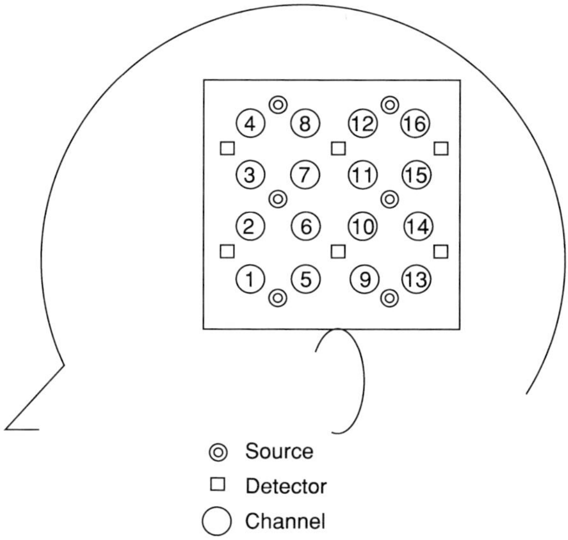

Channel	DPF (759 nm)
1	3.76
2	4.88
3	5.47
4	7.44
5	4.33
6	4.61
7	5.57
8	6.91
9	4.39
10	4.36
11	4.90
12	5.29
13	4.15
14	4.82
15	6.03
16	7.56

FIG. 11. Differential path length factors measured on the left scalp adjacent to the sensorimotor cortex in a 23-year-old healthy adult. The wavelength is 759 nm.

(Bevilacqua *et al.*, 1999; Sassaroli *et al.*, 2000; Simpson *et al.*, 1998). Thus, accurate quantification of concentration change in cerebral Hb remains difficult.

Elimination of the effects of hemodynamic changes in the scalp, which occur on various occasions (for example, during physical exercise and during performance of mental tasks), is also a challenging issue. Methods to solve the two problems simultaneously are being explored. Diffuse optical tomography (DOT), which reconstructs images of Hb concentration changes using multiple light sources and detectors, is a potential technique for quantitative detection of focal changes in cerebral hemodynamics (Arridge, 1999). DOT is not based on the modified Beer–Lambert law and can be performed with a time-domain system (Gao *et al.*, 2002), a frequency-domain system (Pogue *et al.*, 1995), or a CW system (Boas *et al.*, 2004). However, this technique, which requires multichannel instruments, is still in the development stage. TRS is also a potential tool for this purpose. The temporal profile of detected light intensity carries information about depth-dependent attenuation, because light that penetrates into deeper positions in the head is detected later. Several time-domain methods to determine absorption changes with depth resolution have been proposed (Hielscher *et al.*, 1996; Steibrink *et al.*, 2001). Although these methods were successful in certain cases, further investigation must be done in order for them to be applied to human-head measurements.

B. Practical Issues

Except on neonates, it is not possible for NIRS to measure deep brain structures such as the diencephalons noninvasively. It is also difficult to identify the exact brain areas that are beneath the NIRS probes; this would require three-dimensional MRI measurements. However, it has been reported that there is an appropriate relationship between the international 10–20 system of electrode positioning in EEGs and the cortical anatomy (Okamoto *et al.*, 2004; Towle *et al.*, 1993). Thus, we can roughly deduce the measured brain area by using the location of the 10–20 system as a landmark.

Advances in NIRS technology have enabled simultaneous measurements at multiple brain regions with high temporal resolution, which is desirable not only in investigations of regional differences in brain activation but also for the detection of localized brain activation. To improve spatial resolution, various arrangements of source-detector positions are being designed. However, fundamental problems remain to be solved. It takes time and skill to place many source-detector pairs on the hairy scalp and on the heads of neonates and infants. Improvement of NIRS probes will be vital for dissemination of NIRS in various fields of medicine and research.

Data in functional mapping studies with PET and fMRI are usually analyzed with some form of statistical parametric mapping, which refers to the construction of spatially extended statistical processes to test hypotheses about regionally specific effects in neuroimaging data (Friston *et al.*, 1991). However, statistical parametric mapping cannot be directly applied to NIRS studies because there are too few measurement portions, far fewer than the numbers of voxels in PET and fMRI. Standard approaches remain to be developed for the analysis of NIRS data. In addition, although less motion restriction during measurements is one strength of NIRS, head motion easily causes artifacts. Such artifacts are not easily eliminated automatically by a computer; they require inspection by the naked eye.

VIII. Future Prospects

As mentioned in the previous section, NIRS still has a number of technical issues to be explored and practical difficulties to be solved. Nevertheless, NIRS is a tool distinct from other neuroimaging techniques for studies of brain functions and for the diagnosis, assessment, and treatment of psychoneurological diseases. Thus, a variety of novel applications of NIRS, such as NIRS-based brain–computer interface (BCI), are being tried out. BCI provides users with an alternative output channel rather than the normal output path of the brain, i.e., the efferent nervous system and muscles (Donoghue, 2002; Wolpaw *et al.*, 2002). BCI has been given much attention recently as an alternate mode of communication and control for the disabled, including patients suffering from amyotrophic lateral sclerosis (ALS) or those who are "locked in." Most of the current BCI systems rely on the brain's electrical activity to produce scalp EEG signals. However, since the scalp EEG signals are inherently noisy and nonlinear, a more accessible interface that uses a more direct measurement of brain function to control an output device is being explored. NIRS is considered a possible alternative to electrical signals (Ranganatha *et al.*, 2004).

Another optical approach to detect brain activation is also being tested. Conventional NIRS instruments detect signals corresponding to relatively slow hemodynamic responses. In contrast, a much faster signal occurring over a period of tens of milliseconds has been detected by both a frequency-domain system (Gratton *et al.*, 1997; Wolf *et al.*, 2002) and a CW system (Franceschini and Boas, 2004). These fast signals, which are thought to be attributable to scattering changes in neurons, are much weaker than those of hemodynamic origin, and high temporal resolution is required for their detection. The instrumentation and data analysis for the technique have markedly improved over the last few years, making it feasible to detect neuronal activity. This new approach is becoming a powerful clinical tool.

The development of NIRS has taken a long time, comparable to that of fMRI. For the last 25 years, fMRI has established itself as the gold standard of neuroimaging, whereas NIRS has just begun to be used as a neuroimaging technique. However, NIRS is making steady progress, and its strengths and advantages are expected to provide a new direction for functional mapping studies that other neuroimaging techniques cannot handle. Thus, NIRS shows great promise for providing further insights into brain functions and for development of clinical tools.

References

Arridge, S. R., Cope, M., and Delpy, D. T. (1992). Theoretical basis for the determination of optical pathlengths in tissue: Temporal and frequency analysis. *Phys. Med. Biol.* **37,** 1531–1560.

Arridge, S. R. (1999). Optical tomography in medical imaging. *Inverse Problems* **15,** R41–R93.

Auer, L. M., and Sayama, I. (1983). Intracranial pressure oscillations (B-waves) caused by oscillations in cerebrovascular volume. *Acta Neurochir.* **68,** 93–100.

Baddeley, A. D., and Hitch, G. (1974). Working Memory. *In* "The Psychology of Learning and Motivation" (G. A. Bower, Ed.), Vol. 8, pp. 47–89. Academic Press, New York.

Baddeley, A. D. (1986). "Working Memory." Oxford University Press, Oxford.

Baird, A. A., Kagan, J., Gaudette, T., Walz, K. A., Hershlag, N., and Boas, D. A. (2002). Frontal lobe activation during object performance: Data from near-infrared spectroscopy. *NeuroImage* **16,** 1120–1126.

Bartocci, M., Winberg, J., Ruggiero, C., Bergqvist, L. L., Serra, G., and Lagercrantz, H. (2000). Activation of olfactory cortex in newborn infants after odor stimulation: A functional near-infrared spectroscopy. *Pediatr. Res.* **48,** 18–23.

Benaron, D. A., Hintz, S. R., Virrlinger, A., Boas, D., Kleinschmidt, A., Frahm, J., Hirth, C., Obrigh, H., van Houten, J. C., Kermit, E. L., Cheong, W.-F., and Stenvenson, D. K. (2000). Noninvasive functional imaging of human brain using light. *J. Cereb. Blood Flow Metab.* **20,** 469–477.

Bevilacqua, F., Piguet, D., Marquet, P., Gross, J. D., Tromberg, B. J., and Depeursige, C. (1999). *In vivo* local determination of tissue optical properties: Applications to human brain. *Appl. Opt.* **38,** 4939–4950.

Boas, D. A., Culver, J. P., Stott, J. J., and Dunn, A. K. (2002). Three dimensional Monte Carlo code for photon migration through complex heterogeneous media including the adult human head. *Opt. Express* **10,** 159–170.

Boas, D. A., Chen, K., Grebert, D., and Franceschini, M. A. (2004). Improving the diffuse optical imaging spatial resolution of the cerebral hemodynamic response to brain activation in humans. *Opt. Lett.* **29,** 1506–1508.

Born, P., Leth, H., Miranda, M. J., Rostrup, E., Stensgaard, A., Peitersen, B., Larsson, H. B. W., and Lou, H. C. (1998). Visual activation in infants and young children studies by functional magnetic resonance imaging. *Pediatr. Res.* **44,** 578–583.

Brazy, J. E., Lewis, D. V., Mitnick, M. H., and Jöbsis, F. F. (1985). Noninvasive monitoring of cerebral oxygenation in preterm infants: Preliminary observation. *Pediatrics* **75,** 217–225.

Colier, W. N., Quaresima, V., Oeseburug, B., and Ferrari, M. (1999). Human motor-cortex oxygenation changes induced by cyclic coupled movements of hand and foot. *Exp. Brain Res.* **129,** 457–461.

Danen, R. M., Wang, Y., Li, X. D., Thayer, W. S., and Yodh, A. G. (1998). Regional imager for low resolution functional imaging of the brain with diffusing near-infrared light. *Photochem. Photobiol.* **67,** 33–40.

Delpy, D. T., Cope, M., van der Zee, P., Arridge, S., Wray, S., and Wyatt, J. (1988). Estimation of optical pathlength through tissue from direct time of flight measurement. *Phys. Med. Biol.* **33,** 1433–1442.

Donoghue, J. P. (2002). Connecting cortex to machines: Recent advances in brain interfaces. *Nat. Neurosci.* **5,** 1085–1088.

Duncan, A., Whitlock, T. L., Cope, M., and Deply, D. T. (1993). A multiwavelength, wideband, intensity modulated optical spectrometer for near infrared spectroscopy and imaging. *Proc. SPIE* **1888,** 248–257.

Duncan, J. (2001). An adaptive coding model of neural function in prefrontal cortex. *Nat. Rev. Neurosci.* **2,** 820–829.

Eda, H., Oda, I., Ito, Y., Wada, Y., Oikawa, Y., Tsunazawa, Y., Takada, M., Tsuchiya, Y., Yamashita, Y., Oda, M., Sassaroli, A., Yamada, Y., and Tamura, M. (1999). Multi-channel time-resolved optical tomographic imaging system. *Rev. Sci. Instrum.* **70,** 3595–3602.

Fallgatter, A. J., Roesler, M., Sitzmann, L., Heidrich, A., Mueller, T. J., and Strik, W. K. (1997). Loss of functional hemispheric asymmetry in Alzheimer's dementia assessed with near-infrared spectroscopy. *Cogn. Brain Res.* **6,** 67–72.

Fallgatter, A. J., and Strik, W. K. (2000). Reduced frontal functional asymmetry in schizophrenia during a cued continuous performance test assessed with near-infrared spectroscopy. *Schizophr. Bull.* **26,** 913–919.

Firbank, M., Okada, E., and Delpy, D. T. (1998). A theoretical study of the signal contribution of regions of the adult head to near-infrared spectroscopy studies of visual evoked responses. *NeuroImage* **8,** 69–78.

Fox, P., and Raichle, M. E. (1986). Focal Physiological uncoupling of cerebral blood flow and oxidative metabolism during somatosensory stimulation in human subjects. *Proc. Natl. Acad. Sci. USA* **83,** 1140–1144.

Franceschini, M. A., Toronov, V., Filiaci, M. E., Gratton, E., and Fantini, S. (2000). On-line optical imaging of the human brain with 160-ms temporal resolution. *Opt. Express* **6,** 49–57.

Franceschini, M. A., and Boas, D. A. (2004). Noninvasive measurement of neuronal activity with nea-infrared optical imaging. *NeuroImage* **21,** 372–386.

Friston, K. J., Frith, C. D., Liddle, P. F., and Frackowiak, R. S. J. (1991). Comparing functional (PET) images: The assessment of significant change. *J. Cereb. Blood Flow Metab.* **11,** 690–699.

Fukui, Y., Ajichi, Y., and Okada, E. (2003). Monte Carlo prediction of near-infrared light propagation in realistic adult and neonatal head models. *Appl. Opt.* **42,** 2881–2887.

Gao, H., Zhao, H., and Yamada, Y. (2002). Improvement of image quality in diffuse optical tomography using full time-resolved data. *Appl. Opt.* **41,** 778–791.

Germon, T. J., Evans, P. D., Barnet, N. J., Wall, P., Manara, A. R., and Nelson, R. J. (1999). Cerebral near infrared spectroscopy: Emitter-detector separation must be increased. *Br. J. Anaesth.* **82,** 831–837.

Gratton, G., Maier, J. S., Fabiani, M., Mantulinm, W. W., and Gratton, E. (1994). Feasibility of intracranial near-infrared optical scanning. *Psychophysiol.* **31,** 211–215.

Gratton, G., Corballis, P. M., Cho, E., Fabiani, M., and Hood, D. C. (1995). Shades of gray matter: Noninvasive optical images of human brain responses during visual stimulation. *Psychophysiology* **32,** 505–509.

Gratton, G., Fabiani, M., Corballis, P. M., Hood, D. C., Goodman-Wood, M. R., Hirsch, J., Kim, K., Friedman, D., and Gratton, E. (1997). Fast and localized event-related optical signals (EROS) in the human occipital cortex: Comparisons with the visual evoked potential and fMRI. *NeuroImage* **6,** 168–180.

Hebden, J. C., Gibson, A., Yusof, R., Everdell, N., Hillman, E. M. C., Delpy, D. T., Arridge, S. R., Austin, T., Meek, J. H., and Wyatt, J. S. (2002). Three-dimensional optical tomography of the premature infant brain. *Phys. Med. Biol.* **47,** 4155–4166.

Heekeren, H. R., Obrig, H., Wenzel, R., Eberle, K., Ruben, J., Villringer, K., Kurth, R., and Villringer, A. (1997). Cerebral haemoglobin oxygenation during sustained visual stimulation-a near-infrared spectroscopy study. *Philos. Trans. R. Soc. London B Biolo. Sci.* **352,** 743–750.

Hess, A., Stiller, D., Kaulisch, T., Heil, P., and Scheich, H. (2000). New insights into the hemodynamic blood oxygenation level-dependent response through combination of functional magnetic resonance imaging and optical recording in gerbil barrel cortex. *J. Neurosci.* **20,** 3328–3338.

Hielscher, A. H., Liu, H., Chance, B., Tittel, F. K., and Jacques, S. L. (1996). Time-resolved photon emission from layered turbid media. *Appl. Opt.* **35,** 719–728.

Hintz, S. R., Benaron, D. A., van Houten, J. P., Duckworth, J. L., Liu, F. W. H., Spilman, S. D., Stevenson, D. K., and Cheong, W.-F. (1998). Stationary headband for clinical time-of-flight optical imaging at the bedside. *Photochem. Photobiol.* **68,** 361–369.

Hock, C., Villringer, K., Müller-Spahn, F., Wenzel, R., Heekeren, H., Schuh-Hofer, S., Hofman, M., Minoshima, S., Schwaiger, M., Dirnagl, U., and Villringer, A. (1997). Decrease in parietal cerebral hemoglobin oxygenation during performance of a verbal fluency task in patients with Alzheimer's disease monitored by means of near-infrared spectroscopy (NIRS)-correlation with simultaneous rCBF-PET measurements. *Brain Res.* **755,** 293–303.

Horovitz, S. G., and Gore, J. S. (2004). Simultaneous event-related potential and near-infrared spectroscopic studies of semantic processing. *Human Brain Mapp.* **22,** 110–115.

Hoshi, Y., and Chen, S.-J. (2002). Cerebral blood flow changes associated with emotion in children. *Pediatr. Neurol.* **27,** 275–281.

Hoshi, Y., and Tamura, M. (1993a). Detection of dynamic changes in cerebral oxygenation coupled to neuronal function during mental work in man. *Neurosci. Lett.* **150,** 5–8.

Hoshi, Y., and Tamura, M. (1993b). Dynamic multichannel near-infrared optical imaging of human brain activity. *J. Appl. Physiol.* **75,** 1842–1846.

Hoshi, Y., and Tamura, M. (1997). Fluctuations in the cerebral oxygenation state during the resting period in functional mapping studies of the human brain. *Med. Biol. Engin.Comput.* **35,** 328–330.

Hoshi, Y., Hazeki, O., Kakihana, Y., and Tamura, M. (1997). Redox behavior of cytochrome oxidase in the rat brain measured by near-infrared spectroscopy. *J. Appl. Physiol.* **83,** 1842–1848.

Hoshi, Y., Kosaka, S., Xie, Y., Kohri, S., and Tamura, M. (1998). Relationship between fluctuations in the cerebral hemoglobin oxygenation state and neuronal activity under resting conditions in man. *Neurosci. Lett.* **245,** 147–150.

Hoshi, Y., Kohri, S., Matsumoto, Y., Cjo, K., Matsuda, T., Okajima, S., and Fujimoto, S. (2000). Hemodynamic responses to photic stimulation in neonates. *Pediatr. Neurol.* **23,** 323–327.

Hoshi, Y., Kobayashi, N., and Tamura, M. (2001). Interpretation of near-infrared spectroscopy signals: A study with a newly developed perfused rat brain model. *J. Appl. Physiol.* **90,** 1657–1662.

Hoshi, Y., Tsou, B. H., Billock, V. A., Tanosaki, M., Iguchi, Y., Shimada, M., Shinba, T., Yamada, Y., and Oda, I. (2003). Spatiotemporal characteristics of hemodynamic changes in the human lateral prefrontal cortex during working memory tasks. *NeuroImage* **20,** 1493–1504.

Hoshi, Y., Shimada, M., Sato, C., and Iguchi, Y. (2005). Peevaluation of near-infrared light propagation in the adult human head: Implications for functional near-infrared spectroscopy. *J. Biomed. Opt.* (in press).

Ishimaru, A. (1978). Diffusion of a pulse in densely distributed scatters. *J. Opt. Soc. Am.* **68,** 1045–1050.

Isobe, K., Kusaka, T., Nagano, K., Okubo, K., Yasuda, S., Kondo, M., Itoh, S., and Onishi, S. (2001). Functional imaging of the brain in sedated newborn infants using near infrared topography during passive knee movement. *Neurosci. Lett.* **299,** 221–224.

Jöbsis, F. F. (1977). Noninvasive infrared monitoring of cerebral and myocardial oxygen sufficiency and circulatory parameters. *Science* **198,** 1264–1267.

Kato, T., Kamei, A., Takashima, S., and Ozaki, T. (1993). Human visual cortical function during photic stimulation monitoring by means of near-infrared spectroscopy. *J. Cereb. Blood Flow Metab.* **13,** 516–520.

Kienle, A., Patterson, M. S., Dognitz, N., Bays, R., Wagnieres, G., and van den Bergh, H. (1998). Noninvasive determination of the optical properties of two-layered turbid media. *Appl. Opt.* **37,** 779–791.

Kleinschmidt, A., Obrig, H., Requardt, M., Merboldt, K- D., Dirnagl, U., Villringer, A., and Frahm, J. (1996). Simultaneous recording of cerebral blood oxygenation changes during human brain activation by magnetic resonance imaging and near-infrared spectroscopy. *J. Cereb. Blood Flow Metab.* **16,** 817–826.

Lakowicz, J. R., and Berndt, K. (1990). Frequency domain measurement of photon migration in tissues. *Chem. Phys. Lett.* **166,** 246–252.

Lindegaard, K. F., Lundar, T., Wiberg, J., Sjøberg, D., Aaslid, R., and Normes, H. (1987). Variations in middle cerebral artery blood flow investigated with noninvasive transcranial blood velocity measurements. *Stroke* **18,** 1025–1030.

Maki, A., Yamashita, Y., Ito, Y., Watanabe, E., Mayanagi, Y., and Koizumi, H. (1995). Spatial and temporal analysis of human motor activity using noninvasive NIR topography. *Med. Phys.* **22,** 1997–2005.

Matsuo, K., Kato, N., and Kato, T. (2002). Decreased cerebral haemodynamic response to cognitive and physiological tasks in mood disorders as shown by near-infrared spectroscopy. *Psychol Med.* **32,** 1029–37.

McCormick, P. W., Stewart, M., Goetting, M. G., Dujovny, M., Lewis, G., and Ausman, J. I. (1991). Noninvasive cerebral optical spectroscopy for monitoring cerebral oxygen delivery and hemodynamics. *Crit. Care Med.* **19,** 89–97.

Meek, J. K., Firbank, M., Elwell, C. E., Atkinson, J., Braddick, O., and Wyatt, J. S. (1998). Regional hemodynamic response to visual stimulation in wake infants. *Pediatr. Res.* **43,** 840–843.

Miyai, I., Tanabe, H. C., Sase, I., Eda, H., Oda, I., Konishi, I., Tsunazawa, Y., Suzuki, T., Yanagida, T., and Kubota, K. (2001). Cortical mapping of gait in humans: A near-infrared spectroscopic topography study. *NeuroImage* **14,** 1186–1192.

Oda, M., Yamashita, Y., Nakano, T., Suzuki, A., Shimizu, K., Hirano, I., Shimomura, F., Ohmae, E., Suzuki, T., and Tsuchiya, Y. (1999). Nearinfrared time-resolved spectroscopy system for tissue oxygenation monitor. *Proc. Society Photo-Optical Instrumentation Engineers* **3597,** 611–617.

Okada, F., Tokumitsu, Y., Hoshi, Y., and Tamura, M. (1994). Impaired interhemispheric integration in brain oxygenation and hemodynamics in schizophrenia. *Eur. Arch. Psychiatry Clin. Neurosci.* **244,** 17–25.

Okada, E., Firbank, M., Schweiger, M., Arridge, S. R., Cope, M., and Delpy, DT. (1997). Theoretical and experimental investigation of near-infrared light propagation in a model of the adult head. *Appl. Opt.* **36,** 21–31.

Okamoto, M., Dan, H., Sakamoto, K., Takeo, K., Shimizu, K., Kohno, A., Oda, I., Isobe, S., Suzuki, T., Kohyama, K., and Dan, I. (2004). Three-dimensional probabilistic anatomical cranio-cerebral correlation via the international 10–20 system oriented for transcranial functional brain mapping. *NeuroImage* **21,** 99–111.

Patterson, M. S., Chance, B., and Wilson, B. C. (1989). Time resolved reflectance and transmittance for the noninvasive measurement of tissue optical properties. *Appl. Opt.* **28,** 2331–2336.

Peña, M., Maki, A., Kovacic, D., Dehaene-Lambertz, G., Koizumi, H., Bouquet, F., and Mehler, J. (2003). Sounds and silence: An optical topography study of language recognition at birth. *Proc. Natl. Acad. Sci. USA* **100,** 11702–11705.

Pogue, B. W., Patterson, M. S., Jiang, H., and Paulsen, K. D. (1995). Initial assessment of a simple system for frequency domain diffuse optical tomography. *Phys. Med. Biol.* **40,** 1709–1729.

Pryds, O., Greisen, G., and Friis-Hansen, B. (1988). Compensatory increase of CBF supports the cerebral metabolism in preterm infants during hypoglycemia. *Acta Paediatr. Scand.* **77,** 632–637.

Ranganatha, S., Hoshi, Y., and Cuntai, G. (2004). Near Infrared Spectroscopy based Brain-Computer Interface, 3rd International Conference on Experimental Mechanics (ICEM), 29 Nov-1, Dec, 2004, Singapore.

Rea, P. A., Crowe, J., Wickramasinghe, Y., and Rolfe, P. (1985). Non-invasive optical methods for the study of cerebral metabolism in the human newborn: A technique for the future? *J. Med. Eng. Technol.* **9,** 160–165.

Saito, S., Yoshikawa, D., Nishihara, F., Morita, T, Kitani, Y., Amaya, T., and Fujita, T. (1995). The cerebral hemodynamic response to electrically induced seizures in man. *Brain Res.* **673,** 93–100.

Sakatani, K., Chen, S., Lichty, W., Zuo, H., and Wang, Y. (1999). Cerebral blood oxygenation changes induced by auditory stimulation in newborn infants measured by near infrared spectroscopy. *Early Hum. Dev.* **55,** 229–236.

Sato, H., Takeuchi, T., and Sakai, K. L. (1999). Temporal cortex activation during speech recognition: An optical topography study. *Cognition* **73,** B55–B66.

Sassaroli, A., Martelli, F., Tanikawa, Y., Tanaka, K., Araki, R., Onodera, Y., and Yamada, Y. (2000). Time-resolved measurement of *in vivo* optical properties of piglet brain. *Opt. Rev.* **7,** 420–425.

Schmidt, F. E. W., Fry, M. E., Hillman, E. M. C., Hebden, J. C., and Delpy, D. T. (2000). A 32-channel time-resolved instrument for medial optical tomography. *Rev. Sci. Instrum.* **71,** 256–265.

Schroeter, M. L., Zysset, S., Kupka, T., Kruggel, F., and Yves von Cramon, D. (2002). Near-infrared spectroscopy can detect brain activity during a color-word matching Stroop task in an event-related design. *Human Brain Map.* **17,** 61–71.

Shiga, T., Yamamoto, K., Tanabe, K., Nakase, Y., and Chance, B. (1997). Study of an algorithm based on Model experiments and diffusion theory for a portable tissue oximeter. *J. Biomed. Opt.* **2,** 154–161.

Shimada, S., Hiraki, K., Matsuda, G., and Oda, I. (2004). Decrease in prefrontal hemoglobin oxygenation during reaching tasks with delayed visual feedback: A near-infrared spectroscopy study. *Cogn. Brain Res.* **20,** 480–490.

Shinba, T., Nagano, M., Karia, N, Ozawa, N., Shinohara, T., Shimosato, S., and Hoshi, Y. (2004). Near-infrared spectroscopy analysis of frontal lobe dysfunction in schizophrenia. *Bio. Psychiatry* **55,** 154–164.

Simpson, C. R., Kohl, M., Essenpreis, M., and Cope, M. (1998). Near-infrared optical properties of *ex vivo* human skin and subcutaneous tissues measured using the Monte Carlo inversion technique. *Phys. Med. Biol.* **43,** 2465–2478.

Steinbrink, J., Wabnitz, H., Obrig, H., Villringer, A., and Rinneberg, H. (2001). Determining changes in NIR absorption using a layered model of the human head. *Phys. Med. Biol.* **46,** 879–896.

Suto, T., Fukuda, M., Ito, M., Uehara, T., and Mikuni, M. (2004). Multichannel near-infrared spectroscopy in depression and schizophrenia: Cognitive brain activation study. *Biol Psychiatry* **55,** 501–511.

Tanosaki, M., Hoshi, Y., Iguchi, Y., Oikawa, Y., Oda, I., and Oda, M. (2001). Variation of temporal characteristics in human cerebral hemodynamic responses to electric median nerve stimulation: A near-infra-red spectroscopic study. *Neurosci. Lett.* **316,** 75–78.

Toronov, V., Francecshini, M. A., Filiaci, M., Fantini, S., Wolf., M., Michalos, A., and Gratton, E. (2000). Near-infrared study of fluctuations in cerebral hemodynamics during rest and motor stimulation: Temporal analysis and spatial mapping. *Med. Phys.* **27,** 801–815.

Toronov, V., Webb, A., Choi, J. H., Wolf, M., Michalos, A., Gratton, E., and Hueber, D. (2001). Investigation of human brain hemodynamics by simultaneous near-infrared spectroscopy and functional magnetic resonance imaging. *Med. Phys.* **28,** 521–527.

Towle, V. L., Bolanos, J., Suarez, D., Tan, K., Grzeszczuk, R., Levin, D. N., Cakmur, R., Frank, S. A., and Spire, J. (1993). The spatial location of EEG electrodes: Locating the best-fitting sphere relative to cortical anatomy. *Electroencephalogr. Clin. Neurophysiol.* **86,** 1–6.

Villringer, A., Plank, J., Hock, C., Schleikofer, L., and Dirnagl, U. (1993). Near-infrared spectroscopy (NIRS): A new tool to study hemodynamic changes during activation of brain function in human adults. *Neurosci. Lett.* **154,** 101–104.

Wolf, M., Wolf, U., Choi, J. H., Gupta, R., Safonova, L. P., Paunescu, L. A., Michalos, A., and Gratton, E. (2002). Functional frequency-domain near-infrared spectroscopy detects fast neuronal signal in the motor cortex. *NeuroImage* **17,** 1868–1875.

Wolpaw, J. R., Birbaumer, N., McFarland, D. J., Pfurtscheller, G., and Vaughan, T. M. (2002). Brain-computer interface for communication and control. *Cin. Neurophysiol.* **11,** 767–791.

Wyatt, J. S., Cope, M., Delpy, D. T., Wray, S., and Raynolds, E. O. R. (1986). Quantification of cerebral oxygenation and haemodynamics in sick newborn infants by near infrared spectrophotometry. *Lancet* **ii,** 1063–1066.

Zaramella, P., Freato, F., Amigoni, A., Salvadori, S., Marangoni, P., Suppjei, A., Schiavo, B., and Chiandetti, L. (2001). Brain auditory activation measured by near-infrared spectroscopy (NIRS) in neonates. *Pediatr. Res.* **49,** 213–219.

Further Readings

Chance, B., Leigh, J. S., Miyake, H., Smith, D. S., Nioka, S., Greenfeld, R., Finander, M., Kaufmann, K., Levy, W., Young, M., Choen, P., Yoshioka, H., and Boretsky, R. (1988). Comparison of time-resolved and -unresolved measurements of deoxyhemoglobin in brain. *Proc. Natl. Acad. Sci. USA* **85,** 4971–4975.

Chance, B., Andy, E., Nioka, S., Zhou, S., Hong, L., Worden, K., Li, C., Murray, T., Ovetsky, Y., Pidikiti, D., and Thomas, R. (1998). A novel method for fast imaging of brain function, non-invasively, with light. *Opt. Express* **2,** 411–423.

Sevick, E. M., Frisoli, J. K., Burch, C. L., and Lakowicz, J. R. (1990). Localisation of absorbers in scattering media by use of frequency domain measurements of time dependent photon migration. *Appl. Opt.* **33,** 3562–3570.

NEURAL MODELING AND FUNCTIONAL BRAIN IMAGING: THE INTERPLAY BETWEEN THE DATA-FITTING AND SIMULATION APPROACHES

Barry Horwitz* and Michael F. Glabus[†]

*Section on Brain Imaging and Modeling, National Institute on Deafness and Other Communication Disorders, National Institutes of Health, Bethesda, Maryland 20892
[†]Laboratory of Biological Psychiatry, Department of Psychiatry, Louisiana State University Health Sciences Center, Shreveport, Louisiana 71130

This paper discusses the use of computational neural modeling with functional neuroimaging data, especially the data obtained from positron emission tomography and functional magnetic resonance imaging (fMRI) studies of human cognition. We point out that such data require network analysis to most thoroughly interpret them. Two types of modeling methods are discussed. The first is systems-level modeling, which attempts to determine the nodes of the neural networks mediating specific tasks and the strengths of the interactions between the nodes. The second is large-scale neural modeling, which uses large numbers of neurally realistic elements, arranged into different interacting populations and regions, to simulate the tasks of interest, the neural activities in the multiple populations, and the neuroimaging data (e.g., fMRI) in each region of the model for comparison with experimental data. We will lastly show that these two types of modeling can be combined and used to obtain insight into interregional neural interactions.

I. Introduction

Functional brain imaging has, in the last few years, become one of the most powerful tools for investigating the neural bases of cognitive, sensorimotor, and emotional behavior, especially in human subjects. Human functional

INTERNATIONAL REVIEW OF NEUROBIOLOGY, VOL. 66
DOI: 10.1016/S0074-7742(05)66009-6

neuroimaging methods, which include positron emission tomography (PET), functional magnetic resonance imaging (fMRI), and electroencephalography and magnetoencephalography (EEG/MEG), have generated a plethora of new data that have added considerable insight into the functional neuroanatomy of specific cognitive and other high-level functions. However, in spite of the extensive use of these methods, and in spite of the seemingly large number of new findings that have been published, our understanding of many functional-brain-imaging results has been limited in scope, and many studies have had major interpretational difficulties. The reason for this is that functional neuroimaging data possess two characteristics that *require* that computational neural modeling methods be utilized to more fully exploit the data that are generated, and, until relatively recently, such methods have rarely been so used.

The first aspect of these data that requires neural modeling comes about because functional neuroimaging methods, especially PET and fMRI, essentially provide data from most of the brain simultaneously. This is in contrast to the more classical methods that have been used to investigate the neural basis of human cognition: (1) neuropsychological investigations of brain-damaged patients (or lesion analysis in nonhuman animals) that attempt to determine the functional effects of lesions confined to specific brain areas; (2) electrophysiological microelectrode recordings from nonhuman animals (and, to a lesser extent, recordings from neurosurgical patients) that try to ascertain the response properties of neurons in specific brain areas to experimental tasks. These two methods essentially explore the functional specialization of single brain areas. Functional neuroimaging provides the opportunity for researchers to investigate not just what a single brain area does but also how brain regions work together during the performance of individual cognitive tasks. Therefore, functional neuroimaging data encourage network analysis—and network analysis, because of the intrinsic complexity of dealing with interacting elements, necessitates computational neural-modeling methods.

The second characteristic of functional brain-imaging data that calls for neural modeling is their spatiotemporal integrative nature. That is to say, the data generated by functional brain imaging, especially the fMRI and PET methods, represent spatiotemporal scales that are many times larger than those associated with individual neurons; EEG and MEG have temporal scales comparable to that at the neural level, but their spatial resolution is poorer and harder to specify than that of fMRI and PET.

This is a general problem for all areas of biology, but it is particularly acute for neuroscience, where we now have the ability to acquire large quantities of data at multiple spatial and temporal scales of investigation. The spatial scales range all the way from the whole brain to the subcellular or molecular dimension, and the temporal dimensions span the range from days, months, and even years to submillisecond intervals. Because there is no one method that transcends all the different levels, it is difficult for investigators to interpret their data in terms of a single, unified account relating specific behaviors to their underlying neural mechanisms. In order to advance, it will be essential to develop systematic ways to bridge these levels, so that we can integrate neuroscientific information across them and thereby generate coherent and unified accounts of the phenomena that are being investigated. Our contention is that neural modeling provides a way to bridge these levels (Horwitz, 2005; Horwitz *et al.*, 1999, 2000).

In the last few years, there have been various efforts to incorporate computational neural modeling into the analysis and interpretation of functional brain-imaging data. Specifically, a number of explicit network analysis methods, constituting a neural-network approach to functional neuroimaging, have appeared. We have called this type of modeling, with its emphasis on evaluating the interregional effective connectivity, *systems-level modeling* (Horwitz *et al.*, 1999, 2000). A second type of modeling—*large-scale neural modeling*—has recently begun to be used to relate the functional brain imaging signals corresponding to specific tasks to their underlying neural substrates, thus providing an explicit way to integrate neuroscientific data across multiple spatiotemporal scales. The first method stresses fitting experimental functional neuroimaging data so as to obtain the values of the model parameters—the interregional effective connectivities. The second method stresses simulation: experimental functional neuroimaging data are predicted via simulations from a model constructed in terms of neuronal elements.

In this paper, we will present examples of both of these methods in which neural modeling can be used in conjunction with functional neuroimaging data. These examples will emphasize some of the specific neuroscientific questions that these methods enable one to address. Through these examples we hope to demonstrate how computational modeling can be used to more fully exploit the data and how it can be used to lessen some of the interpretational problems that exist using conventional data-analysis methods. In the penultimate section, we will indicate how these two types of modeling can be related to one another.

II. Systems-Level Modeling

Two fundamental assumptions govern how functional neuroimaging data are used to make inferences about which brain processes are involved (Horwitz, 1994). The first, functional specialization, says that different brain regions are engaged in different functions, and it is implemented by comparing the functional signals between two scans (in its most simple formulation), each representing a different experimental condition or group. The loci of the large signal differences between the two scans are assumed to delineate the brain regions differentially involved in the two conditions or groups. The second assumption, functional integration, leads to the covariance paradigm (Horwitz *et al.*, 1992b), which asserts that a task is mediated by a network of interacting brain regions and that different tasks utilize different functional networks (see also McIntosh, 2000). It is assumed that, by examination of some measure of the covariance in brain activities between different pairs of brain areas, one can infer something about which areas are important nodes in the network under study and about how these nodes are functionally connected.

Central to the covariance paradigm is the concept of *functional connectivity*, which is used in the context of neuronal processes to allude to the functional interactions between different areas of the brain, although different authors have simultaneously ascribed different meanings to this term (Horwitz, 2003). Time-series functional connectivity has significance in the study of temporally correlated activity, and Friston (Friston *et al.*, 1993) had this in mind when quoting the term with regard to neurophysiological events occurring between distant, discrete brain regions. Earlier, the term *functional association* was used in the context of PET neuroimaging to describe functional (metabolic or hemodynamic) interactions between regions (for example, in the visual system [Horwitz *et al.*, 1992a]). In all cases, the formal mathematical quantification of these functional relationships depends on the use of interregional covariance or correlation coefficients (e.g., Friston, 1994).

Defining the interaction between two brain regions via their functional connectivity is appealing due to its inherent simplicity. The investigator makes a series of measurements of brain activity (hemodynamic, metabolic, neurophysiological, etc.) at predetermined regions of interest, in response to a cognitive, motor, or sensory demand, and then derives a measure of coherent activity using, for example, the interregional correlation coefficient. Alternatively, one can employ a voxel-based approach, wherein the investigator selects a single seed voxel (or region) and determines the magnitude of the functional connectivity with all other voxels (regions). No assumptions are made, or required, about whether the interaction represents a direct or an indirect influence from other concurrently activated (or deactivated) regions of the brain (Horwitz *et al.*, 1999).

Evaluating functional connectivity can be quite helpful when little is known about the specific networks mediating a task or when the underlying neuroanatomy is not clearly defined. For example, one of the earliest examples of its use was a PET neuroimaging study showing functional disconnection of neocortical areas in Alzheimer's disease (Horwitz *et al.*, 1987). A more recent example revealed the influence of top-down processes in animations of simulated biological motion in a PET study of mental attribution (Castelli *et al.*, 2002). Here, the authors identified a cortical network from other studies of mentalizing that involved three brain regions, and they formed a hypothesis regarding a functional connection between two of them: the extrastriate cortex region V3 and the temporoparietal junction. Castelli *et al.* (2002) were able to demonstrate a clear deficit in the functional connection between these two regions, possibly explaining a behavioral deficit in individuals with autism.

Extending a functional connectivity analysis beyond pairs of regions to include multiple brain regions is superficially straightforward, and it is valuable in helping to identify both patterns of brain activity and the ways these patterns might differ across different conditions or groups of subjects (Bullmore *et al.*, 1996; Friston *et al.*, 1993; Meyer-Lindenberg *et al.*, 2001). Like pairwise correlational analysis, this descriptive approach tells us something about patterns of functional interactions between brain regions and how these patterns differ between experimental conditions or groups, but it doesn't tell us explicitly which regions are directly influencing others (e.g., along specific anatomical pathways); it usually simply shows some kind of indirect interaction. To do this, some type of systems-level neural modeling needs to be employed (Friston, 1994; Horwitz, 1994; McIntosh and Gonzalez-Lima, 1994).

The term *effective connectivity* has come to mean, at the systems level, the direct influence of one neural population on another (Friston, 1994). We think of it as meaning the functional strength of a particular anatomical connection; during one task, an anatomical link could have a stronger effect than during a second condition, and likewise, such a link might be stronger in normal subjects than in a patient group. The evaluation of effective connectivity requires modeling because one needs to select a small group of brain regions to include in the network, and one needs to combine functional neuroimaging data with information about the anatomical linkages between these regions. That is, the validity of effective-connectivity modeling builds on a combination of implied anatomical and functional connections between brain regions. Effective-connectivity modeling techniques such as structural equation modeling (SEM, also called path analysis) are hypothesis-driven methods normally used in a confirmatory rather than an exploratory manner. SEM, when applied to functional neuroimaging data, attempts to account for hypothesized interactions between specified dependent variables (brain regional activities) but can also account for extraneous contributions from regions outside the model. The variance–covariance matrix constructed from the observed measures is tested against an implied model, and minimization of the error term between the two is used as a guide to parsimony of model fit. The model will generate connection strengths for each specified path between the different brain regions in the model. This can be done in response to different experimentally manipulated conditions within one subject, or in response to a single condition across groups of subjects. The crucial types of questions addressed by effective-connectivity analysis focus on how cognitive tasks are mediated by specific anatomical pathways and how the functional strengths of these connections differ between tasks or groups.

SEM was first applied to neuroimaging in a study of the startle response in rats (McIntosh and Gonzalez-Lima, 1991). The first application to a human neuroimaging study used PET data to examine the cortical visual pathways involved in object and spatial vision (McIntosh *et al.*, 1994). Using this method, the authors were able to discriminate between object and spatial vision processes by different values of the effective connections in the occipitotemporal and occipitoparietal areas of the right hemisphere. Perhaps one of the most elegant early demonstrations of SEM applied to PET neuroimaging data was a study of patients with idiopathic Parkinson's disease before and after globus pallidotomy, based on a model of anatomical connections and presumed physiological connections in a model of motor function (Grafton *et al.*, 1994). Here, the authors were able to show how the presumed interregional physiologic connections of the network responded to the removal of one node of the network after surgery. SEM has also shown value in predicting how effective connections change between the dorsal and ventral pathways in response to associative learning of objects and their locations (Buechel *et al.*, 1999). Effective-connectivity modeling has also been used to assess the effect of drugs on object-location learning (Honey *et al.*, 2003) and the effect of different classes of antipsychotic medication on the cortical working-memory network of schizophrenic patients (Schlosser *et al.*, 2003). Though this method of network analysis has usually been applied to groups of subjects, it has also been shown valuable for studying the cortical networks of individual subjects, for example, in an fMRI study of spatial attention (Buechel and Friston, 1997). Differences in cognitive strategy in individual subjects were also found to be reflected in intersubject variability in effective connectivity in a word and pseudo-word reading task (Mechelli *et al.*, 2002). This last fMRI study was notable in that it attempted to model the common influence of experimental variables by connecting a virtual node to an input region in all the single-subject SEM models. This approach therefore allows the modeling of both endogenous variance and exogenous effects induced by experimental design.

A more recent PET neuroimaging study of two-back working memory used single-subject SEM models to determine the neurobiological substrate for task performance and task strategy (Glabus *et al.*, 2003). In this two-back working memory task, subjects were presented with a diamond-shaped figure. A series of single numbers (1, 2, 3 or 4) appeared at random in one of the four corners (changing randomly for each item), but the same number (e.g. 2) always appeared at the same spatial location (i.e. left corner). Subjects were asked to remember the current number and to press one of four buttons on a diamond-shaped button box corresponding to the stimulus presented two-before. Because of its design, this task could be performed using either verbal working memory (keeping track of the numbers) or spatial working memory (keeping track of the spatial location at which each item appears) or some combination of both. Verbal working memory is thought to be mediated primarily by the left hemisphere, whereas spatial working memory has been associated with the right. A group working-memory SEM model was constructed, and individual-subject working-memory models were evaluated and tested against the group model. The authors

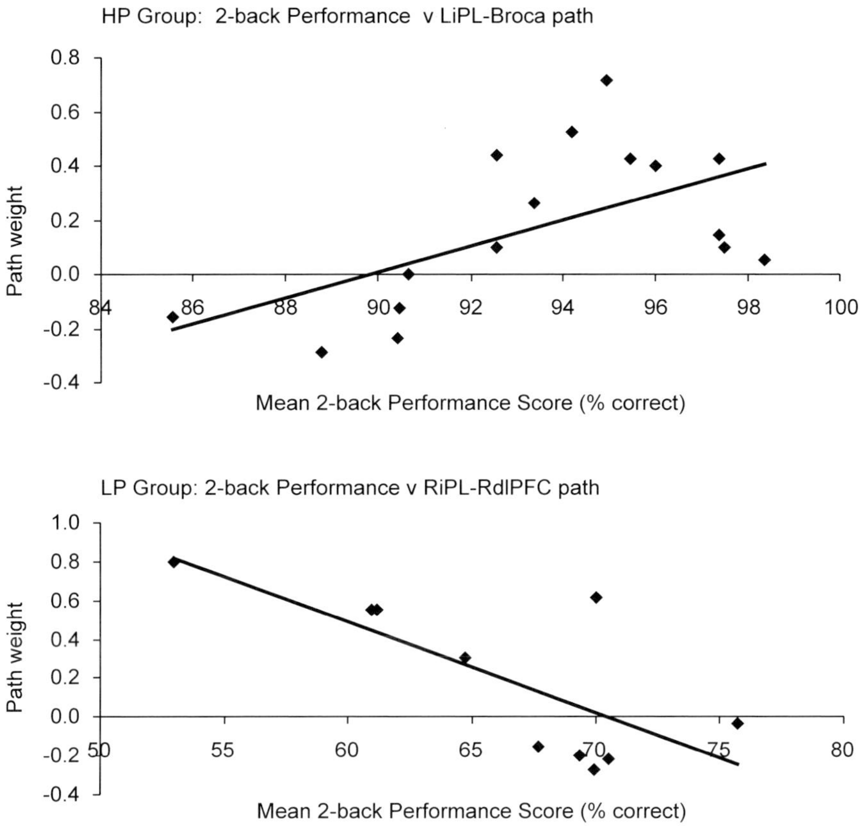

FIG. 1. Significant correlations of working memory performance with path weights for the high-performance group and low-performance group. Upper: correlation of two-back performance score (percent correct) against the path weight between the inferior parietal lobule in the left hemisphere and Broca's region in the high-performance group ($n = 16$, $r = 0.584$, $p = 0.018$). Lower: correlation of two-back performance score (percent correct) against the path weight between right inferior parietal lobule (RiPL) and right dorsolateral prefrontal cortex (Rd1PFC) in the low-performance group ($n = 12$, $r = -0.737$, $p = 0.015$). From Glabus *et al.* (2003).

were able to show that subjects that performed the task well (high performers) had a stronger effective connection between the left inferior parietal lobule and Broca's area and that their performance correlated positively with the strength of the effective connection (Fig. 1, upper part). Individuals that performed the task poorly (low performers), on the other hand, had a strong effective connection between the right inferior parietal lobule and a right dorsolateral prefrontal region (Brodmann area 9/46) and, most importantly, showed a negative correlation between effective connection strength and performance score (Fig. 1, lower part). Furthermore, the analysis of task strategy showed an association between

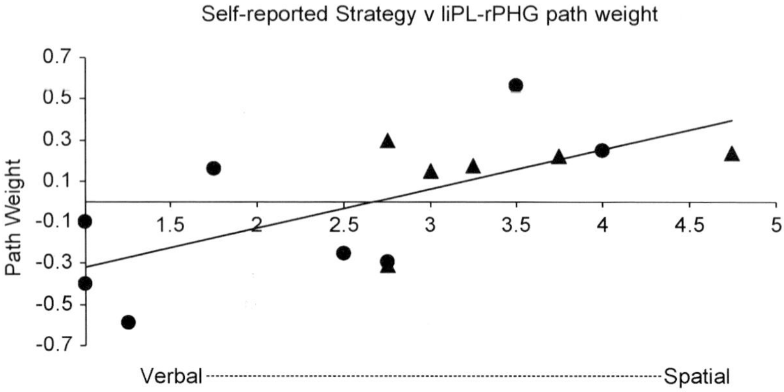

FIG. 2. Correlation of working memory self-reported strategy with the path coefficient weight connecting the parahippocampal gyrus in the right hemisphere (rPHG) to the inferior parietal lobule in the left hemisphere (liPL). Circles denote subjects in the high-performance (HP) group; triangles denote subjects in the low-performance (LP) group. There was no correlation with path weight and strategy for the LP group; for the HP group, $r = 0.755$, $p = 0.03$; for the combined group, $r = 0.66$, $p = 0.01$; for all subjects who reported their strategy, $n = 14$. From Glabus *et al.* (2003).

strategy and the path connecting the right parahippocampal gyrus and the left inferior parietal lobule: subjects who used an increasingly verbal strategy had a negative path weight, whereas subjects who used a spatial strategy showed an increasingly positive path weight (Fig. 2). This analysis also showed that high performers tended to favor a verbal strategy whereas low performers favored a spatial strategy. The implications of these findings demonstrate the power of effective-connectivity analysis to illuminate the neural underpinnings of individual behavior. Specifically, the results from the low performers suggest that the more that subjects used a spatial strategy, the worse their performance on the task, and that the neuroanatomical substrate for this centered on the parietal lobe and its connections.

There are a number of problems with the use of SEM to model functional integration, including some related to measures of "goodness of fit" of a specific model (Bullmore *et al.*, 2000). A particularly important problem is that the technique does not work well with time-varying changes in effective connectivity because SEM reflects a "steady-state" set of correlations between regions, which would be difficult to apply to data from event-related fMRI designs or from studies involving endogenous changes in the state of the subject being scanned. Analysis of time-dependent effective connectivity using SEM would require the construction and testing of numerous models, one for each time point, by the generation of observed versus expected variance–covariance matrices, and this is likely to be impractical. An attempt to address this limitation was made using variable parameter regression in an fMRI study of attentional modulation

(Buechel and Friston, 1998). The authors used this method to model nonlinear effective connectivity in a predictive manner. Another significant limitation is that effective-connectivity models of functional integration model changes in observed hemodynamics (eg., blood oxygenation level–dependent [BOLD] activity with fMRI or regional cerebral blood flow [rCBF] with PET), and not changes in the underlying neural activity, potentially giving rise to problems in interpretation of the results. We will discuss this further in the penultimate section of this article.

One way to deal with some of these issues can be found in dynamic causal modeling (DCM), which is a neuromodeling technique for effective-connectivity evaluation that discriminates between changes at the hemodynamic and neuronal levels (Friston *et al.*, 2003; Penny *et al.*, 2004). DCM is also a confirmatory rather than an exploratory technique, like SEM, with results that relate solely to a specific experiment. DCM relies on a model of known anatomical and presumed physiological connections and a model relating neural activity to the hemodynamic signal. A model of intrinsic connections is constructed that specifies the directionality of interregional connections, and the results of a DCM analysis specify the values of the changes in these effective connections as a function of the changing experimental conditions (which would be equivalent to changes in the path coefficients in SEM, if the experimental design were conducive to an SEM analysis and if we could ignore the effects of the hemodynamic response on the results—see Section IV). Additionally, input connections are specified that signify connections to particular brain regions and modulatory connections are specified that indicate which intrinsic connections can be changed by which inputs (Penny *et al.*, 2004). DCM can be used in the same manner as SEM to answer specific questions relating to changes in effective connectivity in experiments that occur across conditions or across groups of subjects. However, DCM will accommodate both the nonlinear and the dynamic aspects of neuronal interactions. SEM assumes linear interactions that are of steady state and thus are not time-series models.

An elegant demonstration of the power of DCM was shown in a study of top-down and bottom-up processing in visual perception and imagery (Mechelli *et al.*, 2004). Here, the authors developed a model of visual perception and imagery based on three different classes of objects: houses, chairs, and faces. The principal nodes of the model covered brain regions in the fusiform gyrus, prefrontal cortex, inferior occipital cortex (V3), and superior parietal cortex. The authors were able to demonstrate that bottom-up processes dominate image viewing, based on evidence of increased connectivity from the inferior occipital cortex to category-specific brain regions. When subjects imagined the same category of objects, there was increased connectivity between the prefrontal cortex and the same category-specific regions in the inferior temporal and fusiform gyri.

In summary, DCM appears to have several advantages over SEM, especially when applied to fMRI data, and there are also situations in which SEM cannot be used to evaluate data sets but DCM can (Friston *et al.*, 2003). However, there are some limitations on the use of DCM. In the current DCM implementation in the SPM2 software, the number of regions that can be included in a model is limited to eight, and only a single state variable can be used to signify the regional neurodynamics. The computation time for DCM models is also relatively lengthy compared to that required to compute equivalent SEM models. Finally, although DCM attempts to formulate its findings in terms of interactions between neural activities in different brain regions, deconvolution of hemodynamic signals results in only a limited increase in the temporal information that is available. We will see in the next section how large-scale modeling hopes to address problems of this sort.

III. Large-Scale Neural Modeling

Large-scale neural modeling uses computer techniques to generate and simulate neural data at the neuronal level and the systems level that can be compared, respectively, with experimental data obtained by single-unit electrophysiology and by functional neuroimaging methods (e.g., Arbib *et al.*, 1995; Deco *et al.*, 2004; Tagamets and Horwitz, 1998). The neurons that are used in the models are constructed in a neurobiologically realistic way, and the models are configured so that they can perform the same types of tasks as are used experimentally. Generally, a model consists of multiple neuronal populations representing different interconnected brain areas. (For a review, see Horwitz, 2004; Horwitz *et al.*, 2000.)

The principal reason for employing large-scale modeling with functional neuroimaging data is that it provides a method by which to formulate and delineate specific hypotheses, expressed in neurobiological terms, about how interacting neural populations instantiate high-level cognitive and sensorimotor tasks in humans, and to generate simulated data based on these hypotheses that can be compared directly to experimental results, particularly those acquired using functional brain imaging. This is what we will discuss in this section, but it is worth mentioning that this type of modeling can also serve other conceptually distinct purposes: (1) determining how well an experimental design paradigm or analysis method works (e.g., assessing how well DCM expresses the underlying pattern of neural interactions [Lee *et al.*, submitted]); (2) investigating the meaning in neural terms of a macro-level concept (e.g., functional connectivity [Horwitz *et al.*, 2005], which we shall illustrate in Section IV of this paper); (3) combining different types of data with each other (e.g., fMRI and transcranial magnetic stimulation data) (Husain *et al.*, 2002). Note that, in practical terms, these conceptually distinct reasons often merge.

To illustrate the key question that large-scale modeling addresses—how do neurons interact so that specific cognitive and sensorimotor tasks can be implemented?—we will use related models that were developed from visual (Horwitz and Tagamets, 1999; Tagamets and Horwitz, 1998) and auditory (Husain *et al.*, 2004) object processing. These models allow us to simultaneously generate both neuronal and fMRI (or PET) simulated data, as well as to compare both of these to experimentally acquired data. With this approach we are able to formulate specific hypotheses about how auditory and visual objects are processed in the cortex, and to test these hypotheses against data obtained at different spatio-temporal scales. These large-scale neural network models are based on neurophysiological and neuroanatomical data from primate and human studies. Both models perform delayed match-to-sample (DMS) tasks: a visual DMS task for object shape (Horwitz and Tagamets, 1999; Tagamets and Horwitz, 1998) and an auditory DMS task for tonal contours (Husain *et al.*, 2004). A key point about these two models is that they are very similar in their structures, as will be explained in a bit. Because of this, we are essentially positing that visual and auditory (and possibly tactile) object processing all use a set of similar cortical computational mechanisms along each of their respective pathways, although the features on which these mechanisms act depend on the sensory modality. Note also that we are not implying that all sensory features have analogues in the three systems, only that some do.

We will start with the visual object model for which the experimental basis is much stronger than that for the auditory object model. Four major brain areas are included in the model, which collectively represent the ventral object-processing stream (Ungerleider and Mishkin, 1982): (1) primary sensory cortex (V1/V2); (2) secondary sensory cortex (V4); (3) a perceptual integration region (inferior temporal (IT) cortex); and (4) prefrontal cortex (PFC), which plays a key role in short-term memory (Fig. 3). The main feature processed by the neuronal units in the primary and secondary visual cortices is line orientation; other possible stimulus attributes, such as color and texture, are not covered by the model.

An analogous processing stream along the superior temporal gyrus for auditory object processing has been hypothesized by Kaas, Rauschecker and Tian, and others (Kaas *et al.*, 1999; Rauschecker and Tian, 2000). Some examples of auditory objects are words, melodic fragments, and definable environmental sounds. However, experimental support for the auditory object pathway is much weaker than for the visual object pathway. The large-scale model that was generated (Husain *et al.*, 2004) implements this hypothesis by explicitly indicating the neural mechanisms by which auditory object processing would occur in each module comprising the network. Like the visual model, the auditory model contains four major modules: (1) the primary sensory cortex (A1/A2), where neurons respond in a tonotopically organized fashion to up-and-down frequency

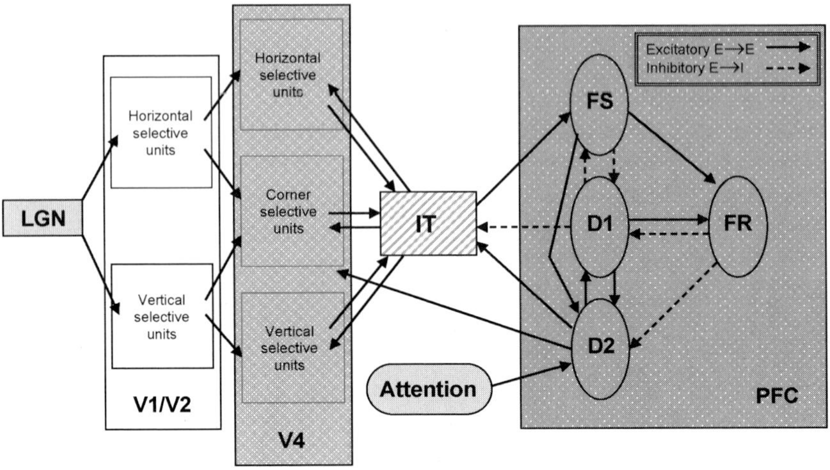

FIG. 3. Network diagram of the visual object–processing model (Tagamets and Horwitz, 1998). The regions of the model (V1/V2, V4, IT, PFC) form a complex network of feedforward and feedback connections; these interregional connections can be either excitatory (excitatory-to-excitatory elements, shown as solid lines) or inhibitory (excitatory-to-inhibitory elements, shown as dashed lines). In the PFC region, there are four submodules: FS contains stimulus-sensitive units, D1 and D2 contain units active during the delay part of a delayed match-to-sample task, and FR contains units whose activity increases if there is a match between the first and second stimuli of a trial. Modified from Horwitz *et al.* (2005), with permission.

sweeps (which are treated as the basic features of the auditory stimuli; other features, such as timbre, are not included in the model); (2) secondary sensory cortex (e.g., lateral and parabelt auditory cortex), where neurons respond to longer up-and-down sweeps, as well as to changes in the direction of frequency sweeps; (3) an area (ST) similar to IT where a complex sound is represented in a distributive manner in the neuronal population; and (4) a prefrontal module analogous to that in the visual model. The experimental evidence for the presence of neurons with these properties has been reviewed by Husain *et al.* (2004).

Every region in the two models is composed of multiple excitatory–inhibitory units, each of which represents a cortical column. Regions are connected by both feedforward and feedback links. In the visual model, as is the case experimentally, there are different scales of spatial integration in the first three stages; the primary sensory region has the neurons with the smallest spatial receptive fields, and the inferior temporal cortex has the neurons with the largest. For the auditory model, it was hypothesized that the temporal receptive field (i.e., the time window of integration) increases as one goes from the primary auditory region to the ST region. The time step used in the models corresponds to approximately 5 msec.

The two models use an identical method to implement the "task instructions" so that the models know whether to perform the DMS task or a control task: a continuous "attention" or "biasing" variable that modulates a subset of prefrontal units by means of diffuse synaptic inputs. The prefrontal units, through feedback connections, modulate activity in posterior areas. The strength of the biasing variable controls whether the stimuli are maintained in short-term memory. The net effect of these arrangements is that activity in each brain area is a mixture of feedforward activity (bottom-up), determined in part by the presence of an input stimulus; feedback activity (top-down), determined in part by the strength of the modulatory attention signal; and local activity within each region.

An important point, indeed, the governing philosophy upon which this type of large-scale modeling rests, is that the assumptions made in constructing these models are about the behavior of elements at the neuronal level. All the macro-level behaviors that the models show are emergent phenomena of the networks. Thus, this approach is very much a "bottom-up" approach. In practice, this means that the parameters controlling the behavior of the model are chosen so that the excitatory elements in each area have simulated neuronal activities resembling, when available, those found in electrophysiological recordings from monkeys performing similar tasks (see, e.g., Funahashi *et al.*, 1993; Kikuchi-Yorioka and Sawaguchi, 2000).

We can simulate functional neuroimaging studies by presenting stimuli to an area of the model representing the lateral geniculate nucleus (LGN) for the visual model and the medial geniculate nucleus (MGN) for the auditory model. Something like 10 such trials might be employed. The fMRI activity is simulated by, first, temporally and spatially integrating the absolute value of the synaptic activity in each region over a time period representing 50–100 msec, which corresponds to the time needed to collect a single slice of fMRI data. Then the time-integrated synaptic activities are convolved with a function representing hemodynamic delay and are subsequently sampled at a time corresponding to the repetition time (TR) of an fMRI scan (Horwitz and Tagamets, 1999). If the simulated results were to correspond to PET data, the absolute synaptic activity would be integrated over the entire set of trials representing a single experimental condition. The absolute value of the integrated synaptic activity is used because it has been found that the fMRI signal (and the rCBF) more closely follows the local field potentials (i.e., the input to neurons) than it does the spiking activity (i.e., the neuronal output) (Lauritzen, 2001; Logothetis *et al.*, 2001).

Both models were initially evaluated by comparing simulated results to experimental PET and fMRI data and to primate (or mammalian) neuronal activity. The simulated neural activities for the visual model were similar to neuronal data found in monkey electrophysiological studies. For the auditory model, primate electrophysiological data were not always available, but Husain *et al.* (2004) showed that the animal literature was compatible with our

TABLE I

COMPARISON OF SIMULATED & EXPERIMENTAL DATA FOR VISUAL AND AUDITORY MODELS

Region-visual model	PSC-simulation	PSC-experiment	Region-auditory model	PSC-simulation	PSC-experiment (right hemisphere)	PSC-experiment (left hemisphere)
V1/V2	3.1%	2.7%	A1/A2	0.6%	30%	18.5%
V4	5.2%	8.1%	Secondary auditory cortex	26.5%	41.2%	40.5%
IT	2.5%	4.2%	ST	50.1%	45.2%	28.8%
PFC	3.5%	4.1%	PFC	103.4%	84.6%	16.7%

Shown are the regions of each model (visual and auditory) and the percent signal change (PSC) between the delayed match-to-sample (DMS) task compared to a control task. For the simulated visual task, the DMS condition corresponded to a matching task for shape and the control task was a passive presentation of degraded shapes (see Tagamets and Horwitz, 1998, for details). Experimental data come from a DMS task for faces compared to a passive viewing task (Haxby *et al.*, 1995). Details as to how PSC values were computed are found in (Tagamets and Horwitz, 1998). For the auditory case, simulated and experimental values for the auditory case are presented in detail in (Husain *et al.*, 2004). Stimuli used for DMS task were tonal contours, each consisting of an up and a down frequency sweep, separated by a constant tone (the duration of each tonal contour was 350 msec, and the delay period was 1 sec). The control task consisted of the same DMS task, except that the stimuli were simple tones. Regions of interest corresponding to the modules of the model at local maxima (for each stimulus type relative to rest) in each hemisphere were selected. These results indicate a close match (except for A1/A2, which includes primary auditory cortex) for regions in the right hemisphere rather than the left. Table modified from (Horwitz, 2004).

hypotheses. Moreover, the simulated PET values for the visual model of the DMS condition, when compared to a control condition, were similar (Tagamets and Horwitz, 1998) to those found in experimental PET studies of face working memory (Haxby *et al.*, 1995), as shown in Table I. For the auditory model, we could generally match quantitatively the simulated fMRI activities to our own fMRI findings in the temporal and frontal cortex during a DMS task for tonal contours (compared to a DMS task for pure tones) (Husain *et al.*, 2004). Thus, both models were able to generate simulated electrophysiological data and simulated PET/fMRI data that generally were in close agreement with experimental data. Importantly, these quantitative agreements strongly argue for the specific hypotheses that were made concerning the neural mechanisms by which multiple interacting brain regions implement these DMS tasks.

Husain *et al.* (2005) went on to test the robustness of the auditory model by applying it to stimuli that were not used in its original construction. They investigated one type of auditory perceptual grouping phenomena, the auditory continuity illusion. Perceptual grouping enables the auditory system to integrate brief, disparate sounds into cohesive perceptual units. This is important for perception because it allows, for example, the separation of attended sounds from environmental noise. The auditory continuity illusion results when a sound object (e.g., pure tone, frequency glide, word) is perceived to continue through occluding noise even when no such signal is physically present in the noise. It serves the especially important purpose of making communication sounds intelligible in a noisy environment. Even though some perceptual grouping phenomena, such as the auditory continuity illusion, have been extensively studied by means of psychophysical experiments, little is known concerning their neural bases.

In the simulations performed by Husain *et al.* (2005), intact stimuli (tonal contours) were matched with fragmented versions (i.e., with inserted silent gaps) of the stimuli. The ability of the model to match fragmented stimuli declined as the duration of the gaps increased (Fig. 4). However, when simulated broadband noise was inserted into these gaps, the matching response was restored, indicating that a continuous stimulus was perceived. The electrical activities of the neuronal units of the model agreed with electrophysiological data (Sugita, 1997), and the behavioral activity of the model matched human behavioral data (e.g., Ciocca and Bregman, 1987; Dannenbring, 1976). In the model, the predominant mechanism implementing temporal induction is the divergence of the feedforward anatomical connections along the auditory processing pathway in the temporal cortex. These simulation results not only attest to the robustness of the model but predict the primary role of the anatomical connectivity of the auditory processing areas in mediating the continuity illusion.

It should be mentioned that another area of modeling that has started to appear is based on efforts to relate cognitive models to functional brain-imaging data. Cognitive models make assumptions about the types of functions that mediate a cognitive task; further assumptions are required to relate each cognitive function to specified brain regions. For example, a well-known cognitive model of reading (Coltheart *et al.*, 2001) includes, among others, modules for visual analysis and for grapheme–phoneme conversion. One of the earliest attempts at combining these types of models with fMRI data can be found in a paper by Just *et al.* (1999), who used a computational model of sentence comprehension called 4CAPS (a production model) to account for the way fMRI activation levels varied as a function of sentence complexity in three brain areas (Broca's, Wernicke's, and the dorsolateral prefrontal cortex). In their computational model, Just *et al.* proposed that resource utilization in a given unit of time in each component of the system may correspond to the amount of activation observed with the neuroimaging measure in the corresponding component during that time interval. Good agreement between the experimental number of activated voxels in Broca's and Wernicke's areas and the prediction of their model for three types of sentences of different complexity was found.

A second example of this style of modeling was provided by Anderson *et al.* (2003). This group examined symbol manipulation tasks using a model

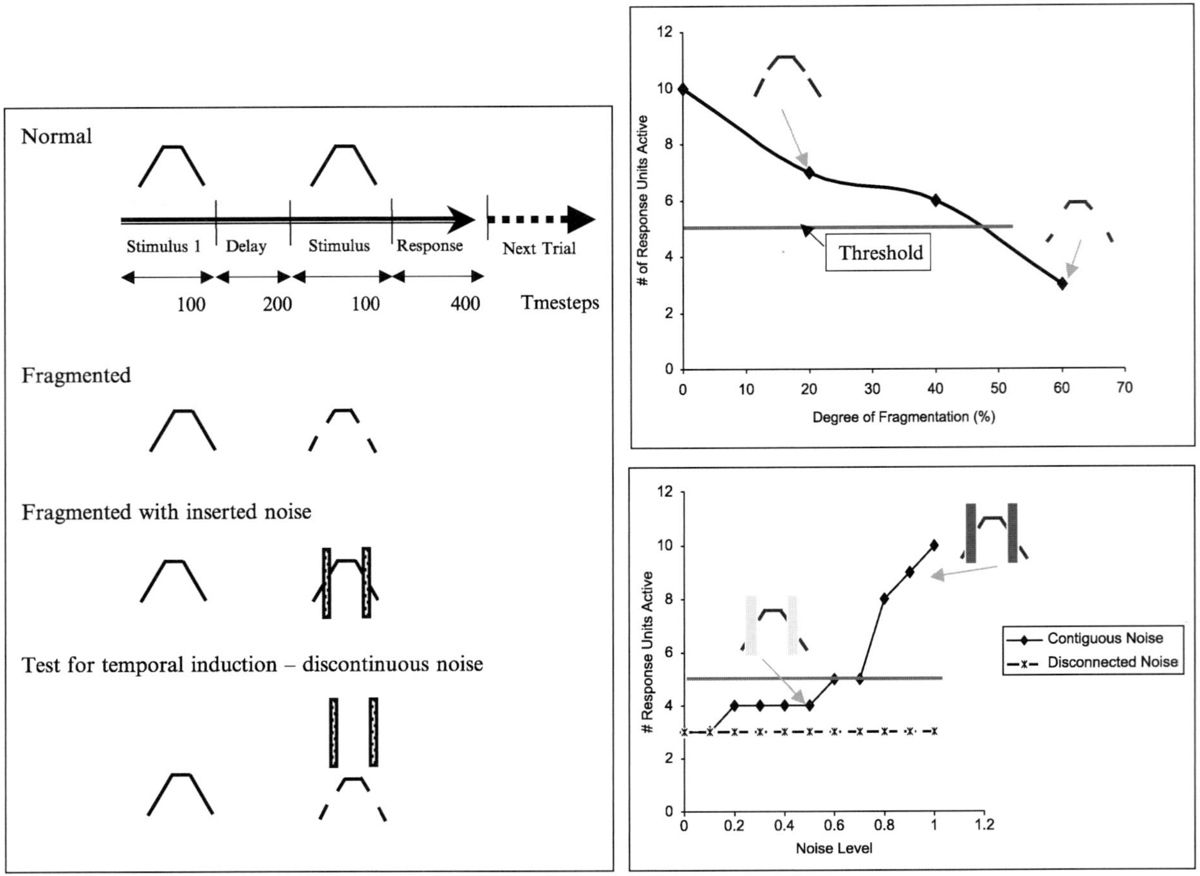
Normal
Stimulus 1
Delay
Stimulus
Response
Next Trial
100
200
100
400
Tmesteps
Fragmented
Fragmented with inserted noise
Test for temporal induction – discontinuous noise
# of Response Units Active
Threshold
Degree of Fragmentation (%)
Response Units Active
Contiguous Noise
Disconnected Noise
Noise Level

called ACT-R, which contains a number of buffers. Somewhat differently from the assumption used by Just *et al.* (1999), Anderson *et al.* suggested that the BOLD response in a brain area represents the integrated duration of time that a buffer is active. They showed that calculated BOLD activity in one buffer of the model (the imaginal buffer) predicted the BOLD response of a left parietal region; that activity in a second buffer (the retrieval buffer) predicted activity in a left prefrontal region; and that activity in a third buffer of the model (the manual buffer) was related to BOLD activity in a motor region.

One difference between the approaches of Just and Anderson is that Anderson *et al.* (2003) assume that each module of their model corresponds to a given brain region. Just *et al.* (1999) also ascribe specific cognitive functions to different brain regions, although they assume that more than one cognitive function can occur in a region and, conversely, that a given cognitive specialization may occur in more than one area, albeit with a different degree of efficiency in each area. Both approaches are different from the large-scale modeling method of Horwitz and colleagues (2004) discussed earlier; in these neurobiologically inspired models, the assumptions about what brain regions do are phrased at the neural, not the cognitive, level. As a result, "cognitive functions" arise as emergent phenomena and are not imposed onto the model. Furthermore, assumptions about the relationship between neural activity and fMRI activity are also expressed in neural, not cognitive or functional, terms. The advantage of this last approach (a bottom-up approach) is that it allows one to relate a cognitive function to its neural implementation. One disadvantage, at least at the present time, is that only relatively low–level cognitive functions can be addressed, since nonhuman models of high-level human cognitive function (e.g., language) do not exist. The use of cognitive models with fMRI data (a top-down approach) allows one to deal with very-high-level cognitive phenomena (e.g., sentence processing, symbol manipulation) but gives little indication as to how such functions are implemented at a neural level. For the time being, both

---

FIG. 4. Left: Illustration of the auditory continuity illusion. Shown at the top is the delayed match-to-sample (DMS) task for a tonal contour (indicated as a frequency-versus-time pattern; each tonal contour is about 350 msec in duration). Fragmented tonal contours, with and without inserted noise, are shown on the bottom. If the noise spans the frequency range of the contour and is loud enough, the fragmented tonal contour is perceived as intact. Right: Performance of the auditory model of Husain *et al.* (2004). Above-threshold activity of 5 or more neurons in the response module (FR) indicates a match between the two stimuli of a DMS trial. The top graph shows that with a short duration gap in the second stimulus, the model (like actual subjects) indicates a match, thus grouping the parts of the tonal contour into a perceptual whole. As the gap widens, a non-match results. If noise is inserted in the gap, and is of weak intensity (yellow), the tonal contour is not considered to continue through the gap (bottom right); if the noise is more intense (red), then perceptual grouping occurs, but only if the band of noise is in the part of frequency space occupied by the tonal contour. See Husain *et al.* (2005) for details. (See Color Insert.)

approaches will be useful and will provide interesting insights, and the cognitive model approach may, when successful, generate a set of target functions at which neurally based large-scale modeling can aim.

IV. Relating Large-Scale Neural Modeling to Systems-Level Modeling

In Section II, we discussed how the concepts of functional and effective connectivity are used with functional neuroimaging data to assess how pairs of brain regions interact during different cognitive tasks. It was mentioned that there are a number of definitions and algorithms that have been used to specify how one computes the functional or effective connectivity (for reviews, see Horwitz, 2003; Horwitz *et al.*, 2005). Most of these definitions and algorithms seem reasonable, even though they often differ as to the particular aspects of region–pair covariability they are accessing. For example, a number of functional and effective connectivity studies using fMRI data have correlated the time courses of the fMRI signal in different brain regions (e.g., Buechel and Friston, 1997; Bullmore *et al.*, 2000), whereas others have correlated some measure of block activity (e.g., Pugh *et al.*, 2000). In Horwitz *et al.* (2005) many other examples of differences in evaluation of functional and effective connectivity are presented.

As was mentioned earlier, one can employ neurally realistic large-scale neural modeling to help identify the neural substrates of particular definitions of functional (and effective) connectivity. This type of neural modeling also can be used to help determine how the evaluated value of the functional/effective connectivity depends on the numerous transformations by which neural activity signals are converted into measurable functional neuroimaging data. Important among these transformations are the following observations: (1) neural activity (i.e., the input to neural populations) results in a spatially and temporally dispersed and delayed hemodynamic response (i.e., the BOLD signal); (2) this hemodynamic signal is sampled at discrete times by the magnetic resonance scanner, yielding the time series that are used to compute functional or effective connectivity. Because both of these steps are implemented in our modeling approach, we can see how they modify the value of the functional and/or effective connectivity (for details, see Horwitz *et al.* [2005]).

Let us consider the visual model discussed earlier. The task, as before, is the DMS for shape, and the control task is "passive viewing" of visual noise patterns. Figure 5 shows the time series of the integrated synaptic activity in V1/V2, IT, and one of the four submodules that are included in the PFC (D1, see Fig. 3). In Fig. 6 we display the time series corresponding to the simulated fMRI in these regions, which, as stated above, we obtain from the integrated synaptic

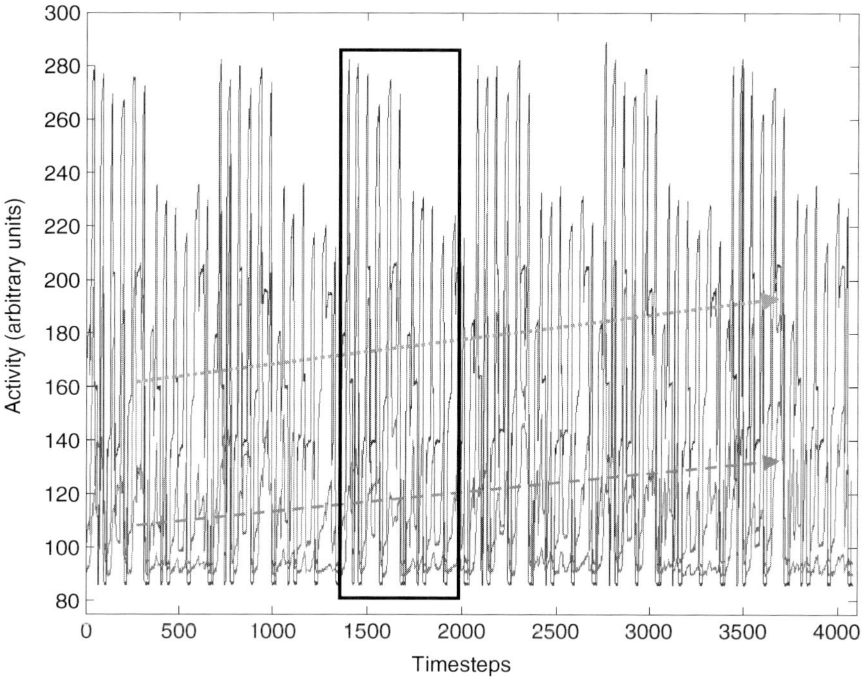

FIG. 5. Graphs of the integrated synaptic activity in the V1/V2 (blue), inferior temporal (IT) (green) and D1 (red) modules. As seen in the V1/V2 graph, there were three DMS trials and three control trials for each of six levels of attention (the lowest is on the left; the highest is on the right). The activities in the IT and D1 modules increase with attention, as indicated by the dashed lines. One block, with three DMS and three control trials, is shown in the box. The temporal integration interval was 50 msec. The values of the activity (y-axis) are in arbitrary units. Modified from Horwitz *et al.* (2005). (See Color Insert.)

activities by convolving them with a hemodynamic response function (whose delay parameter is 6 sec) and then sampling the resulting time series every 2 sec (which would correspond to the TR of the MR scanner). The simplest definition of functional connectivity (within-task correlations of the time series between two brain regions) is used. The findings are reported in detail by Horwitz *et al.* (2005), but we will use one result to illustrate the effect of the hemodynamic delay on functional connectivity. As seen in Fig. 3, we would expect the functional connectivity between the inferior temporal cortex and the D2 module to be large during the task condition, since neurons in the D2 module are active during the delay period of the DMS task (it is the interaction between these neurons and those in D1 that is crucial for maintaining a representation of the first stimulus in short-term memory after the actual stimulus is no longer present). Furthermore,

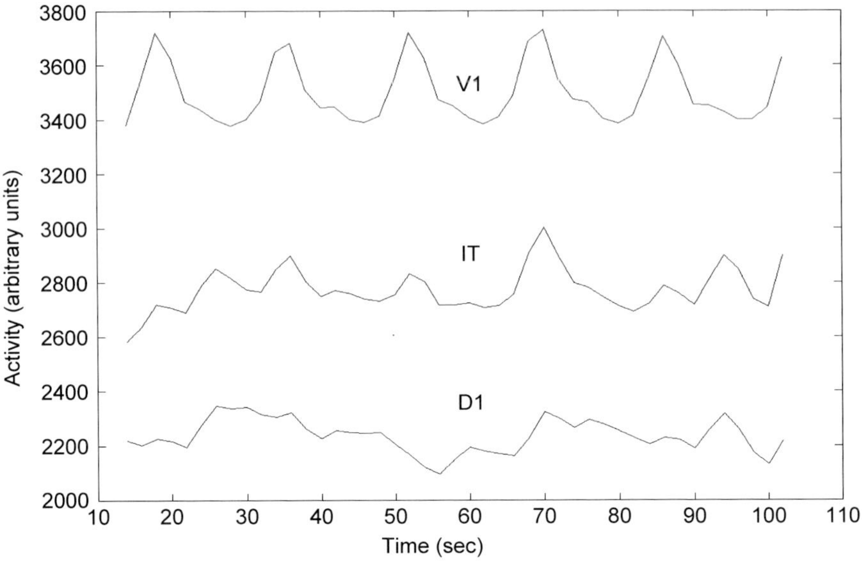

FIG. 6. Graphs of the simulated fMRI signal corresponding to the data shown in Fig. 5. The integrated synaptic activity was convolved with a Poisson function (delay parameter = 6 sec) and sampled with a repetition time of 2 sec. The values along the vertical axis are in arbitrary units. Modified from Horwitz *et al.* (2005).

the functional connectivity would be expected to be reduced during the control task, since nothing needs to be retained in short-term memory. For the integrated synaptic activity (that is, for the neural activity not transformed by a hemodynamic response), we find that the inferior temporal cortex–D2 correlation is 0.80 for the DMS task and 0.43 for the control task, findings that are consistent with our expectations. For the simulated fMRI data, the corresponding values are 0.52 and 0.42. These results suggest that, in this particular case, the effect of convolving the integrated synaptic activities with the hemodynamic response function (and also perhaps temporally sampling the convolved time series relatively slowly) is to blur the differences in functional connectivity between the two conditions. In other cases, other types of alterations may occur.

Our result demonstrates explicitly in the context of a biologically realistic neural model that great care must be taken in the evaluation of interregional functional connectivity using fMRI data. A similar conclusion was reached using simulated data that were generated from a much simpler neural model (two regions, each producing 20 on–off neural events) used by Gitelman and colleagues (2003), who suggested that deconvolution of the fMRI time series should be executed prior to evaluation of interregional functional connectivities. As a result, a deconvolution step has been built into some of the more widely used software packages, such as that supporting DCM (Friston *et al.*, 2003).

V. Conclusions

We have tried to show some of the ways by which computational neural modeling can be applied to interpret functional brain-imaging data, especially those data obtained from the hemodynamic imaging methods of PET and fMRI. We showed that systems-level modeling using techniques such as SEM or DCM can enable us to determine the regions that play key roles in specific cognitive tasks and the strengths of the anatomical links that connect these regions to each other. The main problems with this "top-down" approach are that the results do not provide information about the neural mechanisms operating in the network and that the results have limited spatial and temporal resolution. Indeed, this is why there is great interest within the neuroimaging community in multimodality neuroimaging, where one performs the same tasks using, for example, fMRI and MEG. We also showed that the use of large-scale, neurobiologically realistic neural modeling (a "bottom-up" approach) is a good way to bridge the different spatial and temporal scales. Such models can be used to simulate different types of neural data, including neuroimaging data, at multiple spatial and temporal scales. Finally, we showed that these two types of modeling can be related to each other through the use of the the large-scale models to simulate functional- and effective-connectivity data.

Acknowledgments

The authors wish to thank Drs. Malle Tagamets and Fatima Husain for many useful discussions on computational modeling and Dr. Karen Berman for discussions on working memory. We are also grateful to Drs. James Patterson and Fatima Husain for reading the manuscript.

References

Anderson, J. R., Qin, Y., Sohn, M. H., Stenger, V. A., and Carter, C. S. (2003). An information-processing model of the BOLD response in symbol manipulation tasks. *Psychon. Bull. Rev.* **10,** 241–261.

Arbib, M. A., Bischoff, A., Fagg, A. H., and Grafton, S. T. (1995). Synthetic PET: Analyzing large-scale properties of neural networks. *Hum. Brain Mapp.* **2,** 225–233.

Buechel, C., and Friston, K. J. (1997). Modulation of connectivity in visual pathways by attention: Cortical interactions evaluated with structural equation modeling and fMRI. *Cereb. Cortex* **7,** 768–778.

Buechel, C., and Friston, K. J. (1998). Dynamic changes in effective connectivity characterized by variable parameter regression and Kalman filtering. *Hum. Brain Mapp.* **6,** 403–408.

Buechel, C., Coull, J. T., and Friston, K. J. (1999). The predictive value of changes in effective connectivity for human learning. *Science* **283,** 1538–1541.

Bullmore, E. T., Rabe-Hesketh, S., Morris, R. G., Williams, S. C. R., Gregory, L., Gray, J. A., and Brammer, M. J. (1996). Functional magnetic resonance image analysis of a large-scale neurocognitive network. *Neuroimage* **4,** 16–33.

Bullmore, E., Horwitz, B., Honey, G., Brammer, M., Williams, S., and Sharma, T. (2000). How good is good enough in path analysis of fMRI data? *Neuroimage* **11,** 289–301.

Castelli, F., Frith, C., Happe, F., and Frith, U. (2002). Autism, Asperger syndrome and brain mechanisms for the attribution of mental states to animated shapes. *Brain* **125,** 1839–1849.

Ciocca, V., and Bregman, A. S. (1987). Perceived continuity of gliding and steady-state tones through interrupting noise. *Percept. Psychphys.* **42,** 476–484.

Coltheart, M., Rastle, K., Perry, C., Langdon, R., and Ziegler, J. (2001). DRC: A dual route cascaded model of visual word recognition and reading aloud. *Psychol. Rev.* **108,** 204–256.

Dannenbring, G. L. (1976). Perceived auditory continuity with alternately rising and falling transitions. *Can. J. Psychol.* **30,** 99–114.

Deco, G., Rolls, E. T., and Horwitz, B. (2004). "What" and "where" in visual working memory: A computational neurodynamical perspective for integrating fMRI and singe-cell data. *J. Cogn. Neurosci.* **16,** 683–701.

Friston, K. J. (1994). Functional and effective connectivity in neuroimaging: A synthesis. *Hum. Brain Mapp.* **2,** 56–78.

Friston, K. J., Frith, C. D., Liddle, P. F., and Frackowiak, R. S. J. (1993). Functional connectivity: The principal-component analysis of large (PET) data sets. *J. Cereb. Blood Flow Metab.* **13,** 5–14.

Friston, K. J., Harrison, L., and Penny, W. (2003). Dynamic causal modelling. *Neuroimage* **19,** 1273–1302.

Funahashi, S., Chafee, M. V., and Goldman-Rakic, P. S. (1993). Prefrontal neuronal activity in rhesus monkeys performing a delayed anti-saccade task. *Nature* **365,** 753–756.

Gitelman, D. R., Penny, W. D., Ashburner, J., and Friston, K. J. (2003). Modeling regional and psychophysiologic interactions in fMRI: The importance of hemodynamic deconvolution. *Neuroimage* **19,** 200–207.

Glabus, M. F., Horwitz, B., Holt, J. L., Kohn, P. D., Gerton, B. K., Callicott, J. H., Meyer-Lindenberg, A., and Berman, K. F. (2003). Interindividual differences in functional interactions among prefrontal, parietal and parahippocampal regions during working memory. *Cereb. Cortex* **13,** 1352–1361.

Grafton, S. T., Sutton, J., Couldwell, W., Lew, M., and Waterm, C. (1994). Network analysis of motor system connectivity in Parkinsons's Disease: Modulation of thalamocortical interactions after pallidoctomy. *Hum. Brain Mapp.* **2,** 45–55.

Haxby, J. V., Ungerleider, L. G., Horwitz, B., Rapoport, S. I., and Grady, C. L. (1995). Hemispheric differences in neural systems for face working memory: A PET-rCBF Study. *Hum. Brain Mapp.* **3,** 68–82.

Honey, G. D., Suckling, J., Zelaya, F., Long, C., Routledge, C., Jackson, S., Ng, V., Fletcher, P. C., Williams, S. C., Brown, J., and Bullmore, E. T. (2003). Dopaminergic drug effects on physiological connectivity in a human cortico-striato-thalamic system. *Brain* **126,** 1767–1781.

Horwitz, B. (1994). Data analysis paradigms for metabolic-flow data: Combining neural modeling and functional neuroimaging. *Hum. Brain Mapp.* **2,** 112–122.

Horwitz, B. (2003). The elusive concept of brain connectivity. *Neuroimage* **19,** 466–470.

Horwitz, B. (2004). Relating fMRI and PET signals to neural activity by means of large-scale neural models. *Neuroinformatics* **2,** 251–266.

Horwitz, B. (2005). Integrating neuroscientific data across spatiotemporal scales. *C. R. Biol.* **328,** 109–118.

Horwitz, B., and Tagamets, M.-A. (1999). Predicting human functional maps with neural net modeling. *Hum. Brain Mapp.* **8,** 137–142.

Horwitz, B., Grady, C. L., Haxby, J. V., Schapiro, M. B., and Rapoport, S. I. (1992a). Functional associations among human posterior extrastriate brain regions during object and spatial vision. *J. Cogn. Neurosci.* **4,** 311–322.

Horwitz, B., Soncrant, T. T., and Haxby, J. V. (1992b). Covariance analysis of functional interactions in the brain using metabolic and blood flow data. *In* "Advances in Metabolic Mapping Techniques for Brain Imaging of Behavioral and Learning Functions" (F. Gonzalez-Lima, T. Finkenstaedt, and H. Scheich, Eds.), pp. 189–217. Kluwer Academic Publishers, Dordrecht, The Netherlands.

Horwitz, B., Grady, C. L., Schlageter, N. L., Duara, R., and Rapoport, S. I. (1987). Intercorrelations of regional cerebral glucose metabolic rates in Alzheimer's disease. *Brain Res.* **407,** 294–306.

Horwitz, B., Tagamets, M.-A., and McIntosh, A. R. (1999). Neural modeling, functional brain imaging, and cognition. *Trends Cogn. Sci.* **3,** 91–98.

Horwitz, B., Friston, K. J., and Taylor, J. G. (2000). Neural modeling and functional brain imaging: An overview. *Neural Netw.* **13,** 829–846.

Horwitz, B., Warner, B., Fitzer, J., Tagamets, M.-A., Husain, F. T., and Long, T. W. (2005). Investigating the neural basis for functional and effective connectivity: Application to fMRI. *Phil. Trans. Roy. Soc. Lond. B.* in press.

Husain, F. T., Nandipati, G., Braun, A. R., Cohen, L. G., Tagamets, M.-A., and Horwitz, B. (2002). Simulating transcranial magnetic stimulation during PET with a large-scale neural network model of the prefrontal cortex and the visual system. *Neuroimage* **15,** 58–73.

Husain, F. T., Tagamets, M.-A., Fromm, S. J., Braun, A. R., and Horwitz, B. (2004). Relating neuronal dynamics for auditory object processing to neuroimaging activity. *Neuroimage* **21,** 1701–1720.

Husain, F. T., Lozito, T. P., Ulloa, A., and Horwitz, B. (2005). Investigating the neural basis of the auditory continuity illusion. *J. Cogn. Neurosci.*, in press.

Just, M. A., Carpenter, P. A., and Varma, S. (1999). Computational modeling of high-level cognition and brain function. *Hum. Brain Mapp.* **8,** 128–136.

Kaas, J. H., Hackett, T. A., and Tramo, M. J. (1999). Auditory processing in primate cerebral cortex. *Curr. Opin. Neurobiol.* **9,** 164–170.

Kikuchi-Yorioka, Y., and Sawaguchi, T. (2000). Parallel visuospatial and audiospatial working memory processes in the monkey dorsolateral prefrontal cortex. *Nat. Neurosci.* **3,** 1075–1076.

Lauritzen, M. (2001). Relationship of spikes, synaptic activity, and local changes of cerebral blood flow. *J. Cereb. Blood Flow Metab.* **21,** 1367–1383.

Lee, L., Friston, K. J., and Horwitz, B. Large-scale neural models and dynamic causal modelling: A cross-validation, submitted.

Logothetis, N. K., Pauls, J., Augath, M., Trinath, T., and Oeltermann, A. (2001). Neurophysiological investigation of the basis of the fMRI signal. *Nature* **412,** 150–157.

McIntosh, A. R. (2000). Towards a network theory of cognition. *Neural Netw.* **13,** 861–870.

McIntosh, A. R., and Gonzalez-Lima, F. (1991). Structural modeling of functional neural pathways mapped with 2-deoxyglucose: Effect of acoustic startle habituation on the auditory system. *Brain Res.* **547,** 295–302.

McIntosh, A. R., and Gonzalez-Lima, F. (1994). Structural equation modeling and its application to network analysis in functional brain imaging. *Hum. Brain Mapp.* **2,** 2–22.

McIntosh, A. R., Grady, C. L., Ungerleider, L. G., Haxby, J. V., Rapoport, S. I., and Horwitz, B. (1994). Network analysis of cortical visual pathways mapped with PET. *J. Neurosci.* **14,** 655–666.

Mechelli, A., Penny, W. D., Price, C. J., Gitelman, D. R., and Friston, K. J. (2002). Effective connectivity and intersubject variability: Using a multisubject network to test differences and commonalities. *Neuroimage* **17,** 1459–1469.

Mechelli, A., Price, C. J., Friston, K. J., and Ishai, A. (2004). Where bottom-up meets top-down: Neuronal interactions during perception and imagery. *Cereb. Cortex* **14,** 1256–1265.

Meyer-Lindenberg, A., Poline, J. B., Kohn, P. D., Holt, J. L., Egan, M. F., Weinberger, D. R., and Berman, K. F. (2001). Evidence for abnormal cortical functional connectivity during working memory in schizophrenia. *Am. J. Psy.* **158,** 1809–1817.

Penny, W. D., Stephan, K. E., Mechelli, A., and Friston, K. J. (2004). Modelling functional integration: A comparison of structural equation and dynamic causal models. *Neuroimage* **23** (Suppl. 1), S264–S274.

Pugh, K. R., Mencl, W. E., Shaywitz, B. A., Shaywitz, S. E., Fulbright, R. K., Constable, R. T., Skudlarski, P., Marchione, K. E., Jenner, A. R., Fletcher, J. M., Liberman, A. M., Shankweiler, D. P., Katz, L., Lacadie, C., and Gore, J. C. (2000). The angular gyrus in developmental dyslexia: Task-specific differences in functional connectivity within posterior cortex. *Psychol. Sci.* **11,** 51–56.

Rauschecker, J. P., and Tian, B. (2000). Mechanisms and streams for processing of "what" and "where" in auditory cortex. *Proc. Natl. Acad. Sci. USA* **97,** 11800–11806.

Schlosser, R., Gesierich, T., Kaufmann, B., Vucurevic, G., Hunsche, S., Gawehn, J., and Stoeter, P. (2003). Altered effective connectivity during working memory performance in schizophrenia: A study with fMRI and structural equation modeling. *Neuroimage* **19,** 751–763.

Sugita, Y. (1997). Neural correlates of auditory induction in the cat cortex. *Neuroreport* **8,** 1155–1159.

Tagamets, M.-A., and Horwitz, B. (1998). Integrating electrophysiological and anatomical experimental data to create a large-scale model that simulates a delayed match-to-sample human brain imaging study. *Cereb. Cortex* **8,** 310–320.

Ungerleider, L. G., and Mishkin, M. (1982). Two cortical visual systems. *In* "Analysis of Visual Behavior" (D. J. Ingle, M. A. Goodale, and R. J. W. Mansfield, Eds.), pp. 549–586. MIT Press, Cambridge.

COMBINED EEG AND fMRI STUDIES OF HUMAN BRAIN FUNCTION

V. Menon*,†,‡ and S. Crottaz-Herbette*

*Department of Psychiatry & Behavioral Sciences, Stanford University School of Medicine
Stanford, California 94305

†Program in Neuroscience, Stanford University School of Medicine
Stanford, California 94305

‡Neuroscience Institute at Stanford, Stanford University School of Medicine
Stanford, California 94305

Combined electroencephalography (EEG) and functional magnetic resonance imaging (fMRI) studies show great promise for helping researchers to develop a more comprehensive understanding of the neural basis of behavior, including brain function and dysfunction. The aim of this paper is to review current knowledge and research on the use of combined fMRI and EEG data. We briefly examine the complementary features of the two techniques, and we then describe ways to acquire the two types of data, the relative advantages and disadvantages of acquiring them simultaneously, and strategies and methods for

INTERNATIONAL REVIEW OF
NEUROBIOLOGY, VOL. 66
DOI: 10.1016/S0074-7742(05)66010-2

effectively combining them. We address topics related to both event-related and emergent neural activity.

I. Introduction

Understanding the neural basis of brain functioning requires knowledge about the spatial and temporal aspects of information processing. Functional magnetic resonance imaging (fMRI) and electroencephalography (EEG) are two techniques widely used to noninvasively investigate human brain function. Neither of these technologies alone, however, can provide the information necessary to understand the spatio-temporal aspects of information processing in the human brain. fMRI yields highly localized measures of brain activation, with a good spatial resolution (about 2–3 mm) but a temporal resolution significantly longer than the time needed for most perceptual and cognitive processes. EEG has the necessary temporal resolution to study the dynamics of brain function, but its poor spatial resolution precludes identification of underlying neural sources. fMRI and EEG therefore represent complementary imaging techniques, and combining information from them is a particularly useful way to examine the spatial and temporal dynamics of brain processes (Babiloni *et al.*, 2004; Dale *et al.*, 2000; Liebenthal *et al.*, 2003; Menon *et al.*, 1997). In Section II we review the fundamental features of neural activity as indexed by EEG and fMRI, with a view to better understanding the principles of various methods and their strengths and limitations for combining the two types of data.

II. The Signals

A. EEG

A clear understanding of the basics of EEG signal generation and recording is necessary in order to effectively combine EEG with fMRI. We review some of the relevant issues here; for a detailed review of EEGs, however, the reader is referred to a comprehensive text on this topic (Niedermeyer and Lopes da Silva, 2004). EEG signals recorded on the scalp surface arise from large dendritic currents generated by the quasi-synchronous firing of a large number of neurons (Freeman *et al.*, 2003). At a finer spatial scale, these same currents are also responsible for local field potentials recorded extracellularly *in vivo* in both humans and animals (Steriade, 2001). The local field potential is generated by extracellular currents that pass through the extracellular space in a closed loop.

These currents induce voltage changes (in the uV range) that are smaller than action potentials but that last longer and extend over a larger area of neural tissue. The local field potential reflects the linear sum of overlapping sources (current flows from the intracellular to the extracellular space) and sinks (current flows from the extracellular to the intracellular space). Scalp EEG arises from the passive conduction of currents produced by the summating local field potentials over large neuronal aggregates. The columnar structure of the neocortex facilitates the summation of electrical activity distributed among multiple neuronal groups. EEG activity recorded on a scalp electrode corresponds to the sum of activity from regions near the electrode, but large signals originating from more distal cortical sites can make a significant contribution to the activity observed at a given point on the scalp (Liu *et al.*, 2002). Furthermore, the domains of spatially correlated activity underlying perceptual and cognitive processing are about 2–3 mm in the neocortex; volume conduction by the scalp significantly increases coherence across several cm on scalp recordings (Freeman *et al.*, 2003; Menon *et al.*, 1996). More importantly, the problem of recovering sources of neuronal activity from scalp EEG is fundamentally ill posed (Nunez and Srinivasan, 2006).

Scalp-recorded EEG in healthy adults typically reaches a maximum amplitude of 75 μV (Fig. 1); however, it can reach an amplitude of 1 mV or more in pathological situations such as epileptic seizures (Niedermeyer and Lopes da Silva, 2004). The temporal resolution of EEG is about 1 msec; by this we mean that events of short duration, such as epileptic spikes (which last about 1 msec), can be reliably recorded. Although normal EEG fluctuations have amplitudes of 75 μV or more, the magnitude of useful brain signals buried in these fluctuations is often considerably smaller.

1. *Evoked EEG Activity*

The term "event-related potential" (ERP) refers to the electrical response of the brain to a specific stimulus or cognitive process. Much of the background EEG activity is not time locked to the stimulus. Typically, stimulus-evoked signals are in the range of a few microvolts, which implies that the signal-to-noise ratio is much smaller than 1. Thus, signals cannot often be identified by visual inspection, and signal averaging and other statistical or signal-analysis procedures are therefore needed to recover meaningful brain signals. Averaging the EEG time locked to the stimulus or response increases the signal-to-noise ratio, resulting in reliable detection of useful brain signals (Handy, 2005). For cognitive processes, typically 30–100 stimuli trials are needed to obtain reliable ERP components. The spatial resolution of scalp-recorded ERPs is, however, poor, and this further limits accurate determination of the location of neural sources.

ERPs have been widely used to examine the timing of distributed brain processes involved in perception and cognition (Regan, 1989; Rugg, 1995).

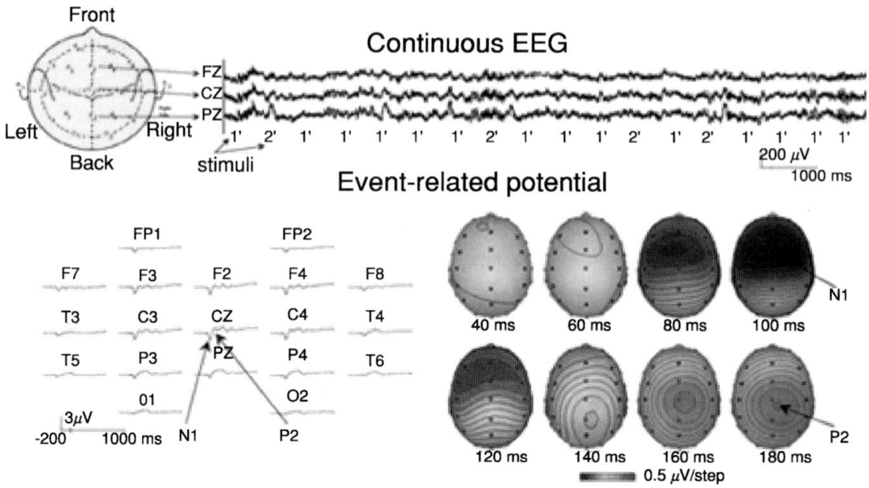

FIG. 1. Top: Raw EEG recorded at electrodes FZ, CZ, and PZ during the presentation of two different auditory tones, labeled "1" and "2." The EEG responses following the presentation of the tones are not visible on the raw EEG. Bottom left: Event-related potentials (ERPs) derived by averaging EEG segments following the presentation of the tones (e.g., stimulus 1 on the raw EEG). The N1 and P2 peaks are negative and positive deflections at 100 msec and 200 msec poststimulus, respectively. Bottom right: Topographical maps every 20 ms, from 40 ms to 180 ms poststimulus, presenting the N1 and P2 peaks of the ERPs. (See Color Insert.)

ERP waveforms can be broken down into several basic components. A "component" is a positive- or negative-going fluctuation that can be visually identified in an ERP waveform. The components that occur prior to 100 ms are thought to reflect information processing in early sensory pathways. Cognitive electrophysiologists have been most interested in the so-called long-latency ERP components, including the P1, P2, P3, N1, N2, and N4 components. These components are named by their polarity (P for positive, N for negative) and either their ordinal position after stimulus onset (P1 is the first positive peak) or their latency after stimulus onset (N4 or N400 is a negative-going component peaking at 400 ms). In general, the mid-latency components occuring between 100 and 200 ms are thought to reflect late sensory and early cognitive processes, while those after 250 ms or so are thought to reflect higher-level cognitive processes (e.g., episodic memory).

2. *Emergent EEG Activity*

Although ERPs have been widely used in human brain research, they have significant limitations. Notably, averaging removes useful brain signals that are not time locked to the stimulus or the response. Cortical activity contains both evoked and emergent patterns of stimulus-related activity (Freeman *et al.*, 2003). Researchers like Walter Freeman have emphasized the importance of investigating both evoked and emergent EEG activity. For example, Ohl *et al.* (2003) examined evoked and emergent patterns in the primary auditory cortex (field AI). They showed that evoked patterns were focally organized at locations corresponding to the thalamically relayed input into the cortical tonotopic map. In contrast, emergent patterns could also discriminate responses to stimuli, but they were broadly distributed and held no apparent relationship to the tonotopic map. More recently, Makeig and colleagues (2002) have extended these ideas to human-scalp EEG data. They have shown that some components of the ERP are generated by stimulus-induced phase resetting of changes in ongoing EEG dynamics. More recent studies have shown unambiguously that averaging filters out much of the information about cortical dynamics available in the unaveraged single trials (Makeig *et al.*, 2004). The combination of unaveraged EEG and fMRI data promises to open new vistas in the study of brain dynamics.

B. fMRI

The basic features of the fMRI signal have been described in Chapter 5 of this volume. Here, we will briefly describe the properties of the fMRI signal as it relates to combining EEG with fMRI. Unlike EEG, fMRI does not directly measure neural activity; instead, it relies on changes in oxygenation, blood volume, and flow (Logothetis and Wandell, 2004). During performance of perceptual and cognitive tasks, regions of the brain that are more active than others will have increased blood flow, resulting in increased oxygen levels. This localized increase in oxygen results in changes in the magnetic properties of the underlying tissue, which we can detect in scanners with high magnetic fields. Thus, we can quantify the amount of activity in given brain regions based on the extent and amount of blood flow to those regions over time. Because of the hemodynamic lag—the amount of time it takes for local blood-oxygen levels to rise and peak—the temporal resolution of fMRI is limited to several (1–6) seconds. One advantage of fMRI over EEG and MEG is that its spatial resolution is an order of magnitude better, allowing researchers to pinpoint where in the brain an operation is occurring, with a resolution approaching 1 mm at high fields (Formisano *et al.*, 2003).

As with EEGs, the baseline fMRI signal shows spontaneous and continuous fluctuations even in the absence of external stimuli. But, unlike with EEGs, the origin of these fluctuations is poorly understood. Human cortical EEG activity has been intensively examined at frequencies ranging from 0.5 Hz to several hundred Hz. Recent studies have, however, demonstrated large-scale very slow fluctuations in the human cortex at frequencies ranging from 0.02 Hz to 0.2 Hz

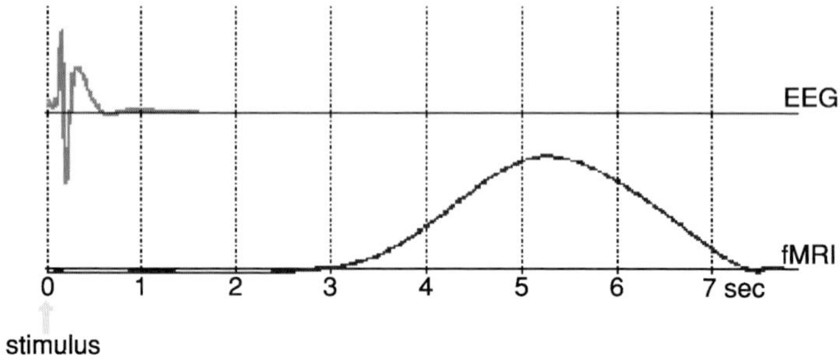

FIG. 2. Relative timing of EEG and fMRI responses following the presentation of a stimulus. The EEG response lasts about 1 second poststimulus, whereas the fMRI (blood oxygen level–dependent) response takes about 2–3 seconds to rise and reaches a maximum at about 5–6 seconds poststimulus. (See Color Insert.)

(Vanhatalo *et al.*, 2004). The relation between these fluctuations and the fMRI baseline is not known, but such oscillations provide useful starting point for further investigations of the relations between the two signals.

Stimulus-evoked and task performance–evoked responses in the fMRI signal are somewhat better understood. When neuronal activity in a given brain region increases, metabolic demands result in an increase of oxygenated blood, which in turn is detected as an increase in fMRI signal intensity (Logothetis and Pfeuffer, 2004). Following onset of task-related neuronal activity, the fMRI signal takes about 2 seconds to increase and 4–6 seconds to reach its peak, and then it recovers to baseline in 12 seconds (Fig. 2). This change in signal, or the response to an impulse, is referred to as the hemodynamic response function (HRF). The form of the HRF has important consequences in terms of experimental design and interpretation of combined EEG and fMRI data acquisition.

III. The Relation Between EEG and fMRI Signals

Combined imaging and physiology experiments in monkeys have indicated that the fMRI signal is better correlated with the local field potential than with multiunit and single-neuron activity (Logothetis and Pfeuffer, 2004). In a seminal study combining microelectrode recording and fMRI in anesthetized monkeys, Logothetis *et al.* (2001) showed a linear correlation between the BOLD response and the stimulus-driven local field potential activity. At a fundamental level, this augurs well for combined EEG and fMRI studies of information processing in the human brain, since the EEG reflects spatially summed and volume-conducted local field potentials.

As we have already noted, EEG signals are directly related to neuronal processing, whereas fMRI responses arise from subsequent changes in blood-oxygenation levels. There are other important differences as well that are relevant for combining the two types of data. For neuronal responses to be recorded on the scalp EEG, it is necessary that sources be located in brain structures that can generate far-field potentials. Brain structures that have a laminar organization, such as the neocortex, can contribute significantly to the scalp EEG. On the other hand, neural sources in structures such as the thalamus and basal ganglia, which have a radial or noncolumnar organization, are less likely to make any significant contributions to the scalp EEG, even when large local field potentials can be recorded from them (Niedermeyer and Lopes da Silva, 2004). fMRI signals, on the other hand, are not directly dependent on the laminar or the radial neuronal organization; they have more to do with the structure of the underlying vascular bed. In the future it will be important to incorporate such constraints into more detailed biophysical modeling.

IV. Nonsimultaneous and Simultaneous EEG and fMRI Data Acquisition

Combined fMRI and EEG studies can be conducted with either simultaneously or nonsimultaneously acquired data. With the former method, EEG and fMRI data are acquired in the scanner; with the latter method, EEG is acquired outside the scanner in a separate session, and the order of the sessions is randomized across subjects.

A. Nonsimultaneous EEG–fMRI Acquisition

A number of brain-imaging studies have combined fMRI and EEG using data recorded in separate sessions (e.g., Ball *et al.*, 1999; George *et al.*, 1995; Heinze *et al.*, 1994; Menon *et al.*, 1997; Opitz *et al.*, 1999; Snyder *et al.*, 1995). An advantage of this approach is that the signal-to-noise ratio of EEG data obtained outside the scanner is usually much better than that of data obtained inside the scanner. Although data are recorded in two separate sessions, differences between the sessions can be minimal, especially those involving ERPs. This is certainly the case for several standard ERP paradigms, such as the oddball ("P300") paradigm and the semantic mismatch ("N400") paradigm, in both of which ERPs may, in fact, be better recorded outside the scanner. Methods for acquiring EEG and fMRI follow standard protocols and so are not discussed further here.

Clearly, a major drawback of this approach is that single-trial EEG and fMRI cannot be combined to examine emergent brain responses which may not be time locked to the stimulus or response. Even for ERPs, there may be significant subjective and experimental differences between the two sessions. For example, there may be differences in subjects' levels of attention, vigilance, motivation, and familiarity with the task. Using simultaneous EEG–fMRI acquisition, it has been shown that arousal levels are an important determinant of brain activation during cognitive tasks (Matsuda *et al.*, 2002). In addition, we must consider that the two separate sessions do not provide the same environment, because during the fMRI session the subject is in a noisy environment, whereas during the EEG session the subject sits generally in a comfortable and quiet room. This is an important issue for combined EEG–fMRI studies of auditory processing.

B. Simultaneous EEG–fMRI Acquisition

The main advantage of acquiring EEG and fMRI data in the scanner is that the two types of data reflect the same neuronal processes. Simultaneous acquisition allows us to ensure that subjects use the same strategy for both kinds of data; this is an issue that is particularly important for tasks involving complex cognitive processing. For epileptic seizure localization, simultaneous EEG and fMRI is clearly important. There are other important clinical problems in which simultaneous EEG and fMRI is critical, particularly those cases where symptoms can change over short durations. For example, some patients with schizophrenia have problems distinguishing self-generated from externally generated percepts during hallucinatory episodes, which can wax and wane unpredictably.

Even outside the clinical domain, simultaneous EEG and fMRI recordings are becoming increasingly useful. Studies of single-trial EEG and fMRI in which researchers seek to better understand neuronal processing that is not necessarily time locked to external events (Makeig *et al.*, 2004) will also benefit from simultaneous recordings. Studies of the resting state and attempts to better understand brain dynamics underlying intrinsic EEG rhythms (Goldman *et al.*, 2002) and the default mode of brain function (Raichle *et al.*, 2001) also rely increasingly on simultaneous EEG and fMRI recordings (Laufs *et al.*, 2003a,b). Simultaneous acquisition is also preferable for clinical and developmental studies, as it cuts down on the total time necessary to acquire data. Finally, for studies where continuous EEG tracks various stages of sleep (e.g., REM/non-REM), simultaneous acquisition of EEG and fMRI data is imperative (Czisch *et al.*, 2004).

There are practical reasons why simultaneous recordings may be necessary. For clinical studies and for studies of children and the aged, nonsimultaneous recordings may not be practical. Most of these subjects are not willing to participate in multiple sessions involving extended periods of time—this has certainly been our experience across many such subject groups. Furthermore, in clinical studies that involve medication, it may not be feasible, reliable, or practical to conduct studies across multiple sessions. Thus, whereas normative studies on college-aged adults can be performed easily in two or more sessions, this is certainly more difficult in clinical and developmental studies. Also, even in normal healthy individuals, particularly in children, memory and learning paradigms involve extended stimulus exposure, which can interfere with encoding and retrieval.

In the past five years, several brain-imaging centers have developed and refined techniques for simultaneous acquisition of EEG and fMRI data (Bonmassar *et al.*, 2001a; Krakow *et al.*, 2000; Lemieux *et al.*, 2001b; Salek-Haddadi *et al.*, 2002) and have used them to detect EEG spikes, characterize resting-state EEG, and investigate the neural basis of ERPs (Bonmassar *et al.*, 1999, 2001b, 2002; Christmann *et al.*, 2002; Lazeyras *et al.*, 2001; Liebenthal *et al.*, 2003; Mulert *et al.*, 2004; Nagai *et al.*, 2004; Thees *et al.*, 2003).

1. *General Considerations*

The procedures used to record EEGs inside the scanner are, for the most part, similar to those used outside the scanner. Figure 3 summarizes a basic setup for data acquisition. EEG data can be recorded in either a referential or a bipolar configuration. In the referential configuration, data for all electrodes are recorded with respect to a neurophysiologically "neutral" electrode that may, for example, be attached to the mastoid (behind the ear lobes). In the bipolar configuration, each electrode is recorded with respect to a neighboring electrode on the scalp. Both kinds of configurations have been used to acquire EEGs in the scanner. One advantage of bipolar electrodes' in a twisted configuration is this makes it possible to significantly reduce gradient noise (Goldman *et al.*, 2000). However, it should be noted that some EEG signals of interest may not be clearly detectable in a bipolar montage. In this case, the bipolar montage can be transformed into a referential montage provided there is appropriate connectivity between electrodes in different head regions. In particular, it is necessary that at least one channel that connects left- and right-side electrodes be included in the bipolar montage (e.g., the C3–C4 channel) (Lagerlund *et al.*, 2003).

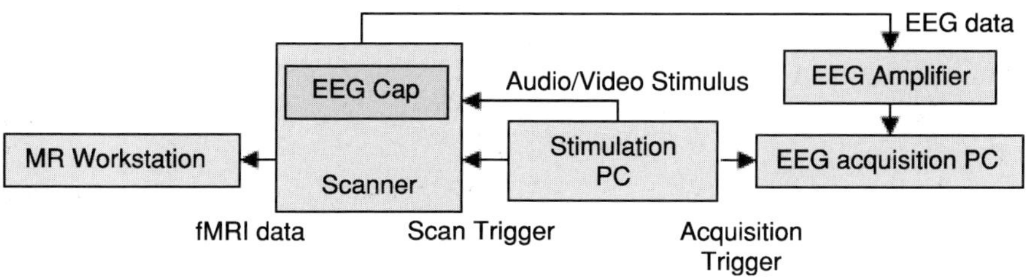

FIG. 3. Schematic diagram showing basic features of a combined EEG–fMRI data acquisition system. (See Color Insert.)

Although 16- to 32-channel recordings have been commonly used, high-density EEG has been recorded with 64 and 128 channels (Scarff *et al.*, 2004). Increasing the number of recording electrodes from 64 to 128 improved the accuracy of the equivalent dipole source localization but decreased the signal-to-noise ratio (SNR) of MR images. This suggests that the 64-electrode setup may be optimal for use in simultaneous recording of EEG and fMRI data. However, spiral fMRI methods may not suffer geometric distortion, as echo-planar imaging methods do, due to off-resonance effects (Pfeuffer *et al.*, 2002).

A number of technical problems must be first overcome before the benefits of simultaneous EEG–fMRI acquisition can be fully realized. Concurrent acquisition of EEG and fMRI has proven to be challenging for a number of reasons, including those related to data quality (Allen *et al.*, 1998, 2000; Goldman *et al.*, 2000; Sijbers *et al.*, 1999, 2000). EEG data acquired in the scanner are contaminated by physiological and imaging artifacts (Ives *et al.*, 1993). These artifacts can reach amplitudes that are several hundred times larger than those of the EEG. Studies using simultaneous EEG–fMRI acquisition can be divided into two categories related to the type of fMRI acquisition: interleaved or fully continuous acquisition. Both of these acquisitions have specific advantages and disadvantages, which will be discussed in the following sections.

2. *Interleaved Acquisition*

Most studies to date have used interleaved EEG–fMRI acquisition protocol, in which EEG data are required continuously and fMRI data are acquired intermittently. fMRI data are acquired using a clustered procedure, whereas EEG data are recorded continuously. Following stimulus presentation, 1 to 2 seconds of the EEG data are first acquired without fMRI scanning (the MRI scanner is quiet during this time interval), followed by a few seconds of fMRI data acquisition (Fig. 4). This is similar to the "clustering" procedure used to acquire fMRI data in auditory experiments (Hall *et al.*, 1999), where it is important that auditory stimuli be heard without interference from scanner noise. Most of the useful stimulus-related EEG is therefore not contaminated by scanner noise, since the neurophysiological response starts a few milliseconds after the stimulus onset and lasts no more than 1 to 2 seconds, depending on the cognitive processes involved in the task. Starting the fMRI data acquisition 1 to 2 seconds after stimulus onset allows us to estimate the BOLD response without contamination from gradient artifacts. The EEG recorded in this manner still has artifacts related to cardiac pulsation. These ballistocardiogram (BCG) artifacts must be removed in order for single-trial EEG data to be useful (see Section V.A for more). The extent to which BCG artifacts are synchronized to stimulus and response is currently not known. It is likely that for tasks that do not involve a strong affective response, the BCG and stimulus-related responses are not synchronized, so that at least for ERPs it is not absolutely necessary to remove the

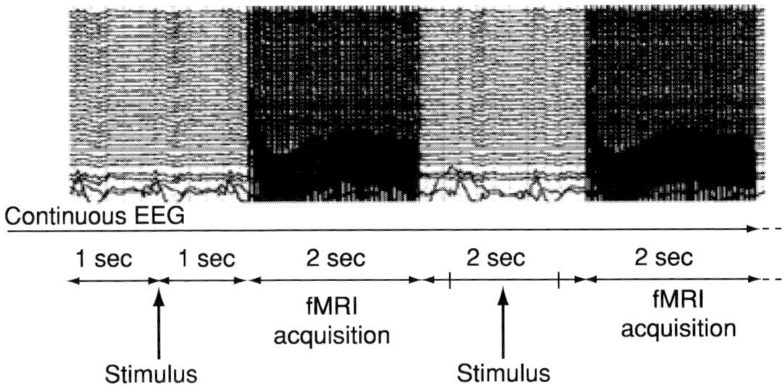

FIG. 4. EEG data from an interleaved EEG–fMRI acquisition in which the fMRI data are is acquired using a clustered procedure, with 2 seconds of EEG acquisition without MRI scanning followed by 2 seconds of fMRI data acquisition (note that the EEG is continuously acquired). In this case, the stimulus (either a tone or a flashing checkerboard) was presented 1 second after the end of each fMRI acquisition interval (TR).

BCG artifacts if a sufficiently large number of trials is used. The main disadvantage of interleaved acquisition is that the rate of stimuli presentation has to be reduced, often almost doubling the length of the experiment (cf. Liebenthal *et al.*, 2003). The interleaved acquisition is currently a good method for EEG–fMRI acquisition, particularly for recording ERPs, since the problem of gradient-artifact removal has not yet been satisfactorily solved. For single-trial EEG, the interleaved acquisition is, however, not optimal.

3. *Fully Simultaneous Acquisition*

In this case, both fMRI and EEG data are acquired continuously. This is useful for both cognitive and clinical studies. For one, rapid event-related studies require the added sampling afforded by simultaneous recordings. For short-lived and unpredictable events such as epileptic spikes, also, this is preferable and can greatly increase the statistical power. However, fully simultaneous acquisition is even more challenging than interleaved acquisition because of the large artifacts in the EEG during fMRI data acquisition (see Section V.B). It is necessary to have EEG amplifiers with large dynamic range so that the amplifiers do not saturate during the fMRI acquisition. Sophisticated algorithms are necessary for removing gradient artifacts. The first studies using fully continuous acquisition sought to determine seizure foci (Benar *et al.*, 2002; Lemieux *et al.*, 2001a). This method has rarely been used for ERP studies (Nagai *et al.*, 2004), partly because ERP signals have low SNR and partly because many experimental paradigms can be successfully implemented with the interleaved acquisition. Procedures to record EEG inside the scanner are similar to those used outside, except that the removal of gradient artifacts requires that

EEG data be recorded at a sampling rate of 10 kHz/channel or higher. This is considerably higher than the 1 kHz/channel sampling typically used in standard EEG recordings. This is necessary both to prevent aliasing of scanner artifacts into the EEG signal and to detect and remove gradient artifacts that contaminate the EEG.

C. fMRI Data Acquisition

Anatomical and functional MRI acquisitions have been performed at scanner strengths of up to 3 Tesla (T) with minimal artifacts from the EEG setup and data acquisition on MRI or fMRI images. Normal imaging protocols are used to acquire fMRI data, except in the case of the interleaved EEG–fMRI acquisition, where standard clustering procedures can be used. These procedures are not described here; for details, the reader is referred to Huettel *et al.* (2004).

V. Artifact Reduction in Simultaneous EEG–fMRI Data Acquisition

A. BCG Artifacts

The BCG artifact is a consequence of the electromotive force (EMF) produced on the EEG electrodes due to small head movements, such as those caused by cardiac pulsation, inside the scanner magnetic field. As pointed out by Sijbers *et al.* (2000), there are three major sources of BCG artifacts: (1) small but firm movement of the electrodes and the scalp due to expansion and contraction of scalp arteries between the systolic and diastolic phases; (2) fluctuation of the Hall voltage due to the pulsatile changes of the blood in the arteries; and (3) small cardiac-related movements of the body (Fig. 5). The cardiac pulse generates artifacts with amplitudes considerably larger than those of EEG signal fluctuations. It is therefore important to develop methods to identify and remove these artifacts in a robust manner. Most methods of eliminating BCG artifacts have focused on either (1) averaged artifact subtraction (AAS), in which a BCG artifact template is estimated by averaging over the intervals of EEG signal that are corrupted by the artifact and subsequent subtraction of the template from the corrupted segments to obtain a clean signal (Allen *et al.*, 1998) or (2) adaptive filtering techniques, which make use of correlations between a reference ECG channel and the EEG channels to estimate the contribution of the BCG artifact in the EEG signals, which is then subtracted to yield clean signals.

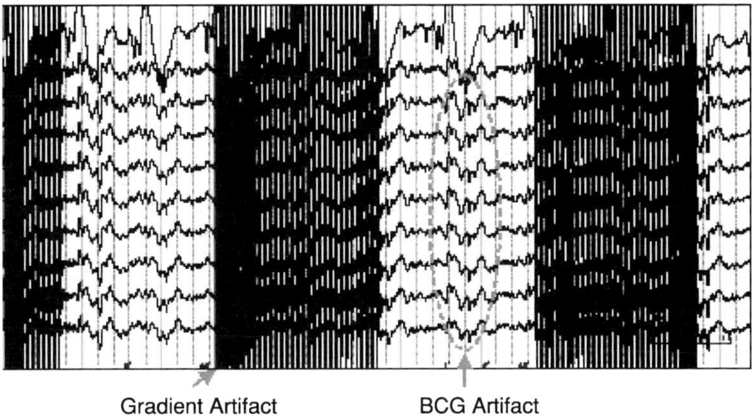

FIG. 5. EEG traces from an interleaved EEG–fMRI data acquisition showing the gradient (imaging) artifact and the ballistocardiography (BCG) (physiological) artifacts. Both artifacts are clearly visible on the EEG data at every electrode.

Currently, the AAS procedure is the most commonly employed method for removing the BCG artifact from EEG data. In this procedure, first, the QRS peaks in the ECG signal are detected, and then EEG activity time locked to these peaks is averaged to give an estimate of the pulse artifact (Allen *et al.*, 1998). The average artifact is then subtracted from the EEG. Goldman *et al.* (2000) have used a method that is conceptually similar to the AAS procedure but that differs in the weights that are applied to data segments prior to averaging. These weights vary inversely with the temporal displacement from the current sample to compensate for the slow changes in the BCG artifact. Along similar lines, Sijbers *et al.* (2000) have used QRS onset detection to create a template of the BCG artifact based on adaptive filtering. They point out that simple averaging would not lead to a satisfactory template, as the ECG is not a stationary signal and hence the rate and duration of BCG artifacts might vary over time. In their approach, median filtering was performed to obtain an artifact template, because it adapts to changes in ECG signals over time. Bonmassar *et al.* (2002) used motion information recorded from a piezoelectric sensor placed on the temporal artery to estimate the motion-artifact noise (originating mostly from BCG), followed by adaptive filtering to subtract the artifact. Srivastava *et al.* (2005) showed that independent components analysis (ICA) is useful for removing BCG artifacts. ICA consistently showed five to six components representing the BCG artifact. Following removal of these components, a significant reduction in spectral power at frequencies associated with the BCG artifact was observed. Preliminary results suggest that ICA-based procedures may be significantly better than other noise-cancellation methods.

B. Gradient Artifacts

The gradient artifact is caused by EMF induction on the EEG leads due to the rapidly switching magnetic-field gradients during fMRI acquisition. This is a periodic artifact with multiple spectral lines in the Fourier spectrum, and the fundamental frequency is governed by scan parameters like repetition time (TR) and number of slices. The standard deviation of this artifact is about 30–50 times as large as that of the EEG signal (Fig. 5), and hence the signal-to-artifact ratio (SAR) is negative: -20 to -30 dB. The current methods to remove this artifact include (1) using short EEG leads for scalp measurements with optical data transmission inside the MR scanner bore to prevent the gradient artifacts from saturating the amplifier; (2) estimating the individual responses of the G_x, G_y, G_z gradients and subtracting the sum of these responses from the actual EEG data to recover just the physiological signal recorded on the scalp (Felblinger *et al.*, 1999); (3) estimating the power spectrum of the gradient artifact through combined median filtering of different noisy sections of the data and using the spectrum as a template for removing the gradient noise through adaptive filtering techniques (Sijbers *et al.*, 1999); (4) straight time averaging of noisy sections of each EEG channel to obtain a template gradient-artifact waveform, subtracting the template from the channel, and cleaning the residual noise using adaptive techniques (Allen *et al.*, 2000).

Most of the commercially available MR-compatible EEG acquisition system do not reduce the gradient artifact to attenuate at the acquisition stage. This leads to small EEG signals and large artifacts. The gradient artifacts have large power in the EEG frequency range, and, hence, conventional low-pass or band-pass filtering cannot be employed to clean the artifact without severely compromising the quality of EEG signals. The power in the gradient artifact is concentrated in very narrow spectral spikes, and, hence, it is useful to use notch filters with very narrow stop bands and large attenuation so that only the power at spike frequencies is suppressed, and therefore there is no significant effect on the power in the neighboring EEG frequency range. However, explicitly designing such notch filters is difficult and, moreover, may result in removal of significant EEG power in the alpha, beta, and mu bands. Advanced signal-analysis procedures such as adaptive noise cancellation will be necessary to satisfactorily remove these artifacts.

VI. Task-Design Issues

Although early fMRI studies used blocked designs, which provide a better SNR, both fMRI and EEG research have converged on event-related designs. In a blocked fMRI design, subjects are presented with alternating task conditions that last about 15–30 seconds each. Blocked designs are efficient for estimating generalized task-related fMRI responses, but they are not optimal for parsing specific component processes and therefore can be combined with EEG only for tasks involving minimal cognitive processing. In event-related fMRI designs, each stimulus (or trial) corresponding to specific task conditions is presented randomly. In slow event-related designs, successive trials are spaced in time so that their evoked BOLD responses do not overlap. In contrast, rapid event-related fMRI designs use shorter intertrial-intervals (ITIs) of about 1–7 seconds. This allows faster data acquisition and also reduces the impact of cognitive processes associated with long ITIs. In this design, the BOLD response to successive trials can overlap, but "jitter" and randomized sequences (Burock *et al.*, 1998; Dale, 1999) can be used to reduce this confound (Friston *et al.*, 1998). Most EEG–fMRI studies, both simultaneous and nonsimultaneous, have used fast event-related designs (Table I). Rapid event-related presentations have been the mainstay of EEG studies over the past 40 years. The performance of most experimental designs for EEG is near optimal, since EEG signals have a rapid decorrelation time and since, if the contribution of slow potentials is small enough, the carryover into trials separated by 2 seconds (the time typically required to acquire whole-brain fMRI data) is quite small. Care, however, must be exercised if the experimental paradigm induces long-lasting slow waves such as the contingent negative variation (Brunia and van Boxtel, 2001). In most studies of perception and cognition, this is not an issue. For fMRI, the task-design issues are much more critical.

The performance of an experimental design for fMRI can be characterized by its estimation efficiency, which is the ability to make an estimate of the hemodynamic response; its detection power, which is the ability to detect brain activation; and its conditional entropy, which is a measure of the randomness of the design (Liu, 2004). The properties of event-related sequences depend on the specific design chosen (e.g., m-sequence designs, clustered m-sequence designs, or mixed designs). There are tradeoffs among efficiency, power, and entropy for each sequence. An event-related design that is high in efficiency may not have the best detection power. For details of the theoretical results and their practical implications for the optimal design of fMRI experiments with multiple trial types, the reader is referred to the excellent papers by Liu and colleagues (Liu and Frank, 2004; Liu *et al.*, 2001). Here, we focus on their relevance to combined EEG–fMRI studies. These considerations apply equally to simultaneously and nonsimultaneously recorded EEG–fMRI data. Whether the efficiency or the detection power is maximized depends on the type of modeling used to integrate EEG and fMRI data. For predictive and correlational analysis of continuous EEG–fMRI data and for more detailed biophysical modeling that relies on current source imaging, estimating the hemodynamic response is important. In such cases, a design that maximizes estimation

TABLE I
SUMMARY OF SOME RECENTLY PUBLISHED EVENT-RELATED EEG–fMRI STUDIES

Reference	Task (no. of subjects)	Sequence parameters	No. of epochs	Scanner and EEG systems (strength in Tesla)	Ballistocardiography artifact removal	Gradient artifact removal	ERP components and analyses	Relation BOLD–ERP
Bonmassar *et al.*, 1999	Visual task, checkerboard reversing at 4 Hz (N unknown)	Interleaved 1-sec fMRI, 1-sec EEG	Not provided	3T, 64 EEG channels, referential montage	Design of filter w that maximizes the ratio of projection of signal and noise onto w	None	Visual N75, P100	Not investigated
Bonmassar *et al.*, 2001b	Visual task, checkerboard reversing at 2 or 4 Hz (2 subjects)	Interleaved 30-sec fMRI, 30-sec EEG (15 sec of checkerboard, 15 sec of uniform gray field)	1200 epochs acquired, average on 500 epochs	1.5T, 32 EEG channels, bipolar montage	Short epoch (125–250 ms) compared to BCG pulsatility (~1 sec) leading to a low probability of corrupted epochs	None	Visual N75, P100 fMRI constraint EEG source localization	Not investigated
Bonmassar *et al.*, 2002	Visual task, checkerboard at 4 Hz (15 subjects)	Interleaved 1-sec fMRI, 1-sec EEG	Average 100, 250 epochs	1.5T, 32 EEG channels, bipolar montage	Adaptive filtering using a piezoelectric transducer motion detector	None	N75, P1, comparison outside, static 1.5T field and interleaved acquisition	Not investigated
Christmann *et al.*, 2002	Somato-sensory task (6 subjects)	Interleaved 2.5-sec fMRI, 2.5-sec EEG	2000 epochs acquired	1.5T, 26 EEG channels, referential montage	Not specified	None	N20–P30–P60, (ERPs not shown) fMRI seeded dipole modeling	Not investigated

Mulert *et al.*, 2002	Auditory oddball task (10 subjects)	Interleaved 2-sec fMRI, 1-sec EEG	not provided	1.5T, 29 EEG channels, referential montage	No specific BCG artifact removal	None	N1 and P3	Not investigated
Liebenthal *et al.*, 2003	Passive auditory oddball task (7 subjects)	Interleaved 2-sec fMRI, 8-sec EEG (blocks of 16 stimuli)	Average on at least 190 epochs	1.5T, 30 EEG channels, referential montage	Averaged pulse artifact subtraction followed by adaptive filtering	None	Mismatch negativity (MMN)	Correlation between BOLD and MMN
Thees *et al.*, 2003	2 somato-sensory tasks (6 subjects)	Interleaved 1-sec fMRI, 1-sec EEG	125, 175 epochs acquired, average 100, 140 epochs	1.5T, 32 EEG channels, referential montage	None	None	Dipole modeling	Not investigated
Foucher *et al.*, 2003	Visual oddball task (5 subjects)	Interleaved 2.4-sec fMRI, 1.6-sec EEG	35 epochs acquired (rare stimuli), average on 26 trials	2T, 10 EEG channels, referential montage	Orthogonalization of the EEG data with respect to ECG signal	None	Visual N200 and P300, event-related gamma oscillations	Discussed but not directly tested
Mulert *et al.*, 2004	Auditory oddball task (10 subjects)	Interleaved 2-sec fMRI, 1-sec EEG	Average on at least 40 epochs	1.5T, 27 EEG channels, referential montage	No specific BCG artifact removal	None	Auditory N1, P2 and P3; comparison inside/outside scanner; current density sources	Not investigated
Nagai *et al.*, 2004	Auditory CNV task (5 subjects)	Continuous	Average on 40 epochs	2T, 10 EEG channels, referential montage	Average pulse artifact subtraction followed by adaptive filtering	Average artifact subtraction, adaptive filtering	Auditory CNV	Correlation ERP–fMRI signals

efficiency should be emphasized. For dipole modeling and group-wise correlational analysis, on the other hand, the detection power is more important. In these cases, it is also important to verify that the task design is optimal for detection of the specific effect of interest. A design that is optimal for estimating the differential activation to two stimulus types, for example, may not be optimal for estimating the activation to each stimulus individually.

VII. Integrating fMRI and ERP Data

Table I summarizes the ERP components examined in several recent EEG–fMRI studies. Currently, two methods are widely used to integrate ERP and fMRI data; one relies on using ERPs as predictors of fMRI response, and the other uses dipole-based biophysical modeling to integrate ERPs and fMRI data. These methods are equally applicable to simultaneously and nonsimultaneously acquired EEG–fMRI data. Several researchers are now developing newer, more biophysically rigorous methods for integrating ERP and fMRI data. One such approach involves calculating three-dimensional volume currents at each voxel in the cortex from the surface measurements. In these distributed source models, the current density is estimated along the cortical surface and thresholded using statistical parametric analysis to generate dynamic maps of cortical response. A discussion of these methods is beyond the scope of this chapter, the interested reader is referred to Dale *et al.* (2000) and Kiebel and Friston (2004).

A. Correlation of fMRI and ERP Data

One method for combining fMRI and ERP data is to use correlation analyses. In this approach, fMRI activation is correlated with the amplitude of a specific ERP component. The analysis can be performed either at the group level or at the individual-subject level. In the former, the amplitude of the ERP response and one task-related contrast image per subject are subjected to a correlational analysis. The amplitudes of the ERPs are measured either at the maximum peak of the component or in a time window that can include one or several ERP peaks. This method has the advantage of mapping brain responses that are specifically correlated with a specific and reliable ERP signal, but the correlational analysis provides little information about the temporal profile or temporal order of responses that led to the specific ERP component.

The individual subject–level analysis is a more powerful approach to combining event-related EEG and fMRI data. Here, windowed ERP amplitudes from a parametric manipulation are used as covariates in a general linear model–based analysis of the fMRI data. For example, Liebenthal *et al.* (2003) used such an approach to examine brain generators underlying mismatch negativity (MMN) (Naatanen, 1995). The MMN is elicited when subjects listen passively to a sequence of two different tones, one presented less frequently than the other. The presentation of the infrequent tone evokes an increased negative deflection at about 150 ms following stimulus onset. This negativity is thought to index automatic detection of stimulus deviancy. Liebenthal *et al.* (2003) used ERP amplitudes from three conditions—standards, small deviants, and large deviants—in a correlational analysis of fMRI data, and they found that restricted regions of the left and right superior temporal gyri, Heschl's gyri, and planum temporale showed significant correlation with the amplitude of the MMN. This approach and its extensions, using unaveraged EEG data, are likely to be widely used in the next few years as the first stage in the integration of fMRI and EEG data. It is expected that these methods will eventually be supplanted by more detailed biophysical modeling, once they are sufficiently validated and tested.

B. fMRI-Constrained Dipole Modeling

A more general method for combining EEG and fMRI data has been to use spatial information based on structural and functional MRI data to constrain the location of ERP sources (Fig. 6). One common approach uses dipoles to model the neural activity; this method is useful and reliable when the approximate locations of the ERP sources are known (Scherg and Von Cramon, 1985). Dipoles are a good model of the distal electrical fields generated by specific neural sources that are active above the background EEG. Single or bilaterally symmetric dipoles are often used when neuronal activity is confined to primary sensory regions. Beyond this, the main problem is to determine the location of the cortical sources and to then model the time course of activity in each location source.

fMRI activations, if used with the appropriate neuroanatomical and neurofunctional constraints, can be extremely effective in specifying initial dipole locations. These models require information about the geometry and conductivities of the different layers of the head, including the cortex, the skull, and the scalp. The contribution of each cortical source to the scalp-recorded signals can then be computed, and best-fitting dipole orientations and strengths can be estimated from a linear inverse solution (Babiloni *et al.*, 2003; Liu *et al.*, 1998; Scherg and Von Cramon, 1985). This analysis also yields the expected time course for each dipole. To the extent that the signals are high at latency windows that correspond to the peak latency of given ERP components and are low outside of those windows, this technique provides strong evidence for localizing

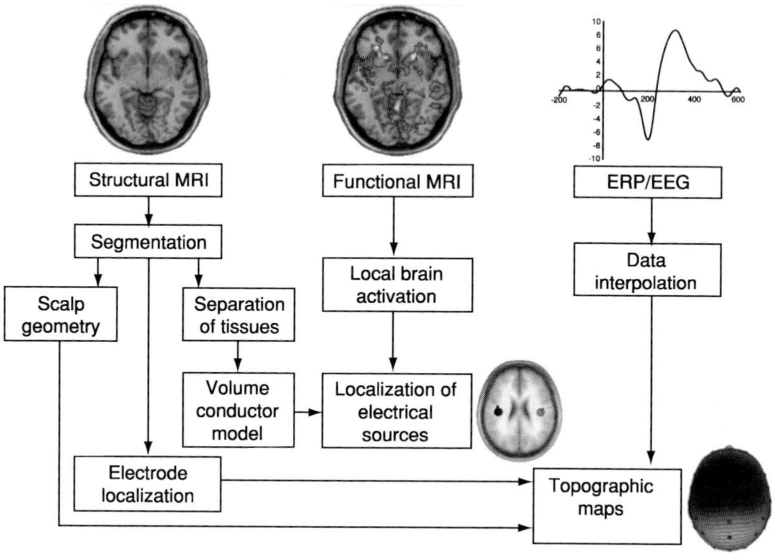

FIG. 6. Schematic overview of some of the steps involved in biophysical modeling of EEG data using constraints derived from structural and functional MRI data. (See Color Insert.)

the source of the scalp-recorded ERP component. The orientation and strength of the dipoles can be fitted on the entire ERP window or on a window of specific components of the ERP. Finally, to investigate the relationship between the ERP and fMRI activation, correlations between dipole strength and the height and extent of fMRI activation can be examined. For further details on dipole modeling, the reader is referred to a recent review by Darvas *et al.* (2004).

Various approximations are used to integrate fMRI activations into the source-localization procedures. One approach is to use Talairach coordinates of fMRI activation to be entered into a three-shell model of the cortex, skull, and scalp (Zanow and Peters, 1995). This approach obviates complex problems with fMRI and EEG coregistration and has been widely used to localize ERP sources (Fuchs *et al.*, 2004; Liotti *et al.*, 2000; Menon *et al.*, 1997; Opitz *et al.*, 1999; Pouthas *et al.*, 2000; Tarkka *et al.*, 1995; Wang *et al.*, 1999; Woldorff *et al.*, 2002). The use of individual subject MRIs to construct more accurate head models is, however, becoming increasingly common (Fuchs *et al.*, 2004).

Regardless of the models used, prior information about possible and impossible sources needs to be incorporated into the analysis. For example, brain areas that do not have the layered organization (including elements such as the thalamus and basal ganglia) needed to generate far fields that can be recorded on the scalp (Nunez, 1999; Pedley and Traub, 1990) should not be included in the dipole model.

We now describe two examples from our research that help explain these ideas. In one study, we used fMRI and ERP with an auditory oddball task to investigate the neural bases of directed attentional processing (Menon *et al.*, 1997). During this task, subjects were asked to detect deviant target stimuli presented randomly within trains of standard stimuli. Event-related fMRI and EEG data were acquired from the same subjects during performance of identical tasks. We combined the two types of data to investigate the neural generators of the P300. Target detection elicited significantly greater activation bilaterally in the temporal–parietal cortex, thalamus, and anterior cingulate cortex (ACC). Spatio-temporal modeling of ERPs based on dipole locations derived from the event-related fMRI indicated that bilateral sources in the temporal–parietal cortex are the main generators of the P300 (Fig. 7). The findings provide convergent fMRI and EEG evidence of significant activation of the temporal–parietal cortex in the interval from 285–610 ms after stimulus onset during target detection. The methods developed in the study provided a novel multimodal neuroimaging technique by which to investigate the spatio-temporal aspects of processes underlying brain functioning.

In a recent study, we extended this analysis and used combined ERP and fMRI to investigate the timing of early, modality-specific, attentional modulation by the ACC. The ACC has been reported to modulate responses in higher-order-association cortices; however, it is not known whether the ACC has a modulatory effect on early sensory processing regions. We used fMRI and effective-connectivity analyses to examine activation and connectivity of the ACC using fMRI data along with fMRI-constrained dipole modeling of ERPs obtained from subjects who performed auditory and visual oddball attention tasks. fMRI activation showed that the ACC response was similar for auditory and visual tasks; however, effective-connectivity analyses showed modality-specific effects with increased ACC influence on the precuneus during the visual task and on the Heschl's and superior temporal gyri during the auditory task (Fig. 8). ERP data recorded using the same oddball tasks showed large N2 and P3 components in response to the target stimuli in both modalities. Dipole modeling of the whole ERP segment was based on source locations determined from fMRI activations; we created one dipole model for the auditory task and another for the visual task. When the fMRI activations were similar in the two tasks, the dipole was placed at the same coordinates in both. This analysis showed that the ACC is the major generator of the N2b–P3a attention-related components in both modalities and that the primary sensory regions generate a small mismatch signal about 50 msec prior to feedback from the ACC. Responses in these sensory regions are significantly enhanced 60 msec after feedback from the ACC. Taken together, these results provide converging neuroimaging and electrophysiological evidence of top-down attentional modulation of early sensory processing by the ACC. Our findings suggest a model of attentional control based on dynamic

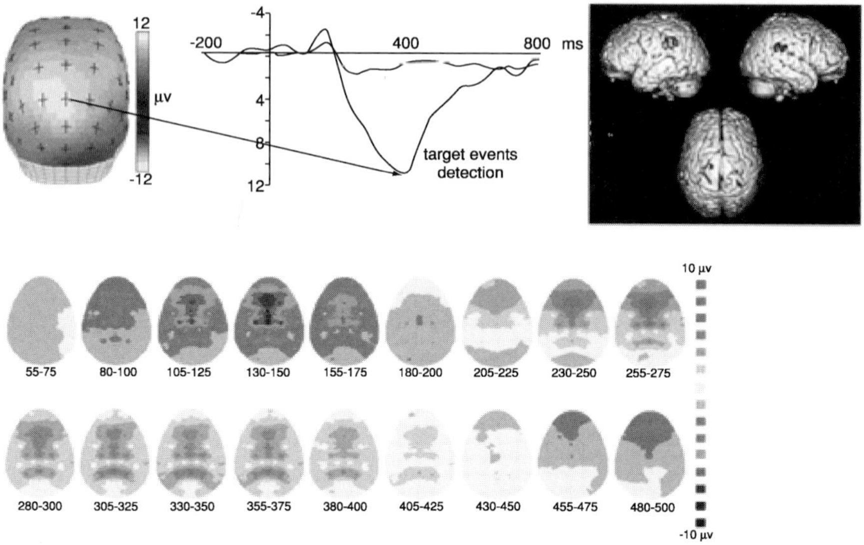

FIG. 7. Top left: Topography and time course of event-related potentials (ERPs) from normal adults during an "oddball" target detection task. The P300 component elicited during target detection has maximum amplitude over the parietal cortex approximately 300–400 ms after onset of the deviant stimulus. Top right: Event-related fMRI responses during target detection. Activated regions include the left and right temporo-parietal cortex 1. Bottom: Spatio-temporal dynamics of brain potentials acquired during target detection. (See Color Insert.)

bottom-up and top-down interactions between the ACC and primary sensory regions.

VIII. Integrating fMRI and (Single-Trial) EEG Data

There is growing interest in combining single trial, or non-phase-locked, EEG with fMRI. These signals are related to perceptual and cognitive processing, but they may or may not be locked to an external stimulus. Combined EEG–fMRI techniques have been used to identify the neural correlates of clinically or behaviorally important spontaneous EEG activity, such as interictal spikes, the alpha rhythm, and sleep waves. The area where the greatest progress has been made is epileptic-seizure localization. The common method of integrating EEG and fMRI data here is to transform EEG data into a physiologically meaningful covariate to be used in a voxel-based general linear model (Salek-Haddadi *et al.*, 2003).

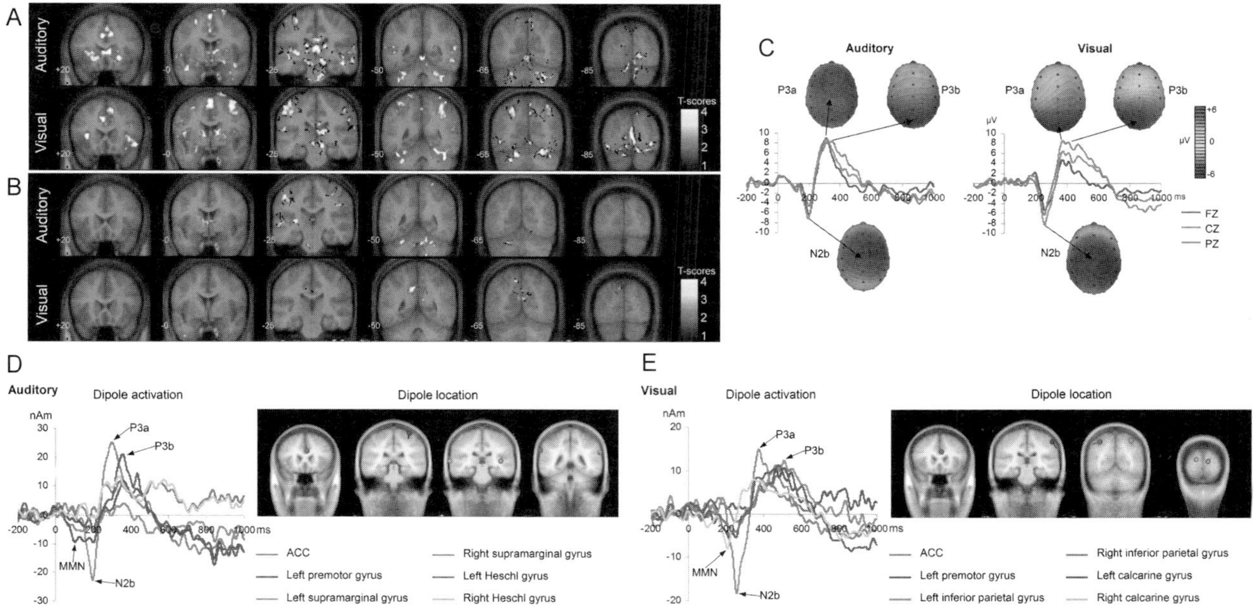

FIG. 8. (A) Brain areas that showed significantly greater activation during auditory (top) and visual (bottom) target detection. (B) Brain regions that showed significantly greater connectivity with the anterior cingulate cortex (ACC) during target detection. (C) Difference waveforms (targets minus standards) of the auditory and visual ERPs at electrodes FZ, CZ, and PZ. The topographical maps shown for the N2 and P3 components, however, reveal very little useful information. (D and E) The fMRI-derived dipole model of the ERPs, however, shows a large contribution by the ACC to the N2b–P3a components in both auditory and visual modalities. (See Color Insert.)

A. EEG and Epilepsy Seizure Localization

The recording of EEG during fMRI has opened up new avenues in epilepsy research. In fact, initial applications of single-trial EEG and fMRI have, until very recently, been limited to seizure localization. Interictal discharges underlying epileptogenic neural activity can be readily detected on scalp EEG; however, the origin of this activity cannot be inferred except in terms of lobes and hemispheres. This work differs from most applications of EEG–fMRI in that each patient has a different spatial and temporal distribution of the epileptogenic activity, and signatures of this activity need to be determined with relatively high accuracy and efficiency.

In the past few years, the simultaneous measurement of EEG and fMRI has been used by several groups to study interictal activity in patients with epilepsy (Aghakhani *et al.*, 2004; Archer *et al.*, 2003; Krakow *et al.*, 2001; Lazeyras *et al.*, 2000; Lemieux *et al.*, 2001a). Although early studies used the EEG spikes to trigger fMRI acquisition (Krakow *et al.*, 1999; Lazeyras *et al.*, 2000; Seeck *et al.*, 1998), use of continuous EEG and fMRI data is now common (Lemieux *et al.*, 2001b). In either case, these studies have benefited from the relatively high amplitude of epileptogenic activity, which, at about 100 μV, is visible on scalp EEG—standing in contrast to perceptual and cognitive ERPs, which have amplitudes of about 5–10 μV after averaging over several trials. However, in some patients, the events are usually not very frequent and of limited duration (from a fraction of a second to [rarely] more than 10 seconds) (Bagshaw *et al.*, 2005), making the study of interictal epileptiform events extremely challenging in these cases. Major applications of fMRI in epilepsy include the localization of task-correlated language and memory function and the localization of ictal and paroxysmal phenomena (Detre, 2004). For example, research from several laboratories has shown that language lateralization by fMRI provides results comparable to those from intracarotid amobarbital testing (Waites *et al.*, 2005). Recent studies have also shown that widespread regions of the cerebral cortex and thalamus underlie generalized spike and wave (GSW) or polyspike and wave bursts in patients with idiopathic generalized epilepsy (Aghakhani *et al.*, 2004). Interestingly, both activation and deactivation were observed in relation to the GSW; activation predominated over deactivation in the thalamus, whereas the opposite was seen in the cerebral cortex. The presence of a thalamic BOLD response in most human patients converges on the thalamic involvement seen in animal models. Findings such as these provide further assurance of the usefulness of the combined EEG–fMRI approach to better understand the brain systems underlying various forms of epilepsy. For further details, the reader is referred to Lemieux (2004).

B. EEG and Resting-State fMRI

fMRI studies in humans have revealed task-specific increases in brain activity that are associated with various mental activities. Many of these studies also show task-independent decreases, especially when tasks with high-level cognitive load have been compared to a passive state, such as simple fixation or closed eyes. These decreases have raised the possibility that there might be a baseline or resting state of brain function involving a specific set of mental operations (Gusnard *et al.*, 2001). The "default-mode" hypothesis of brain function (Raichle *et al.*, 2001) proposes that there is a specific neural network whose activity predominates in the relaxed, resting state and that activity in this default-mode neural network decreases during the performance of a demanding externally cued task. This hypothesis is supported by fMRI studies of the conscious resting state (Greicius *et al.*, 2003) and cognitively undemanding tasks (Greicius and Menon, 2004), which have shown strong temporal coherence of neural activity among brain regions implicated in the default mode. Changes in ongoing EEG activity unrelated to any specific externally mediated tasks have also been well known. For example, the alpha rhythm in the EEG increases when subjects stay relaxed with their eyes closed, and the beta rhythms are prominent when individuals are alert and attentive (Niedermeyer and Lopes da Silva, 2004). Resting-state fMRI and EEG have been used to examine the neural bases of these dynamic changes in baseline state as well as to examine their relation to default-mode networks.

Here, the analytic approach is to correlate ongoing EEG fluctuations with changes in the resting-state fMRI signal. For example, the spectral power of the EEG in the alpha range can be used as a covariate in a standard general linear model. Using this approach, several studies have examined the neural bases of the alpha rhythm using data from simultaneous EEG–fMRI recordings (Goldman *et al.*, 2000, 2002; Laufs *et al.*, 2003a,b). In these studies, the amplitude of the EEG in the alpha band (8–12 Hz) is computed over a moving window that steps about every 30 seconds across a total recording interval of 5 to 8 minutes. The resulting waveform is convolved with the HRF and used as a covariate in a conventional analysis of fMRI data. Goldman *et al.* (2002) found that the alpha rhythm was correlated with fMRI response in the thalamus. On the other hand, Laufs *et al.* (2003a) reported that there was little positive correlation of resting-state brain activity with alpha power (8–12 Hz), but strong and widespread negative correlation was observed in the lateral frontal and parietal cortices, which are known to support attention and working memory. Power in a 17- to 23-Hz range of beta activity was positively correlated with activity in retrosplenial, temporo-parietal, and dorsomedial prefrontal cortices, regions that overlap to some extent with the default-mode network identified in fMRI studies. These findings raise the possibility that beta rhythms may index spontaneous cognitive operations during conscious rest. It is likely that simultaneously recorded, multichannel beta- and gamma-band EEG might assist in the interpretation of task-related fMRI data, since high beta and gamma EEG amplitudes both imply high rates of energy utilization (Freeman, 2004a,b).

IX. Conclusion

In the past 5 years, considerable progress has been made on several fronts in combined EEG and fMRI studies of human brain function. These include optimal study design, data acquisition, and data analysis. As better methods for removing artifacts in MRI scanning become available, it is to be expected that fully continuous EEG–fMRI recordings will become the mainstay of multimodal functional brain imaging. An area that clearly needs further research is the development, and validation, of procedures and algorithms for biophysical modeling of EEG and fMRI signals. Such models will help researchers to better integrate spatial and temporal information in fMRI and EEG. Validation of procedures for artifact reduction is also necessary; this can be easily accomplished using computer simulations, but very little research has been performed in this area. Finally, we note that although clinical applications have thus far been limited to epilepsy, in the future combined EEG and fMRI studies will also provide new insights into the dynamical bases of psychiatric, neurological. and neurodevelopmental disorders.

Acknowledgments

It is a pleasure to thank Dr. Gary Glover and Gaurav Srivastava for useful discussions. This chapter is dedicated to the memory of Achatil Bhaskara Menon.

References

Aghakhani, Y., Bagshaw, A. P., Benar, C. G., Hawco, C., Andermann, F., Dubeau, F., and Gotman, J. (2004). fMRI activation during spike and wave discharges in idiopathic generalized epilepsy. *Brain* **127,** 1127–1144.

Allen, P. J., Josephs, O., and Turner, R. (2000). A method for removing imaging artifact from continuous EEG recorded during functional MRI. *Neuroimage* **12,** 230–239.

Allen, P.J., Polizzi, G., Krakow, K., Fish, D. R., and Lemieux, L. (1998). Identification of EEG events in the MR scanner: The problem of pulse artifact and a method for its subtraction. *Neuroimage* **8,** 229–239.

Archer, J. S., Abbott, D. F., Waites, A. B., and Jackson, G. D. (2003). fMRI "deactivation" of the posterior cingulate during generalized spike and wave. *Neuroimage* **20,** 1915–1922.

Babiloni, F., Babiloni, C., Carducci, F., Romani, G. L., Rossini, P. M., Angelone, L. M., and Cincotti, F. (2003). Multimodal integration of high-resolution EEG and functional magnetic resonance imaging data: A simulation study. *Neuroimage* **19,** 1–15.

Babiloni, F., Mattia, D., Babiloni, C., Astolfi, L., Salinari, S., Basilisco, A., Rossini, P. M., Marciani, M. G., and Cincotti, F. (2004). Multimodal integration of EEG, MEG and fMRI data for the solution of the neuroimage puzzle. *Magn. Reson. Imaging* **22,** 1471–1476.

Bagshaw, A. P., Hawco, C., Benar, C. G., Kobayashi, E., Aghakhani, Y., Dubeau, F., Pike, G. B., and Gotman, J. (2005). Analysis of the EEG-fMRI response to prolonged bursts of interictal epileptiform activity. *Neuroimage* **24,** 1099–1112.

Ball, T., Schreiber, A., Feige, B., Wagner, M., Lucking, C. H., and Kristeva-Feige, R. (1999). The role of higher-order motor areas in voluntary movement as revealed by high-resolution EEG and fMRI. *Neuroimage* **10,** 682–694.

Benar, C. G., Gross, D. W., Wang, Y., Petre, V., Pike, B., Dubeau, F., and Gotman, J. (2002). The BOLD response to interictal epileptiform discharges. *Neuroimage* **17,** 1182–1192.

Bonmassar, G., Anami, K., Ives, J., and Belliveau, J. W. (1999). Visual evoked potential (VEP) measured by simultaneous 64-channel EEG and 3T fMRI. *Neuroreport* **10,** 1893–1897.

Bonmassar, G., Hadjikhani, N., Ives, J. R., Hinton, D., and Belliveau, J. W. (2001a). Influence of EEG electrodes on the BOLD fMRI signal. *Hum. Brain Mapp.* **14,** 108–115.

Bonmassar, G., Purdon, P. L., Jaaskelainen, I. P., Chiappa, K., Solo, V., Brown, E. N., and Belliveau, J. W. (2002). Motion and ballistocardiogram artifact removal for interleaved recording of EEG and EPs during MRI. *Neuroimage* **16,** 1127–1141.

Bonmassar, G., Schwartz, D. P., Liu, A. K., Kwong, K. K., Dale, A. M., and Belliveau, J. W. (2001b). Spatiotemporal brain imaging of visual-evoked activity using interleaved EEG and fMRI recordings. *Neuroimage* **13,** 1035–1043.

Brunia, C. H., and van Boxtel, G. J. (2001). Wait and see. *Int. J. Psychophysiol.* **43,** 59–75.

Burock, M. A., Buckner, R. L., Woldorff, M. G., Rosen, B. R., and Dale, A. M. (1998). Randomized event-related experimental designs allow for extremely rapid presentation rates using functional MRI. *Neuroreport* **9,** 3735–3739.

Christmann, C., Ruf, M., Braus, D. F., and Flor, H. (2002). Simultaneous electroencephalography and functional magnetic resonance imaging of primary and secondary somatosensory cortex in humans after electrical stimulation. *Neurosci. Lett.* **333,** 69–73.

Czisch, M., Wehrle, R., Kaufmann, C., Wetter, T. C., Holsboer, F., Pollmacher, T., and Auer, D. P. (2004). Functional MRI during sleep: BOLD signal decreases and their electrophysiological correlates. *Eur. J. Neurosci.* **20,** 566–574.

Dale, A. M. (1999). Optimal experimental design for event-related fMRI. *Hum. Brain Mapp.* **8,** 109–114.

Dale, A. M., Liu, A. K., Fischl, B. R., Buckner, R. L., Belliveau, J. W., Lewine, J. D., and Halgren, E. (2000). Dynamic statistical parametric mapping: Combining fMRI and MEG for high-resolution imaging of cortical activity. *Neuron* **26,** 55–67.

Darvas, F., Pantazis, D., Kucukaltun-Yildirim, E., and Leahy, R. M. (2004). Mapping human brain function with MEG and EEG: Methods and validation. *Neuroimage* **23**(Suppl. 1), S289–S299.

Detre, J. A. (2004). fMRI: Applications in epilepsy. *Epilepsia* **45**(Suppl. 4), 26–31.

Felblinger, J., Slotboom, J., Kreis, R., Jung, B., and Boesch, C. (1999). Restoration of electrophysiological signals distorted by inductive effects of magnetic field gradients during MR sequences. *Magn. Reson. Med.* **41,** 715–721.

Formisano, E., Kim, D. S., Di Salle, F., van de Moortele, P. F., Ugurbil, K., and Goebel, R. (2003). Mirror-symmetric tonotopic maps in human primary auditory cortex. *Neuron* **40,** 859–869.

Freeman, W. J. (2004a). Origin, structure, and role of background EEG activity. Part 1. Analytic amplitude. *Clin. Neurophysiol.* **115,** 2077–2088.

Freeman, W. J. (2004b). Origin, structure, and role of background EEG activity. Part 2. Analytic phase. *Clin. Neurophysiol.* **115,** 2089–2107.

Freeman, W. J., Holmes, M. D., Burke, B. C., and Vanhatalo, S. (2003). Spatial spectra of scalp EEG and EMG from awake humans. *Clin. Neurophysiol.* **114,** 1053–1068.

Friston, K. J., Josephs, O., Rees, G., and Turner, R. (1998). Nonlinear event-related responses in fMRI. *Magn. Reson. Med.* **39,** 41–52.

Fuchs, M., Ford, M. R., Sands, S., and Lew, H. L. (2004). Overview of dipole source localization. *Phys. Med. Rehabil. Clin. N. Am.* **15,** 251–262.

George, J. S., Aine, C. J., Mosher, J. C., Schmidt, D. M., Ranken, D. M., Schlitt, H. A., Wood, C. C., Lewine, J. D., Sanders, J. A., and Belliveau, J. W. (1995). Mapping function in the human brain with magnetoencephalography, anatomical magnetic resonance imaging, and functional magnetic resonance imaging. *J. Clin. Neurophysiol.* **12,** 406–431.

Goldman, R. I., Stern, J. M., Engel, J., Jr., and Cohen, M. S. (2000). Acquiring simultaneous EEG and functional MRI. *Clin. Neurophysiol.* **111,** 1974–1980.

Goldman, R. I., Stern, J. M., Engel, J., Jr., and Cohen, M. S. (2002). Simultaneous EEG and fMRI of the alpha rhythm. *Neuroreport* **13,** 2487–2492.

Gusnard, D. A., Raichle, M. E., and Raichle, M. E. (2001). Searching for a baseline: Functional imaging and the resting human brain. *Nat. Rev. Neurosci.* **2,** 685–694.

Hall, D. A., Haggard, M. P., Akeroyd, M. A., Palmer, A. R., Summerfield, A. Q., Elliott, M. R., Gurney, E. M., and Bowtell, R. W. (1999). "Sparse" temporal sampling in auditory fMRI. *Hum. Brain Mapp.* **7,** 213–223.

Handy, T. C. (2005). "Event-related potentials: A methods handbook," MIT Press, Cambridge, Mass.

Heinze, H. J., Mangun, G. R., Burchert, W., Hinrichs, H., Scholz, M., Munte, T. F., Gos, A., Scherg, M., Johannes, S., Hundeshagen, H.., *et al.* (1994). Combined spatial and temporal imaging of brain activity during visual selective attention in humans. *Nature* **372,** 543–546.

Huettel, S. A., Song, A. W., and McCarthy, G. (2004). "Functional magnetic resonance imaging." Sinauer Associates, Publishers, Sunderland, Mass.

Ives, J. R., Warach, S., Schmitt, F., Edelman, R. R., and Schomer, D. L. (1993). Monitoring the patient's EEG during echo planar MRI. *Electroencephalogr. Clin. Neurophysiol.* **87,** 417–420.

Kiebel, S. J., and Friston, K. J. (2004). Statistical parametric mapping for event-related potentials: I. Generic considerations. *Neuroimage* **22,** 492–502.

Krakow, K., Allen, P. J., Symms, M. R., Lemieux, L., Josephs, O., and Fish, D. R. (2000). EEG recording during fMRI experiments: Image quality. *Hum. Brain Mapp.* **10,** 10–15.

Krakow, K., Messina, D., Lemieux, L., Duncan, J. S., and Fish, D. R. (2001). Functional MRI activation of individual interictal epileptiform spikes. *Neuroimage* **13,** 502–505.

Krakow, K., Woermann, F. G., Symms, M. R., Allen, P. J., Lemieux, L., Barker, G. J., Duncan, J. S., and Fish, D. R. (1999). EEG-triggered functional MRI of interictal epileptiform activity in patients with partial seizures. *Brain* **122**(Pt 9), 1679–1688.

Lagerlund, T. D., Sharbrough, F. W., and Busacker, N. E. (2003). A general method for remontaging based on a singular value decomposition algorithm. *J. Clin. Neurophysiol.* **20,** 179–187.

Laufs, H., Kleinschmidt, A., Beyerle, A., Eger, E., Salek-Haddadi, A., Preibisch, C., and Krakow, K. (2003a). EEG-correlated fMRI of human alpha activity. *Neuroimage* **19,** 1463–1476.

Laufs, H., Krakow, K., Sterzer, P., Eger, E., Beyerle, A., Salek-Haddadi, A., and Kleinschmidt, A. (2003b). Electroencephalographic signatures of attentional and cognitive default modes in spontaneous brain activity fluctuations at rest. *Proc. Natl. Acad. Sci. USA* **100,** 11053–11058.

Lazeyras, F., Blanke, O., Perrig, S., Zimine, I., Golay, X., Delavelle, J., Michel, C. M., de Tribolet, N., Villemure, J. G., and Seeck, M. (2000). EEG-triggered functional MRI in patients with pharmacoresistant epilepsy. *J. Magn. Reson. Imaging* **12,** 177–185.

Lazeyras, F., Zimine, I., Blanke, O., Perrig, S. H., and Seeck, M. (2001). Functional MRI with simultaneous EEG recording: Feasibility and application to motor and visual activation. *J. Magn. Reson. Imaging* **13,** 943–948.

Lemieux, L. (2004). Electroencephalography-correlated functional MR imaging studies of epileptic activity. *Neuroimaging Clin. N. Am.* **14,** 487–506.

Lemieux, L., Krakow, K., and Fish, D. R. (2001a). Comparison of spike-triggered functional MRI BOLD activation and EEG dipole model localization. *Neuroimage* **14,** 1097–1104.

Lemieux, L., Salek-Haddadi, A., Josephs, O., Allen, P., Toms, N., Scott, C., Krakow, K., Turner, R., and Fish, D. R. (2001b). Event-related fMRI with simultaneous and continuous EEG: Description of the method and initial case report. *Neuroimage* **14,** 780–787.

Liebenthal, E., Ellingson, M. L., Spanaki, M. V., Prieto, T. E., Ropella, K. M., and Binder, J. R. (2003). Simultaneous ERP and fMRI of the auditory cortex in a passive oddball paradigm. *Neuroimage* **19,** 1395–1404.

Liotti, M., Woldorff, M. G., Perez, R., and Mayberg, H. S. (2000). An ERP study of the temporal course of the Stroop color-word interference effect. *Neuropsychologia* **38,** 701–711.

Liu, A. K., Belliveau, J. W., and Dale, A. M. (1998). Spatiotemporal imaging of human brain activity using functional MRI constrained magnetoencephalography data: Monte Carlo simulations. *Proc. Natl. Acad. Sci. USA* **95,** 8945–8950.

Liu, A. K., Dale, A. M., and Belliveau, J. W. (2002). Monte Carlo simulation studies of EEG and MEG localization accuracy. *Hum. Brain Mapp.* **16,** 47–62.

Liu, T. T. (2004). Efficiency, power, and entropy in event-related fMRI with multiple trial types. Part II: Design of experiments. *Neuroimage* **21,** 401–413.

Liu, T. T., and Frank, L. R. (2004). Efficiency, power, and entropy in event-related FMRI with multiple trial types. Part I: Theory. *Neuroimage* **21,** 387–400.

Liu, T. T., Frank, L. R., Wong, E. C., and Buxton, R. B. (2001). Detection power, estimation efficiency, and predictability in event-related fMRI. *Neuroimage* **13,** 759–773.

Logothetis, N. K., Pauls, J., Augath, M., Trinath, T., and Oeltermann, A. (2001). Neurophysiological investigation of the basis of the fMRI signal. *Nature* **412,** 150–157.

Logothetis, N. K., and Pfeuffer, J. (2004). On the nature of the BOLD fMRI contrast mechanism. *Magn. Reson. Imaging* **22,** 1517–1531.

Logothetis, N. K., and Wandell, B. A. (2004). Interpreting the BOLD signal. *Annu. Rev. Physiol.* **66,** 735–769.

Makeig, S., Delorme, A., Westerfield, M., Jung, T. P., Townsend, J., Courchesne, E., and Sejnowski, T. J. (2004). Electroencephalographic brain dynamics following manually responded visual targets. *PLoS Biol.* **2,** e176.

Makeig, S., Westerfield, M., Jung, T. P., Enghoff, S., Townsend, J., Courchesne, E., and Sejnowski, T. J. (2002). Dynamic brain sources of visual evoked responses. *Science* **295,** 690–694.

Matsuda, T., Matsuura, M., Ohkubo, T., Ohkubo, H., Atsumi, Y., Tamaki, M., Takahashi, K., Matsushima, E., and Kojima, T. (2002). Influence of arousal level for functional magnetic resonance imaging (fMRI) study: Simultaneous recording of fMRI and electroencephalogram. *Psychiatry Clin. Neurosci.* **56,** 289–290.

Menon, V., Ford, J. M., Lim, K. O., Glover, G. H., and Pfefferbaum, A. (1997). Combined event-related fMRI and EEG evidence for temporal-parietal cortex activation during target detection. *Neuroreport* **8,** 3029–3037.

Menon, V., Freeman, W. J., Cutillo, B. A., Desmond, J. E., Ward, M. F., Bressler, S. L., Laxer, K. D., Barbaro, N., and Gevins, A. S. (1996). Spatio-temporal correlations in human gamma band electrocorticograms. *Electroencephalogr. Clin. Neurophysiol.* **98,** 89–102.

Mulert, C., Jager, L., Schmitt, R., Bussfeld, P., Pogarell, O., Moller, H. J., Juckel, G., and Hegerl, U. (2004). Integration of fMRI and simultaneous EEG: Towards a comprehensive understanding of localization and time-course of brain activity in target detection. *Neuroimage* **22,** 83–94.

Naatanen, R. (1995). The mismatch negativity: A powerful tool for cognitive neuroscience. *Ear Hear.* **16,** 6–18.

Nagai, Y., Critchley, H. D., Featherstone, E., Fenwick, P. B., Trimble, M. R., and Dolan, R. J. (2004). Brain activity relating to the contingent negative variation: An fMRI investigation. *Neuroimage* **21,** 1232–1241.

Niedermeyer, E., and Lopes da Silva, F. H. (2004). "Electroencephalography: Basic principles, clinical applications, and related fields," 5th Ed. Lippincott Williams & Wilkins, Philadelphia.

Nunez, P. L. (1999). "Electric Fields of the Brain: The Neurophysics of EEG." Oxford University Press.

Nunez, P. L., and Srinivasan, R. (2006). "Electric fields of the brain: The neurophysics of EEG," 2nd Ed. Oxford University Press, New York.

Ohl, F. W., Deliano, M., Scheich, H., and Freeman, W. J. (2003). Analysis of evoked and emergent patterns of stimulus-related auditory cortical activity. *Rev. Neurosci.* **14,** 35–42.

Opitz, B., Mecklinger, A., Von Cramon, D. Y., and Kruggel, F. (1999). Combining electrophysiological and hemodynamic measures of the auditory oddball. *Psychophysiology* **36,** 142–147.

Pedley, T. A., and Traub, R. D. (1990). "Physiological Basis of the EEG." Raven Press, New York.

Pfeuffer, J., Van de Moortele, P. F., Ugurbil, K., Hu, X., and Glover, G. H. (2002). Correction of physiologically induced global off-resonance effects in dynamic echo-planar and spiral functional imaging. *Magn. Reson. Med.* **47,** 344–353.

Pouthas, V., Garnero, L., Ferrandez, A. M., and Renault, B. (2000). ERPs and PET analysis of time perception: Spatial and temporal brain mapping during visual discrimination tasks. *Hum. Brain Mapp.* **10,** 49–60.

Raichle, M. E., MacLeod, A. M., Snyder, A. Z., Powers, W. J., Gusnard, D. A., and Shulman, G. L. (2001). A default mode of brain function. *Proc. Natl. Acad. Sci. USA* **98,** 676–682.

Regan, D. (1989). "Human Brain Electrophysiology: Evoked Potentials and Evoked Magnetic Fields in Science and Medicine." Elsevier Science, New York.

Rugg, N. (1995). ERP studies of memory. *In* "Electrophysiology of Mind: Event-Related Potentials and Cognition" (M. Rugg and M. Coles, Eds.), pp. 113–127. Oxford University Press, Oxford.

Salek-Haddadi, A., Friston, K. J., Lemieux, L., and Fish, D. R. (2003). Studying spontaneous EEG activity with fMRI. *Brain Res. Rev.* **43,** 110–133.

Salek-Haddadi, A., Merschhemke, M., Lemieux, L., and Fish, D. R. (2002). Simultaneous EEG-Correlated Ictal fMRI. *Neuroimage* **16,** 32–40.

Scarff, C. J., Reynolds, A., Goodyear, B. G., Ponton, C. W., Dort, J. C., and Eggermont, J. J. (2004). Simultaneous 3-T fMRI and high-density recording of human auditory evoked potentials. *Neuroimage* **23,** 1129–1142.

Scherg, M., and Von Cramon, D. (1985). Two bilateral sources of the late AEP as identified by a spatio-temporal dipole model. *Electroencephalogr. Clin. Neurophysiol.* **62,** 32–44.

Seeck, M., Lazeyras, F., Michel, C. M., Blanke, O., Gericke, C. A., Ives, J., Delavelle, J., Golay, X., Haenggeli, C. A., de Tribolet, N., and Landis, T. (1998). Non-invasive epileptic focus localization using EEG-triggered functional MRI and electromagnetic tomography. *Electroencephalogr. Clin. Neurophysiol.* **106,** 508–512.

Sijbers, J., Michiels, I., Verhoye, M., Van Audekerke, J., Van der Linden, A., and Van Dyck, D. (1999). Restoration of MR-induced artifacts in simultaneously recorded MR/EEG data. *Magn. Reson. Imaging* **17,** 1383–1391.

Sijbers, J., Van Audekerke, J., Verhoye, M., Van der Linden, A., and Van Dyck, D. (2000). Reduction of ECG and gradient related artifacts in simultaneously recorded human EEG/MRI data. *Magn. Reson. Imaging* **18,** 881–886.

Snyder, A. Z., Abdullaev, Y. G., Posner, M. I., and Raichle, M. E. (1995). Scalp electrical potentials reflect regional cerebral blood flow responses during processing of written words. *Proc. Natl. Acad. Sci. USA* **92,** 1689–1693.

Srivastava, G., Crottaz-Herbette, S., Lau, K. M., Glover, G. H., and Menon, V. (2005). ICA-based procedures for removing ballistocardiogram artifacts from EEG data acquired in the MRI scanner. *Neuroimage* **24,** 50–60.

Steriade, M. (2001). "The Intact and Sliced Brain." Mit Press, Cambridge, Massachusetts.

Tarkka, I. M., Stokic, D. S., Basile, L. F., and Papanicolaou, A. C. (1995). Electric source localization of the auditory P300 agrees with magnetic source localization. *Electroencephalogr. Clin. Neurophysiol.* **96,** 538–545.

Thees, S., Blankenburg, F., Taskin, B., Curio, G., and Villringer, A. (2003). Dipole source localization and fMRI of simultaneously recorded data applied to somatosensory categorization. *Neuroimage* **18,** 707–719.

Vanhatalo, S., Palva, J. M., Holmes, M. D., Miller, J. W., Voipio, J., and Kaila, K. (2004). Infraslow oscillations modulate excitability and interictal epileptic activity in the human cortex during sleep. *Proc. Natl. Acad. Sci. USA* **101,** 5053–5057.

Waites, A. B., Shaw, M. E., Briellmann, R. S., Labate, A., Abbott, D. F., and Jackson, G. D. (2005). How reliable are fMRI-EEG studies of epilepsy? A nonparametric approach to analysis validation and optimization. *Neuroimage* **24,** 192–199.

Wang, J., Zhou, T., Qiu, M., Du, A., Cai, K., Wang, Z., Zhou, C., Meng, M., Zhuo, Y., Fan, S., and Chen, L. (1999). Relationship between ventral stream for object vision and dorsal stream for spatial vision: An fMRI + ERP study. *Hum. Brain Mapp.* **8,** 170–181.

Woldorff, M. G., Liotti, M., Seabolt, M., Busse, L., Lancaster, J. L., and Fox, P. T. (2002). The temporal dynamics of the effects in occipital cortex of visual-spatial selective attention. *Brain Res. Cogn.* **15,** 1–15.

Zanow, F., and Peters, M. J. (1995). Individually shaped volume conductor models of the head in EEG source localisation. *Med. Biol. Eng. Comput.* **33,** 582–588.

INDEX

C

G

O

P

Q

R

CONTENTS OF RECENT VOLUMES

Volume 39

Volume 40

Volume 41

Volume 42

Volume 43

Volume 44

Volume 45

Volume 46

Volume 47

Volume 48

Volume 49

Volume 50

Volume 51

Volume 52

Volume 53

Volume 54

Volume 55

Volume 56

Volume 57

Volume 58

Volume 59

Volume 60

Volume 61

Volume 62

Volume 63

Volume 64

Volume 65

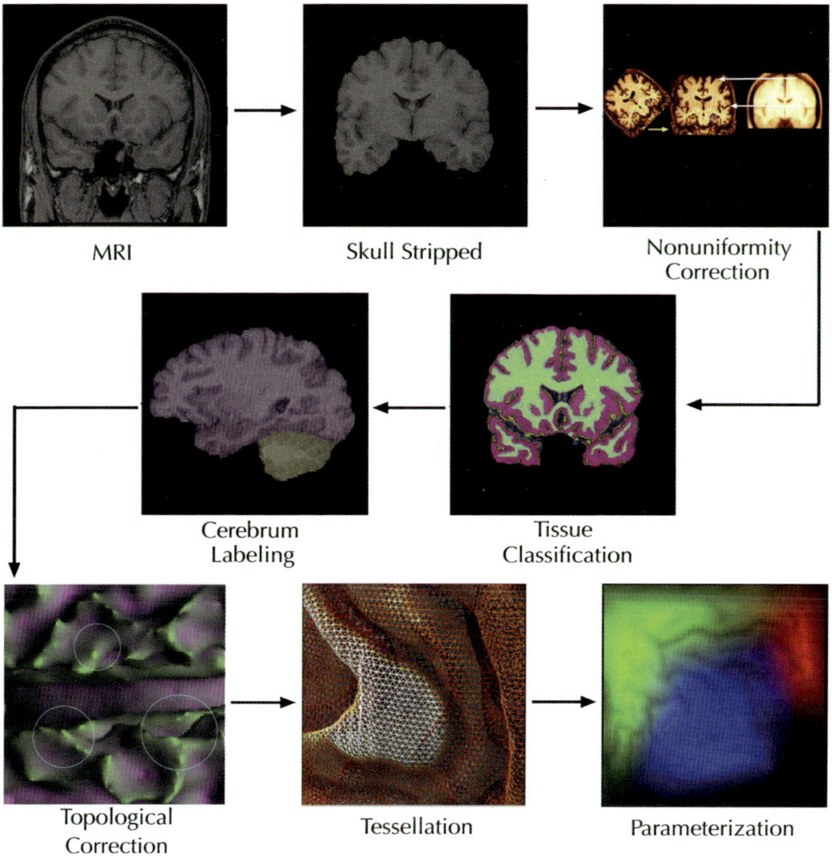

TOGA AND THOMPSON, FIG. 1. Analyzing cortical data. This schematic shows a sequence of image-processing steps that can be applied to MRI scans to map how development and disease affect the cortex. The steps include manual or automated removal of the scalp and other nonbrain tissues from the image (to create 'skull stripped' data), alignment of the data to a standard coordinate space, correction for intensity nonuniformity, tissue classification, parcellation, and creation of cortical surface models (These procedures are detailed in the main text). To help in the comparison of cortical anatomy across subjects, a three-dimensional (3D) cortical surface model can be reconstructed from the scan, defining the interface between gray and white matter. If these interfaces are defined using intensity isosurfaces in the images, additional steps may be needed in order to ensure that the reconstructed surface is topologically equivalent to a sphere (a process called topology correction, which removes small bridges and connections). After this step, the surface can be tessellated using a triangular or rectilinear mesh structure (in tessellation, a small region including the gneu of the corpus callosum is shown). Finally, the surface can be flattened into a square, disc, or sphere, which enables points on the 3D surface to be referenced using 2D coordinates (a process called parameterization). When surfaces are flattened, information on the 3D locations of cortical points can be retained through a color-coding scheme that stores *x,y,z* locations using the color channels red, green, and blue, respectively. Flat maps can then be compared across subjects, as shown in Fig. 2.

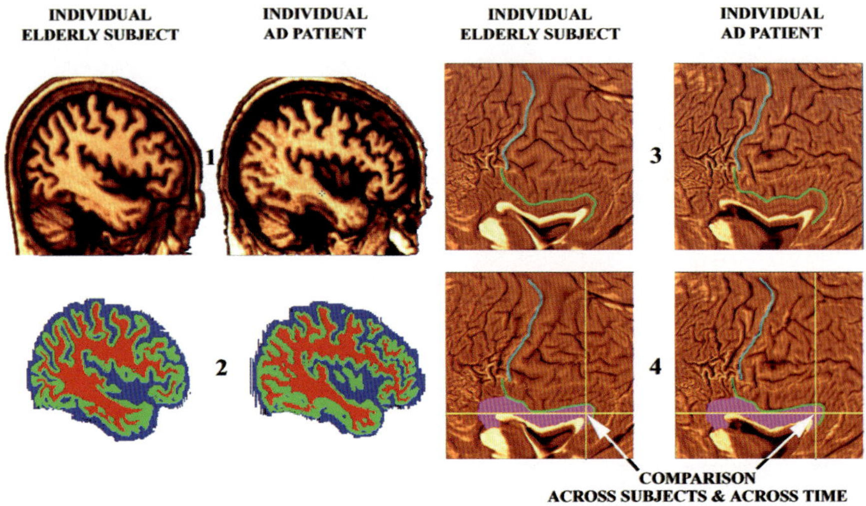

TOGA AND THOMPSON, FIG. 2. Comparing gray matter across subjects. Gray matter is easier to compare across subjects if adjustments are first made for the gyral patterning differences across subjects. This adjustment can be made using cortical pattern matching, which is illustrated here on example brain MRI datasets from a healthy control subject (left column) and from a patient with Alzheimer's disease (right column). First, the MRI images have extracerebral tissues deleted from the scans (stage 1), and individual voxels are classified as gray matter, white matter, or cerebrospinal fluid (CSF) (shown here in green, red, and blue colors) (stage 2). After a 3D geometric model of the cortex is flattened (stage 3), features such as the central sulcus (light blue curve) and cingulate sulcus (green curve) can be reidentified (*http://www.loni.ucla.edu/~esowell/edevel/new_sulcvar.html*). An elastic warp is applied (stage 4) that moves these features, and entire gyral regions (pink colors), into the same reference position in "flat space." After alignment of sulcal patterns from all individual subjects, group comparisons can be made at each 2D pixel (yellow cross-hairs) that effectively compare gray-matter measures across corresponding cortical regions. The cortical measure that is compared, across groups or over time, may be cortical thickness (Fig. 3) or gray-matter density, which is the amount of gray matter (stage 2) that lies within a fixed radius (typically 10–15 mm) of each cortical point. The results of these statistical comparisons can then be plotted back onto an average three-dimensional (3D) cortical model made for the group, and significant findings can be visualized as color-coded maps. Such algorithms bring gray-matter maps from different subjects into a common anatomical reference space, overcoming individual differences in gyral patterns and shape by matching locations point by point throughout the cortex. This enhances the precision of intersubject statistical procedures that detect localized changes in gray matter.

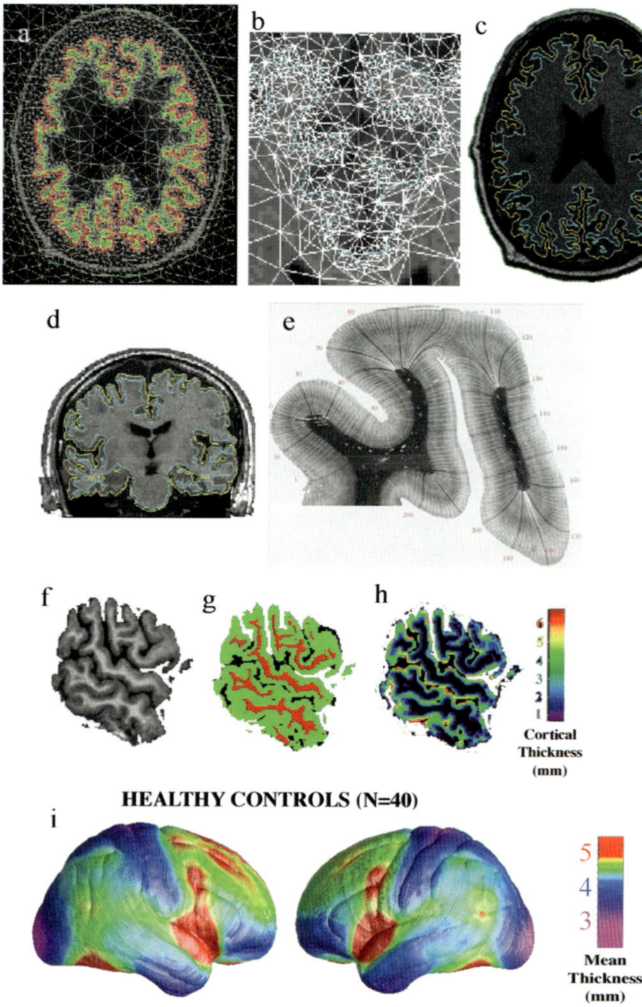

TOGA AND THOMPSON, FIG. 3. (*Continued*)

Toga and Thompson, Fig. 3. Mapping cortical thickness. Several algorithms have been developed to measure the thickness of the cortical gray matter, which is affected in a variety of diseases and changes throughout development. In one approach (Xu *et al.*, 2003, 2005), we used adaptive finite element meshes (a) (shown close up in b), with refinement in a *narrow band* around the cortex (colors, a), to segment the inner and outer cortical surfaces from MRI data (c, d). A parallel line of work measured cortical thickness in two-dimensional (2D) histology (Annese *et al.*, 2002) by solving the heat equation and integrating the gradient of the solution (e). In related work (Thompson *et al.*, 2004, 2005), we computed average maps of cortical thickness for several populations. MRI scans (f) are classified as gray matter, white matter, CSF, or a background class (represented by green, red, black, and white, respectively, in g). To quantify cortical gray-matter thickness, we used the 3D distance measured from the cortical white matter–gray matter boundary in the tissue-classified brain volumes to the cortical surface (gray matter–CSF boundary) in each subject (h). We first resampled tissue-classified brain volumes to the cortical surface (gray matter–CSF boundary) in each subject (h). We first resampled tissue-classified brain volumes to 0.33-mm isotropic voxels to obtain distance measures indexing gray-matter thickness at subvoxel spatial resolution. Gray-matter thickness, measured at thousands of homologous cortical locations in each subject, is then compared across subjects and averaged at each cortical surface location, which provides spatially detailed maps of local thickness differences within or between groups. The thickness data may also be smoothed by a surface-based kernel to enhance the signal-to-noise ratio, before cross-subject comparisons are made. Frame i shows the mean cortical thickness in a group of 40 healthy young adults, ranging from low values in primary sensorimotor and visual cortices (2–3 mm; yellow colors) to the highest values on the medial wall in cingulate areas (up to 6 mm; purple colors). The regional variations in these maps agree with those found in the classical cortical thickness maps derived postmortem by von Economo (Sowell *et al.*, 2004).

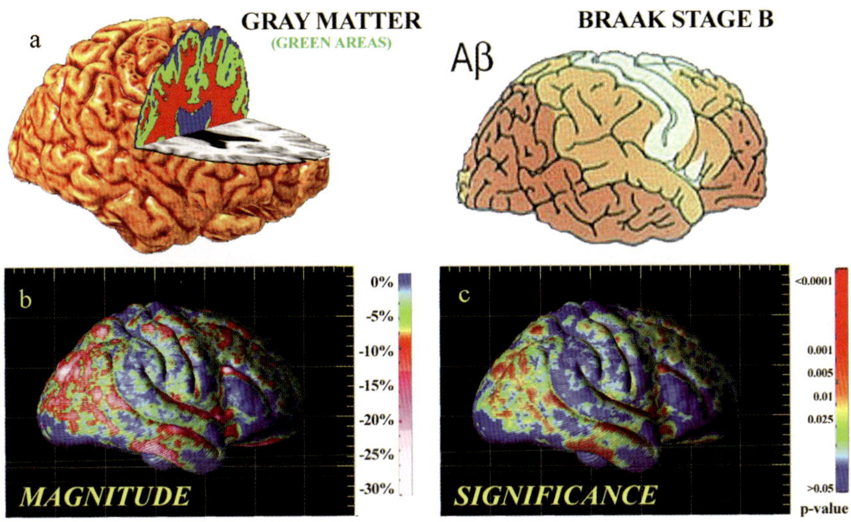

TOGA AND THOMPSON, FIG. 4. Gray-matter deficits in early Alzheimer's disease. Here, the local amount of cortical gray matter (shown in green colors; a) is compared across 26 patients with mild-to-moderate Alzheimer's (age: 75.8 ± 1.7 years; MMSE score: 20.0 ± 0.9) and 20 matched elderly controls (72.4 ± 1.3 years). At this stage of Alzheimer's disease, there is a reduction in gray matter that reaches 30% in the temporal–parietal regions (b). Frame c shows a map of the statistical significance of these deficits (Thompson *et al.*, 2003). Intriguingly, the pattern of temporal-lobe gray-matter loss, seen with MRI, spatially matches the pattern of beta-amyloid (Aβ) deposition seen postmortem (Braak and Braak, 1991, 1997; Braak *et al.*, 2000). The inset panel (Braak Stage B) is adapted from data reported by Braak and Braak (1997). It shows regions with minimal (white), moderate (orange), and severe (red) beta-amyloid deposition. Because amyloid deposition and gray-matter loss may not be synchronized, these maps can represent different stages of Alzheimer's disease; however, there is a clear spatial agreement between MRI and beta-amyloid maps in the severity of the deficits. Intriguingly, both maps indicated that *primary sensorimotor regions* (white in the amyloid map) and the *superior temporal gyrus* (blue colors in c) showed little damage relative to other temporal-lobe gyri. These overall MRI patterns are similar to those in independent studies by Baron *et al.* (2001), O'Brien *et al.* (2001), and Burton *et al.* (2002).

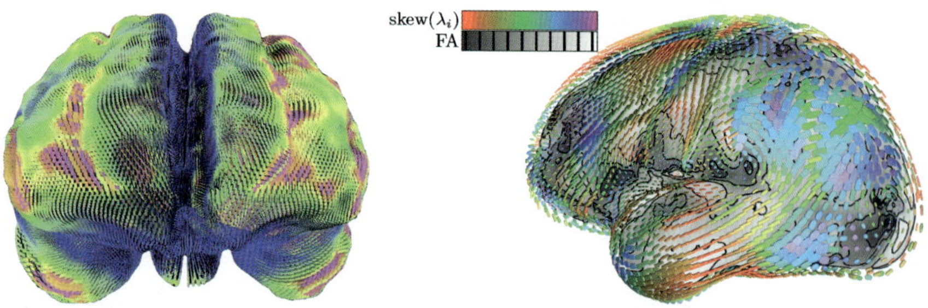

TOGA AND THOMPSON, FIG. 5. Mapping directional patterns of brain variation. This image shows a tensor map of variability for normal subjects, created after 20 elderly subjects' data were mapped into Talairach space (all right-handed; 10 males, 10 females). Ellipsoidal glyphs indicate the principal directions of variation—they are most elongated along directions where anatomic variation is greatest across subjects. Each glyph represents the covariance tensor of the vector fields that map individual subjects onto their group average. Because gyral patterns constrain the mappings, the fields reflect variations in cortical organization at a more local level than can be achieved by matching only global cortical geometry. Note the elongated glyphs in the anterior temporal cortex and the very low variability (in any direction) in the entorhinal and inferior frontal areas. By better defining the parameters of allowable normal variations, we can use the resulting information to distinguish normal anatomical variants from abnormal. Frame b shows a rendering of the same data using superquadric glyphs, which help to visualize the principal axes of variation (Kindlmann *et al.*, 2004). More subtle aspects of anatomical variation can be visualized, including the skew of the eigenvalues of variation (shown in color in b) and the fractional anisotropy (FA) of the variation (shown in grayscale). The anisotropy indicates whether there is a strong directional preference in the population variability.

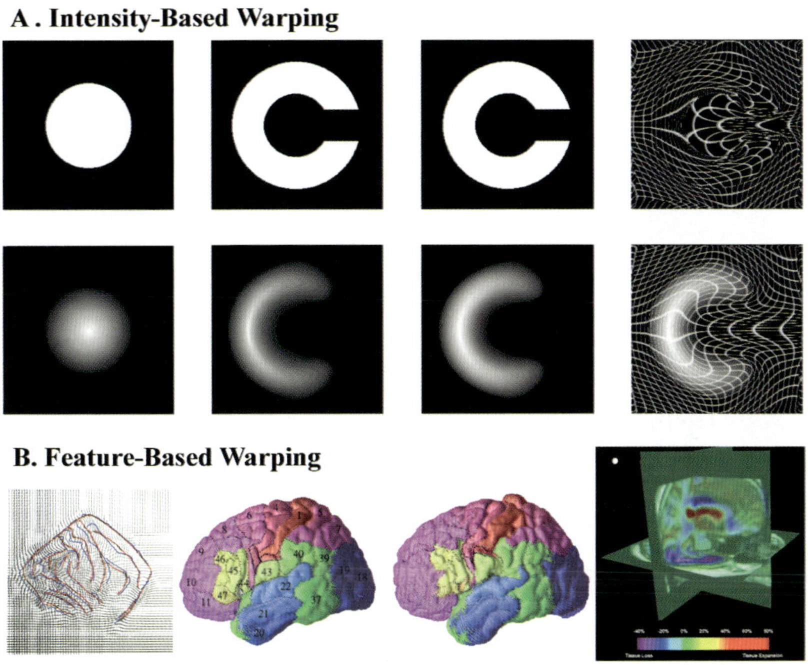

TOGA AND THOMPSON, FIG. 6. Warping algorithms used in brain imaging. As illustrated here, brain image–warping algorithms (Toga, 1998) fall into two major categories: intensity-based algorithms (A, first two rows), and feature-based algorithms (B, bottom row), depending on the information used to align one image with the other. The top row shows a circle and a C shape, and then shows the result of deformation of the circle to the C shape using an intensity-based fluid warping algorithm (data from Chiang *et al.*, 2005). The final column shows the complex pattern of fluid deformation required to transform the circle to the C. Note that this algorithm does not require any user intervention; it automatically finds the deformation field, or "flow," that best aligns one image-intensity distribution with the other. The second row shows a similar experiment in which the circle and the C shape have a range of intensities; the circle is then deformed so that its intensity profile matches that of the C shape. The final row shows examples of feature-based warping. In the first panel, the flattened sulcal pattern of an individual is aligned with an average template of the cortical sulci; in the next two panels, one cortical surface (labeled in color with Brodmann areas) has its labels transferred to another subject's cortical surface with surface-based warping (Rasser *et al.*, 2004; Thompson *et al.*, 2000). In this mapping, three-dimensional (3D) sulcal curves lying in the cortex drive the warped mapping of one surface onto the other, and the Brodmann labels are carried along by the mapping. The final panel shows an application of warping to measure rates of annual tissue loss in semantic dementia patients scanned serially with MRI (data from Leow *et al.*, 2005). A combination of surface-based and intensity-based approaches (i.e., a hybrid approach) is used to derive a 3D elastic deformation mapping in which blue colors denote tissue loss and red colors denote regional expansion (observed here in the lateral ventricles).

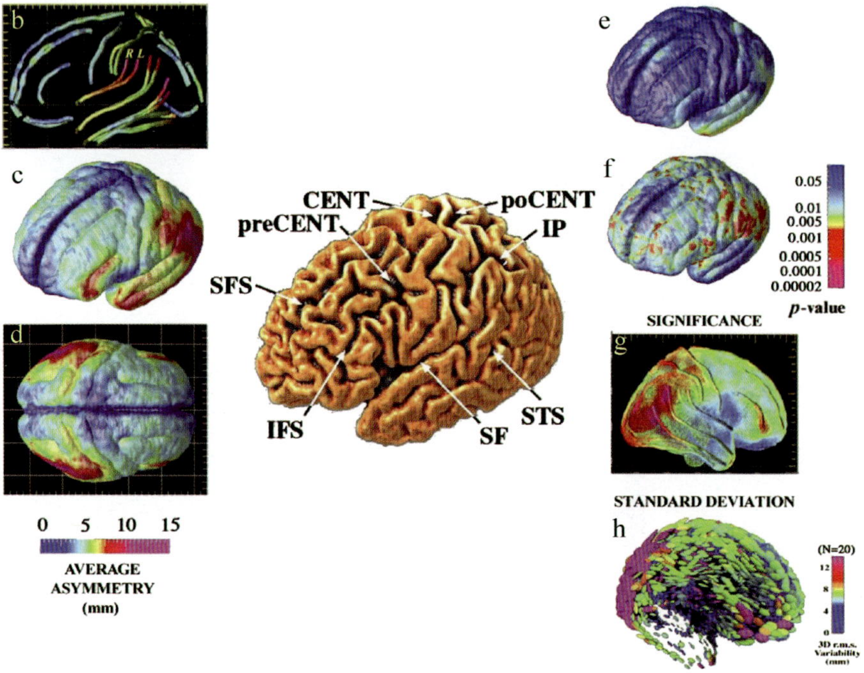

TOGA AND THOMPSON, FIG. 7. Multisubject maps of brain asymmetry. Image-analysis techniques make it possible to distinguish systematic asymmetries in a population or a specific group of subjects from random fluctuations in anatomy. After individual MRI scans are aligned and scaled into a standard three-dimensional (3D) space, 3D curves representing the primary sulcal pattern are digitized (a; center image). (Sulci include central [CENT], precentral [preCENT], postcentral [poCENT], intraparietal [IP], superior frontal [SFS], inferior frontal [IFS], superior temporal, and Sylvian fissures [SF]). The magnitude of asymmetry in the average anatomy, from the average of these curves across 20 normal subjects (b), is shown in color (red colors denote greater asymmetry). Extension of these methods to surfaces (c, d) reveals prominent asymmetries in Broca's anterior speech area and in language regions surrounding the Sylvian fissure. By comparison of the average magnitude of these asymmetries to their standard error (e), regions of significant asymmetry are identified (f). Asymmetries are greatest in brain regions with greatest gyral pattern variability across subjects (g, h). The tensor map (h; see also Fig. 5) shows that the preferred directions of intersubject anatomical variability are also approximately aligned with the direction of interhemispheric asymmetry.

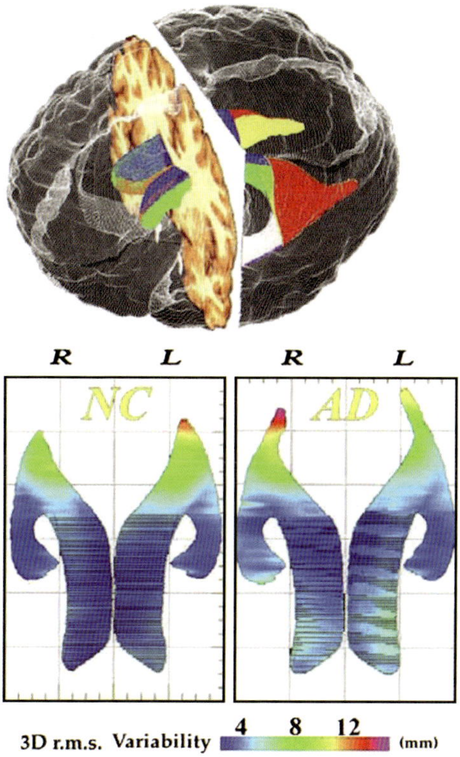

TOGA AND THOMPSON, FIG. 8. Population-based maps of average ventricular anatomy in normal aging and in Alzheimer's disease. In patients and controls, three-dimensional (3D) parametric surface meshes were used to model 14 ventricular elements, and meshes representing each surface element were averaged by hemisphere in each group. An average model for Alzheimer's patients (red) is superimposed on an average model for matched normal controls (blue). Mesh averaging reveals enlarged occipital horns in the Alzheimer's patients and high stereotaxic variability in both groups. Extreme variability at the occipital horn tips also contrasts sharply with the stability of septal and temporal ventricular regions. A top view of these averaged surface meshes reveals localized asymmetry, variability, and displacement within and between groups. These subcortical asymmetries emerge only after averaging of anatomical maps in large groups of subjects.

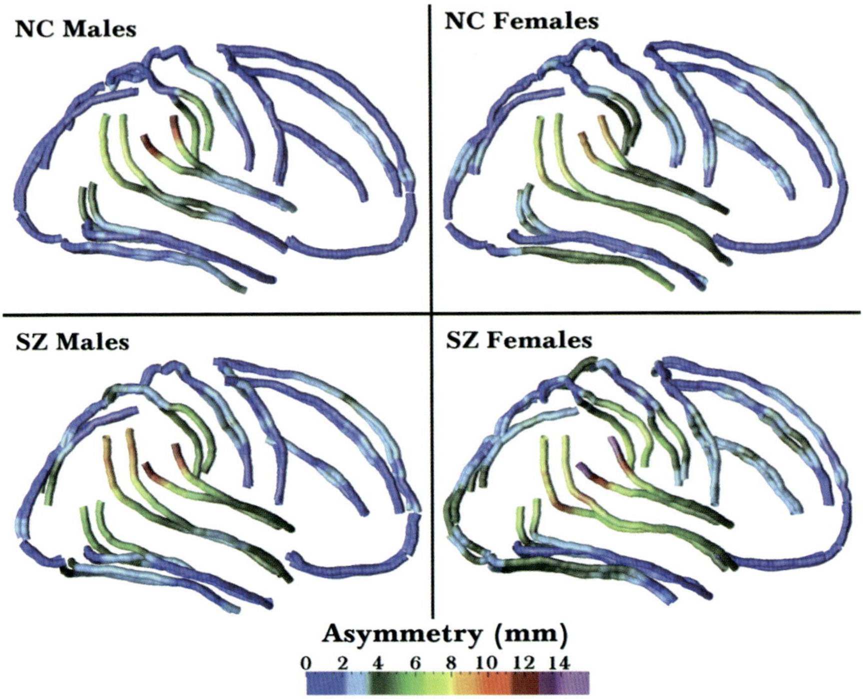

TOGA AND THOMPSON, FIG. 9. Brain asymmetry in schizophrenia. To map group variability and asymmetry in sulci and surrounding anatomy, we obtain three-dimensional (3D) cortical extractions from magnetic resonance (MR) scans and bring them into register using surface-warping algorithms that enable the averaging of equivalent cortical regions that can be delineated through the tracing of major hemispheric sulci across subjects. The atlas data can be stratified by gender and disease group (SZ, schizophrenic patients; NC, normal controls) in order to determine whether asymmetries are altered, on average, in each group. Here the color code denotes the magnitude of the asymmetry (in millimeters) in the average sulcal pattern for each group.

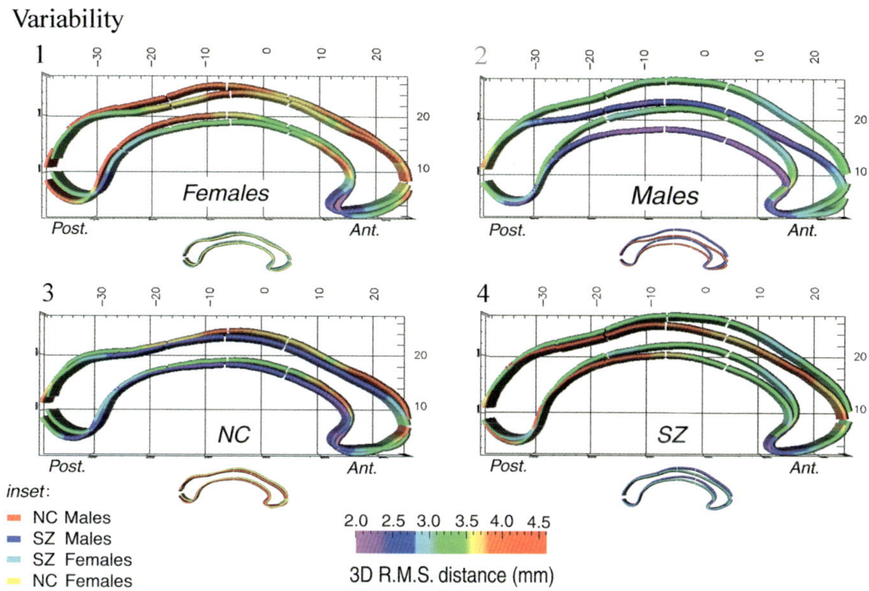

TOGA AND THOMPSON, FIG. 10. Average corpus callosum shapes in schizophrenia. Midsagittal corpus callosum boundaries were averaged from 25 patients with chronic schizophrenia (DSM-III-R criteria; 15 males, 10 females; age: 31.1 ± 5.6 years) and from 28 control subjects matched for age (30.5 ± 8.7 years), gender (15 males, 13 females), and handedness (one left-handed subject per group). Profiles of anatomic variability around the group averages are also shown (in color) as a root-mean-squared (RMS) deviation from the mean. Anatomical averaging reveals a much more pronounced and significant bowing effect in the schizophrenic patients than in the nonschizophrenic controls. Male patients show a significant increase in curvature for superior and inferior callosal boundaries ($p < 0.001$), with a highly significant sex-by-diagnosis interaction ($p < 0.004$). The sample was stratified by sex and diagnosis, and separate group averages show that the disease induces less bowing in females (panel 1) than in males (panel 2). While gender differences are not apparent in controls (panel 3), a clear gender difference is seen in the schizophrenic patients (panel 4). Abnormalities localized in a disease-specific atlas can therefore be analyzed to reveal interactions between disease and demographic parameters. Data from Narr *et al.*, 2000.

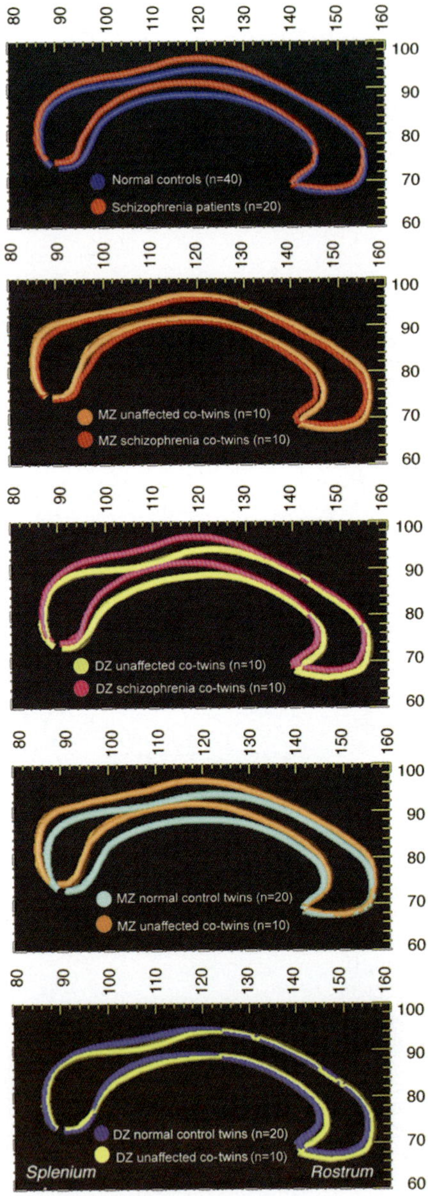

TOGA AND THOMPSON, FIG. 11. Callosal surface averages mapped in groups defined by a biological risk for schizophrenia. Average anatomical mesh models of the corpus callosum are shown in different colors to illustrate differences between (1) monozygotic (MZ) and dizygotic (DZ) co-twins discordant for schizophrenia and (2) healthy monozygotic and dizygotic twin pairs. From the top, midsagittal callosal averages are mapped in (1) schizophrenia patients and controls; (2) unaffected and affected MZ co-twins; (3) unaffected and affected DZ co-twins; (4) unaffected MZ co-twins of the schizophrenia probands and MZ control twin pairs; and (5) unaffected DZ co-twins of the schizophrenia probands and DZ control twin pairs. Data from Narr *et al.*, 2003.

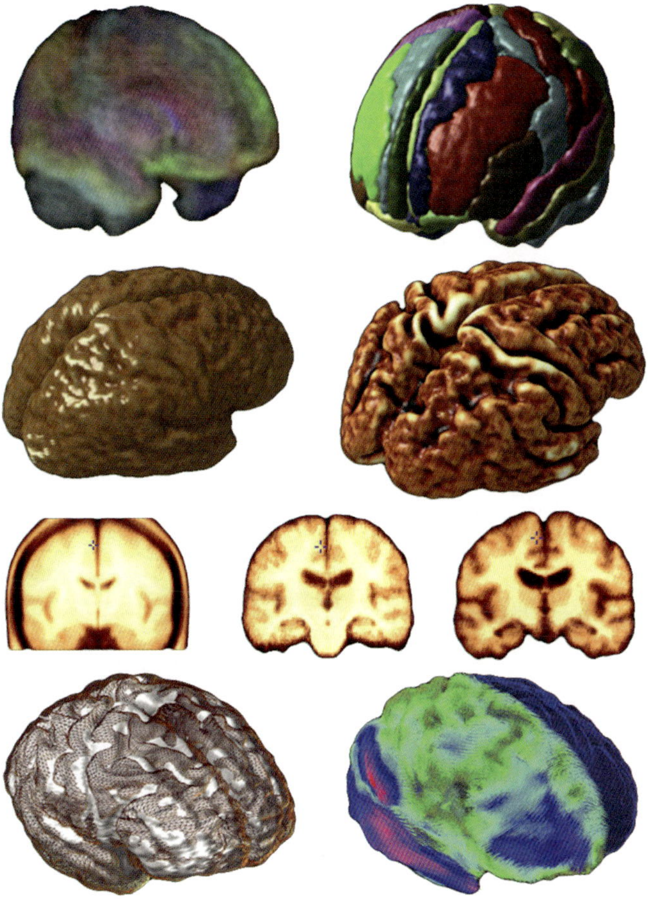

TOGA AND THOMPSON, FIG. 12. Methods for averaging brains. (a) In a widely used average brain image template (ICBM305; row 3, panel 1) based on voxel-wise intensity averaging of 305 young normal subjects' scans (Evans *et al.*, 1994), anatomical features are not well resolved at the cortex. Cortical variability is represented by probability clouds (top left) that describe the frequency of incidence for each gyrus at each stereotaxic voxel, after linear registration and voxel-by-voxel comparison. These statistical/probabilistic maps of anatomy, or SPAMs, can be used to parcellate anatomy in new subjects, in order to help derive automated, but approximate, measures of volumes for brain substructures. In an affine brain template (row 2, first panel, and row 3, second panel), similarly constructed from Alzheimer's disease patients' scans, the cortical average is also poorly resolved. By contrast, anatomical features are highly resolved, even at the cortex, in the *Continuum–Mechanical Brain Template* (row 3, final panel), which applies a continuum–mechanical transformation to each brain before intensity averaging is done. Scans are elastically reconfigured into a group mean configuration, using surface-based warping to match 84 surface models (including gyral pattern elements) across all subjects. Reconfigured scans are then averaged voxel by voxel, after intensity normalization, to produce a group image template with the average geometry *and* average image intensity for the group. Vector-field transformations of extremely high spatial dimension (bottom row) are required in order to resolve cortical features, in their mean configuration, after scans are averaged together.

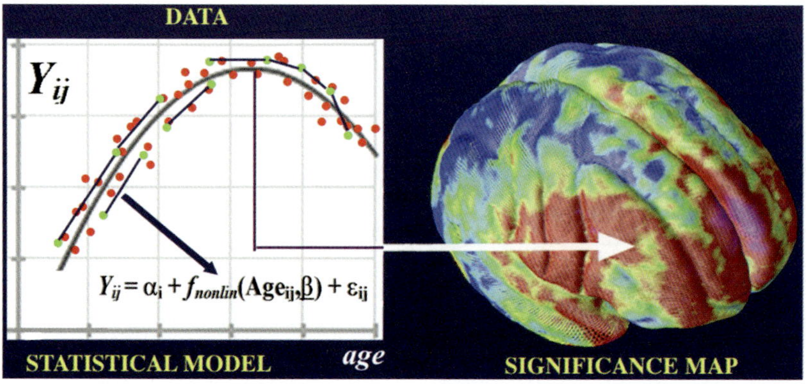

TOGA AND THOMPSON, FIG. 13. Data, statistical models, and maps. This schematic (Thompson *et al.*, 2004) shows some of the steps used in mapping cortical change. First, measures (Y_{ij}) are defined that can be obtained longitudinally (green dots) or only once (red dots) in a group of subjects at different ages. Fitting of statistical models to these data (Statistical Model, left panel) produces estimates of parameters that can be plotted onto the cortex, using a color code. These parameters can include age at peak (see the arrow at the peak of the curve), significance values, estimated statistical parameters such as rates of change, or effects of drug treatment or risk genes.

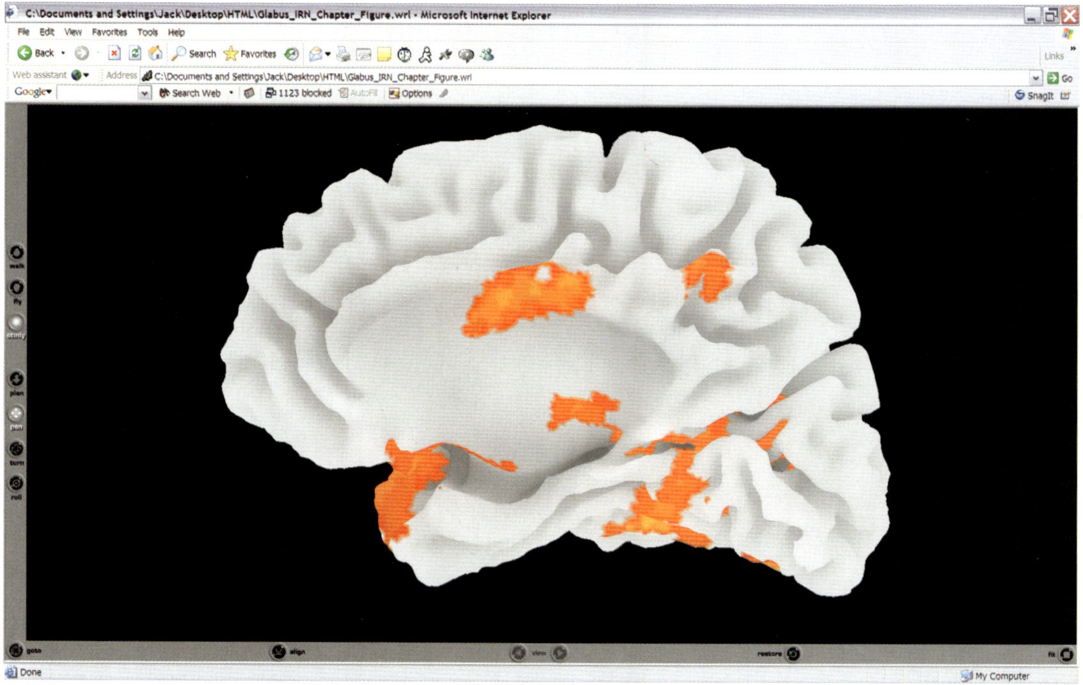

HORN *ET AL.*, FIG. 2. The novel visualization of fMRI results can enhance or emphasize areas not fully appreciated in the original published article. This figure shows a VRML rendering, exported from the Caret software package, of the memory encoding statistical main effects from the study of Druzgal and D'Esposito (2001) viewed as interactive object using the Cortona Client for MS Windows XP, Version 4.2.R93, from Parallel Graphics, Inc. (*http://www.parallelgraphics.com*). This rendering highlights the robust activation of the fusiform gyrus, the superior portion of the anterior cingulate, inferior parietal lobule, as well as the temporal pole in the right cerebral hemisphere. Viewed as a virtual reality object would enable a student or researcher to interactively translate, rotate, or zoom in on regions of particular interest and to view the data as a whole, rather than as a collection of 2D overlay graphics.

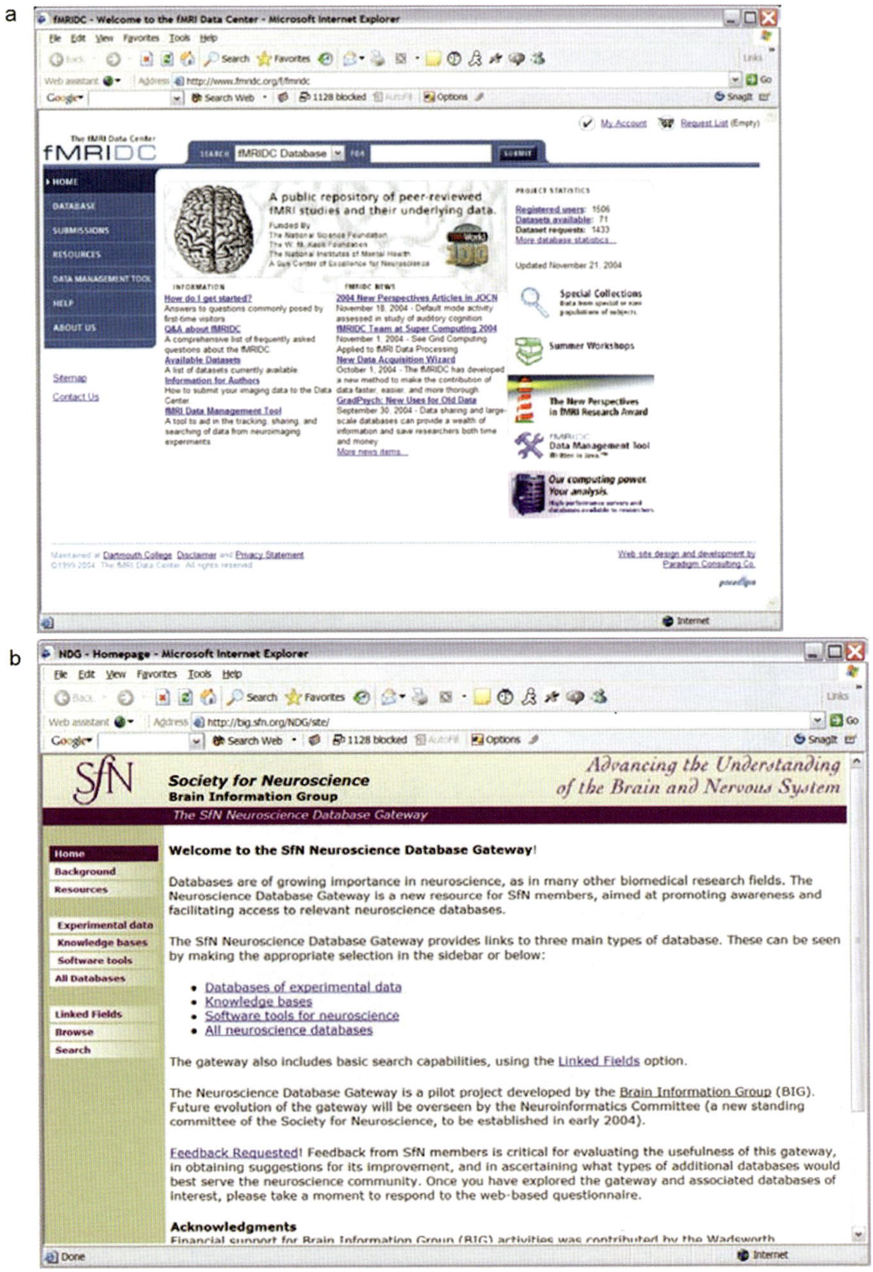

HORN *ET AL.*, FIG. 3. (*Continued*)

c

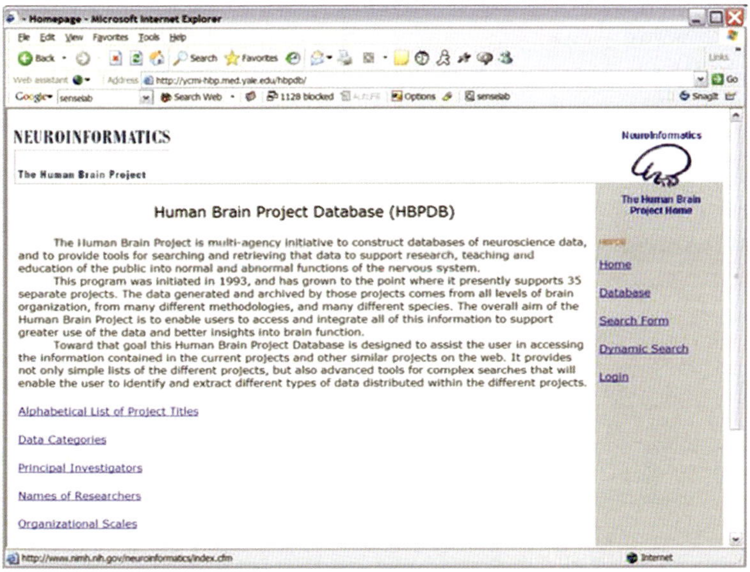

HORN *ET AL.*, FIG. 3. (a) The fMRI Data Center web site permits researchers to browse and request complete fMRI data sets from peer reviewed, published journal articles; (b) The Society for Neuroscience provides an ever increasing portal to online neuroscience data archives (http://web.sfn.org/content/Programs/NeuroscienceDatabaseGateway/index.html), from molecular- to systems-levels, in which workers may obtain data for novel analysis and visualization; and (c) The NIMH Human Brain Project actively promotes the sharing of primary research data and lists the database efforts of investigators funded under its program (http://ycmi-hbp.med.yale.edu/hbpdb/).

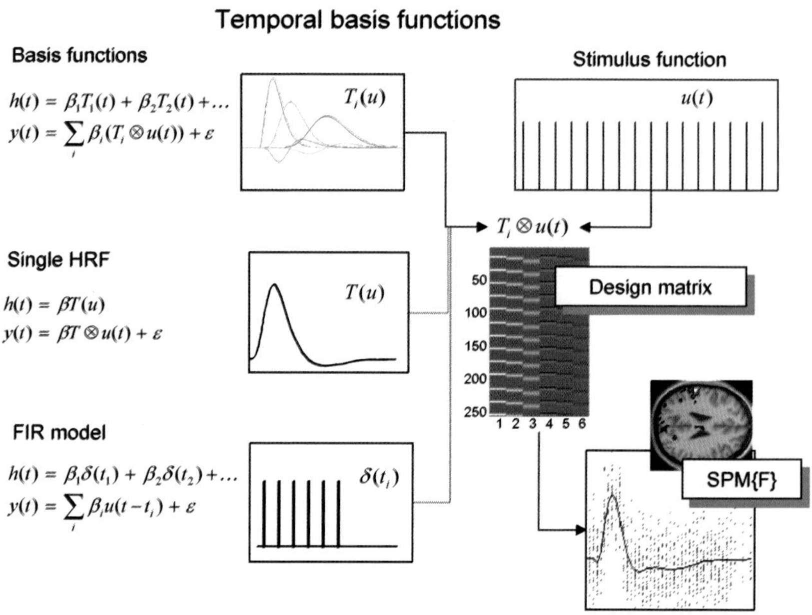

FRISTON *ET AL.*, FIG. 4. Temporal basis functions offer useful constraints on the form of the estimated response that retain the flexibility of finite impulse-response (FIR) models and the efficiency of single-regressor models. The specification of these constrained FIR models involves setting up stimulus functions $u(t)$ that model expected neuronal changes (e.g., boxcars of epoch-related responses or spikes [delta functions] at the onset of specific events or trials). These stimulus functions are then convolved with a set of basis functions $T_i(t)$ of peristimulus time that, in some linear combination, model the hemodynamic impulse-response function (HRF). The ensuing regressors are assembled into the design matrix. The basis functions can be as simple as a single canonical HRF (middle) or as complicated as a series of delayed delta functions (bottom). The latter case corresponds to a FIR model, and the coefficients constitute estimates of the impulse-response function at a finite number of discrete sampling times. Selective averaging in event-related fMRI (Dale and Buckner 1997) is mathematically equivalent to this limiting case.

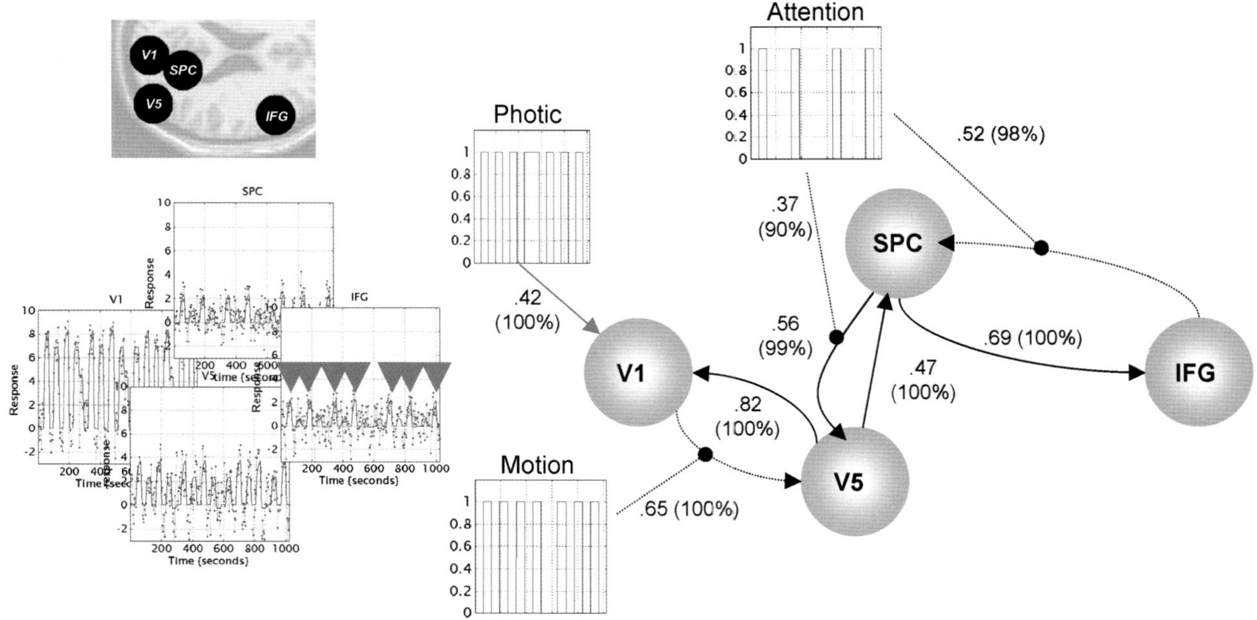

FRISTON *ET AL.*, FIG. 7. (*Continued*)

FRISTON *ET AL.*, FIG. 7. Results of a dynamic causal model analysis of attention to visual motion with fMRI. Right panel: Functional architecture based upon the conditional estimates shown alongside their connections, with the percent confidence that they exceeded a non-trivial threshold shown in brackets. The most interesting aspects of this architecture involve the role of motion and attention in exerting bilinear effects. Critically, the influence of motion is to enable connections from V1 to the motion-sensitive area V5. The influence of attention is to enable backward connections from the inferior frontal gyrus (IFG) to the superior parietal cortex (SPC). Furthermore, attention increases the influence of the SPC on V5. Dotted arrows connecting regions represent significant bilinear effects in the absence of a significant intrinsic coupling. Left panel: Fitted responses based upon the conditional estimates and the adjusted data are shown for each region in the dynamic causal model. The insert (upper left) shows the location of the regions. Subjects were studied with fMRI under identical stimulus conditions (visual motion subtended by radially moving dots) while the attentional component of the task (detection of velocity changes) was manipulated. The data were acquired from a normal subject at 2 Tesla using a Magnetom VISION (Siemens, Erlangen, Germany) whole-body MRI system, equipped with a head volume coil. Contiguous multislice T2*-weighted fMRI images were obtained with a gradient echo-planar sequence (TE = 40 ms; TR = 3.22 seconds; matrix size = 64 × 64 × 32; voxel size 3 × 3 × 3 mm). Each subject had four consecutive 100-scan sessions comprising a series of 10-scan blocks under 5 different conditions, D F A F N F A F N S. The first condition, D, was a dummy condition that allowed for magnetic saturation effects. F (fixation) corresponded to a low-level baseline where the subjects viewed a fixation point at the center of a screen. In condition A (attention), subjects viewed 250 dots moving radially from the center at 4.7 degrees per second and were asked to detect changes in radial velocity. In condition N (no attention), the subjects were asked simply to view the moving dots. In condition S (stationary), subjects viewed stationary dots. The order of A and N was swapped for the last two sessions. In all conditions, subjects fixated on the center of the screen. There were no speed changes during scanning. No overt response was required in any condition.

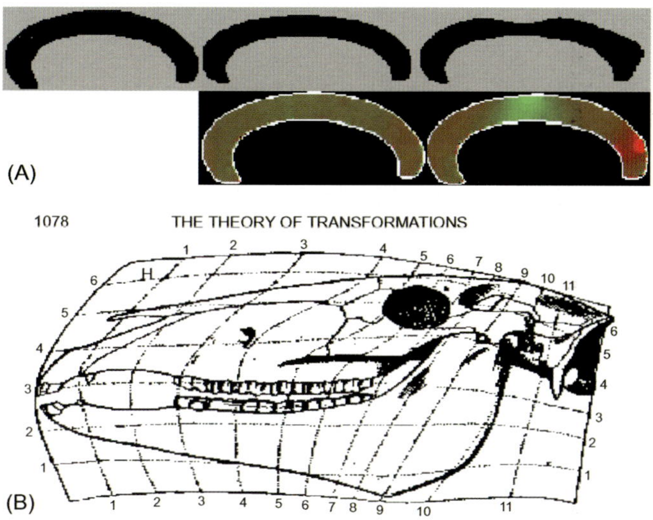

DAVATZIKOS, FIG. 1. Using a shape transformation for morphometric measurements. (A) (Top left) A callosal template. (Top middle and right) Two individual shapes. (Bottom) Respective color-coded maps of the determinant of the Jacobian of the shape transformation mapping the template to the two shapes. Contraction is colored green and expansion is colored red. Voxel-wise comparison of these images reveals local shape differences of the respective shapes. (B) Seminal work by D'Arcy Thompson in 1917 using shape transformations to make comparisons among species.

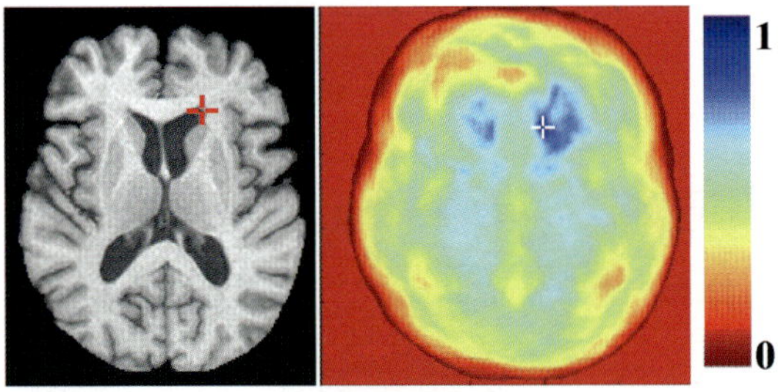

DAVATZIKOS, FIG. 2. The point marked by a cross has a relatively distinctive GMI-based attribute vector. The color-coded image on the right shows the degree of similarity between the attribute vector of the marked (by crosses) point and the attribute vector of every other point in the brain. 1 is maximum similarity and 0 is minimum similarity.

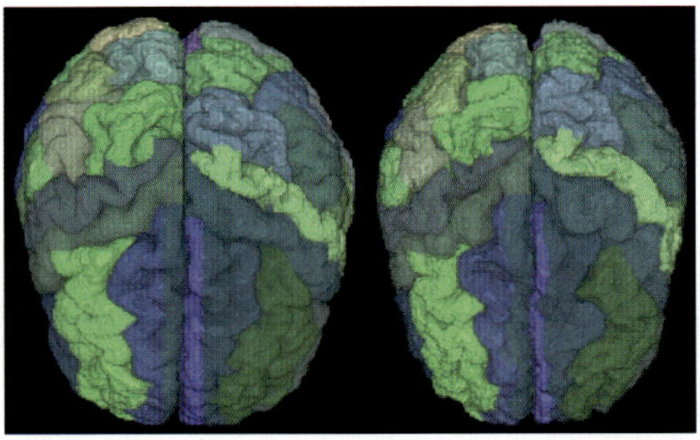

DAVATZIKOS, FIG. 4. Representative example of automated definition of regions of interest, by warping a pre-labeled atlas (left) to an individual's MR images (the warped atlas is shown on the right as a color-coding of a volume rendering of the target brain). This automated ROI definition makes it possible to apply the method to studies with large sample sizes in a streamlined way.

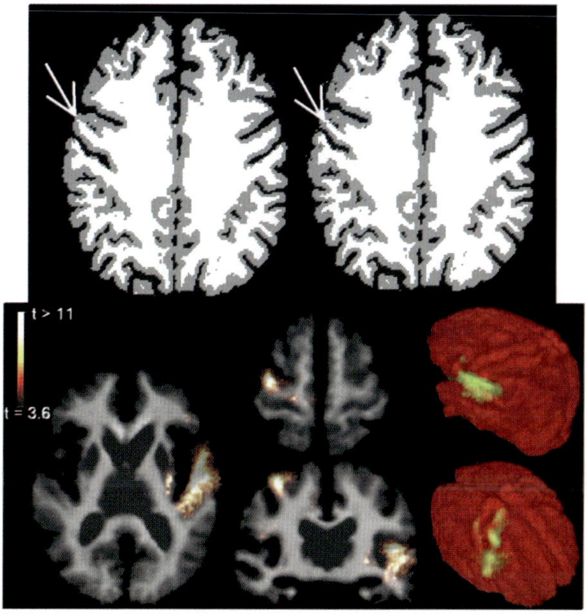

DAVATZIKOS, FIG. 6. (Top) Representative slices from the level of the precentral gyrus, with simulated atrophy indicated by the arrows (left is before and right is after uniform 30% atrophy within the gyrus was applied). (Bottom) Regions detected by the RAVENS analysis, overlaid on the average WM RAVENS maps of the 24 individuals. The two detected regions were exactly where atrophy was simulated. Reprinted from (Davatzikos *et al.*, 2001).

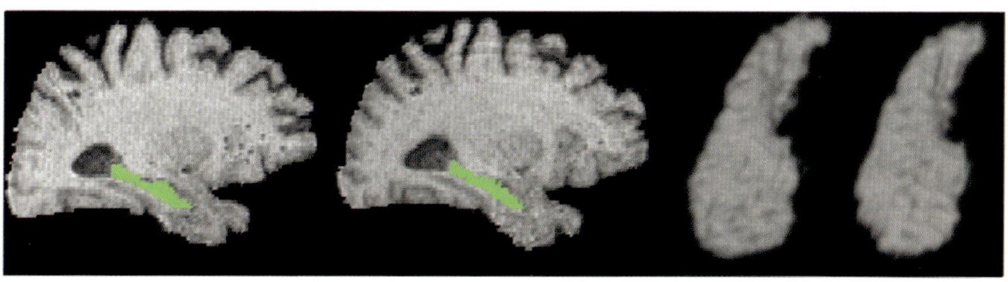

DAVATZIKOS, FIG. 8. Automated segmentation results using 3D HAMMER for subject AD, years 3 and 4. The 5% difference in volumes, in this case, is not readily appreciated visually from these images. (The sections are only approximately corresponding, since the scans were at slightly different orientations. 3D renderings are shown on the right.)

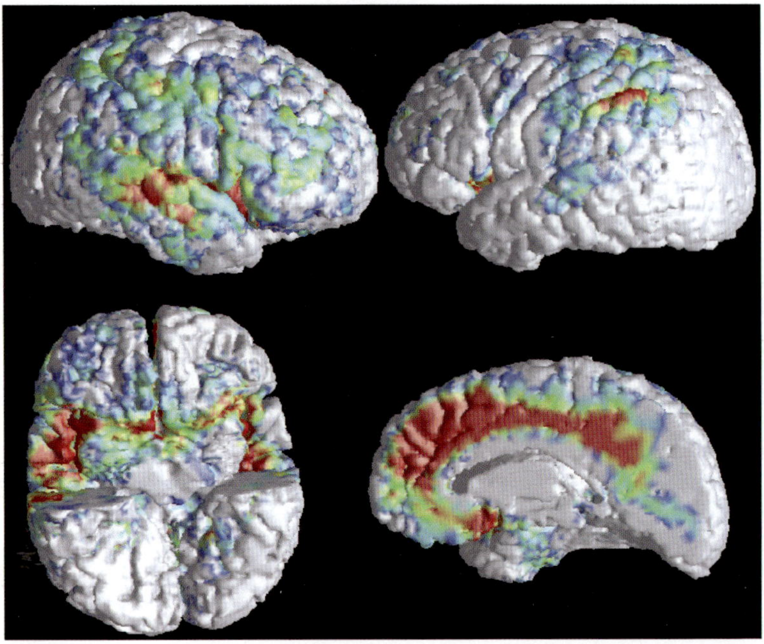

DAVATZIKOS, FIG. 9. Regions displaying significant longitudinal grey matter atrophy over a 4-year period. Estimates of longitudinal atrophy were determined by segmentation into GM, WM and CSF, then applying the mass-preserving RAVENS methodology described in the text, which deforms each individual's brain into alignment with a template brain, while preserving tissue mass by converting it to density. Voxel-based analysis of the resultant tissue density maps is equivalent to voxel-based volumetric analysis and therefore of atrophy quantification. Reprinted from (Resnick *et al.*, 2003).

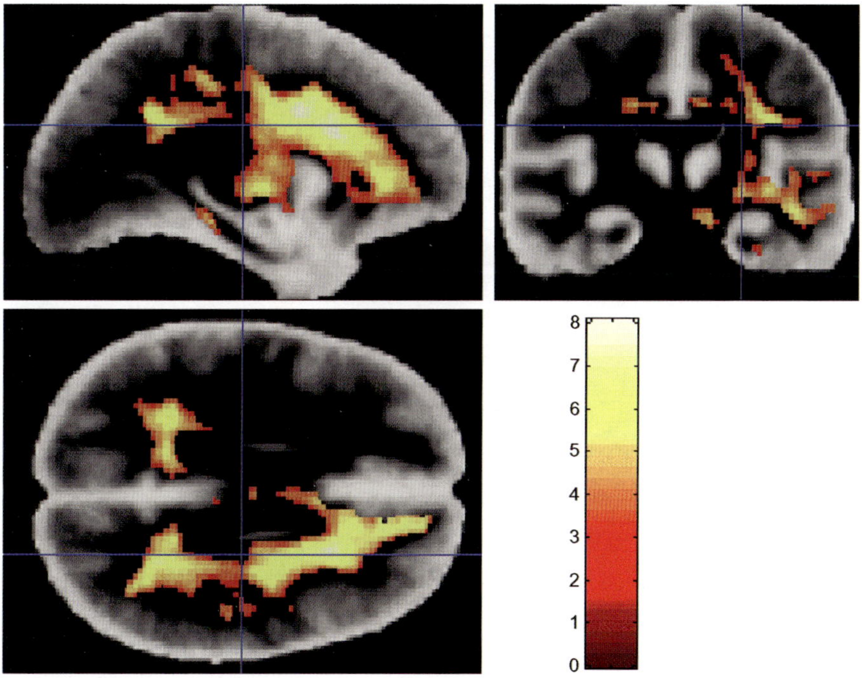

DAVATZIKOS, FIG. 10. Statistical parametric maps of T1 signal darkening with normal aging, obtained using the RAVENS methodology for elastic registration. T1 darkening might be due to a variety of factors, such as vascular disease, demyelination, mineral deposition or other degenerative processes. Reprinted from (Davatzikos and Resnick, 2002).

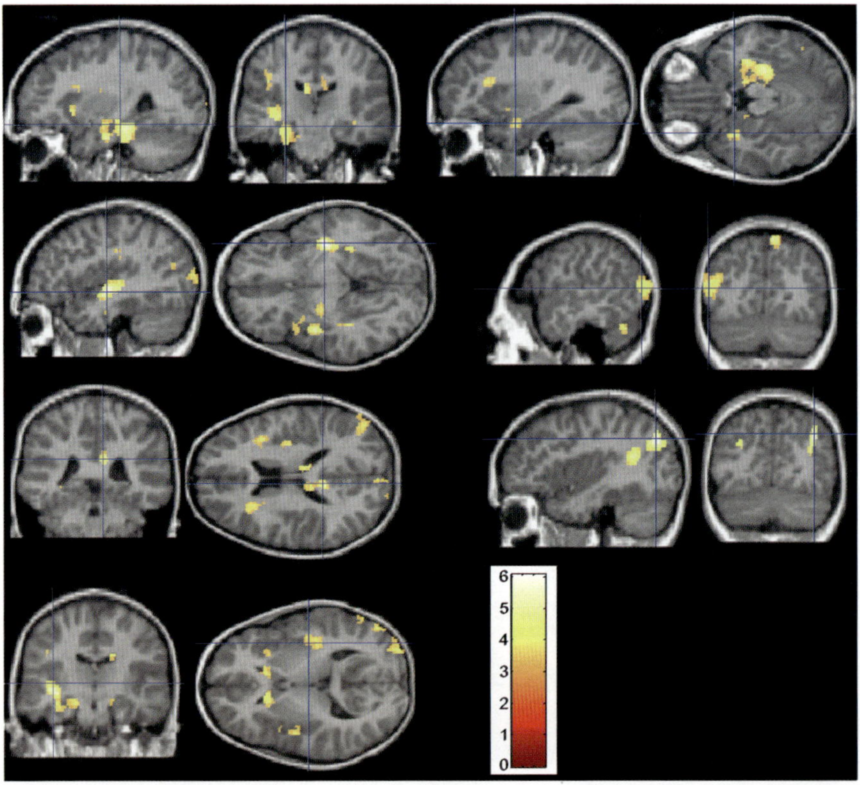

DAVATZIKOS, FIG. 11. Visual summary of several detected regions with significant group differences between XXY brains and normal control brains. The underlying image is the template that we used to normalize individual brains. Color-coding was based on the values of the t-statistic. Only the voxels with significant group differences, i.e., the corrected p-values exceeding a significance threshold 0.005, are shown. (A) *Left* hippocampal formation; (B) *Left* superior temporal gyrus; (C) Cingulate region; (D) *Left* Insula; (E) *Right* Amygdala; (F) *Left* middle temporal gyri; (G) *Right* parietal lobe WM. Reprinted from (Shen *et al.*, 2004).

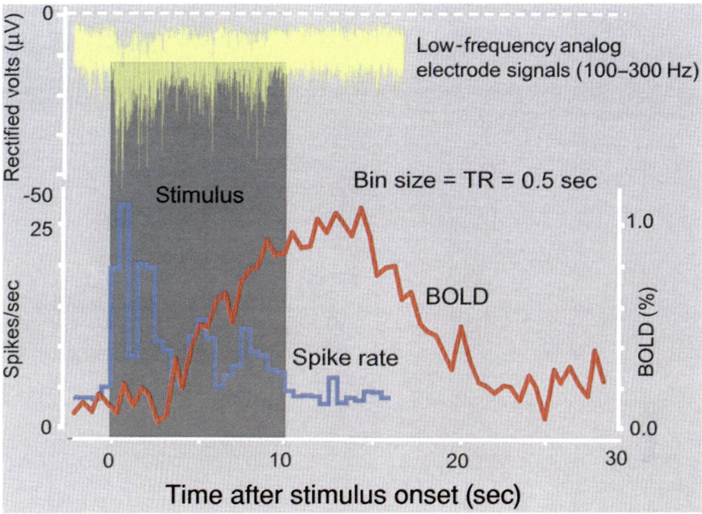

KIM, FIG. 1. The time course of BOLD and single-unit recordings from the same cortical location. Identical visual stimuli were used for fMRI and subsequent single-unit recording sessions. Blue trace: peristimulus histogram of the spike activity. Bin size for the histogram = 0.5 sec = TR (scan repeat time) for fMRI. Red trace: BOLD percent changes during visual stimulation. X-axis: time after stimulus onset. Left Y-axis: Spikes per second. Right Y-axis: BOLD percent changes. Gray box: stimulus duration. The black trace above indicates the original low-frequency analog signals (100–300 Hz) underlying the depicted spike counts.

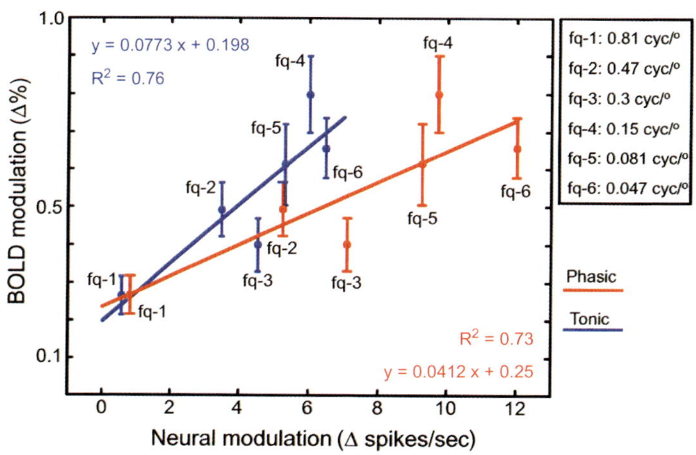

KIM, FIG. 2. Results of direct comparison between BOLD and single-unit recordings across all sites (n = 58). X-axis: neural modulation in spikes per second. Y-axis: % BOLD modulation. The data points fq-1 through fq-6 indicate the six spatial frequencies used to elicit neural responses. Coefficient of determination of the regression line $R^2 = 0.85$. Correlation data are plotted for tonic (blue) and phasic (orange) data separately.

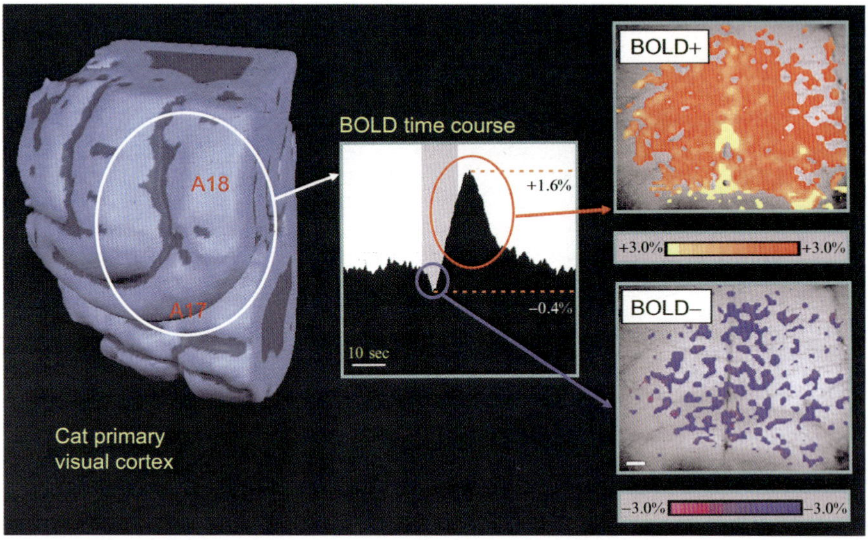

KIM, FIG. 3. Improvement of BOLD spatial specificity by use of nonconventional functional MRI signals. Left side: three-dimensional rendering of the cat visual cortex from which the fMRI signals were acquired. The time course on the right side shows biphasic evolution of MR signals, resulting in the early deoxygenation contrast. If used, such deoxygenation signals produce high-resolution images of exceedingly high functional specificity (termed BOLD−) that contrast with conventional BOLD fMRI signals (termed BOLD+).

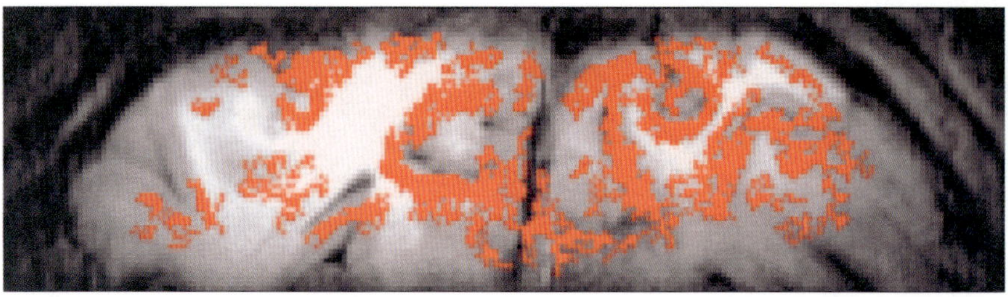

KIM, FIG. 4. $0.5\text{-} \times 0.5\text{-} \times 2\text{-mm}^3$ Hahn spin echo (HSE) images of the visual activation in the human visual cortex obtained at 7 Tesla (T), demonstrating that high-spatial-resolution images can be obtained robustly in the human brain at 7 T by the normally weak but more accurate HSE imaging signals single-shot spin echo planar imaging with field-of-view reduction to 3.2 cm × 12.8 cm (64 × 256) (partial Fourier) TE/TR: 70 msec/2 s. Adapted from Ugurbil *et al.* (2003).

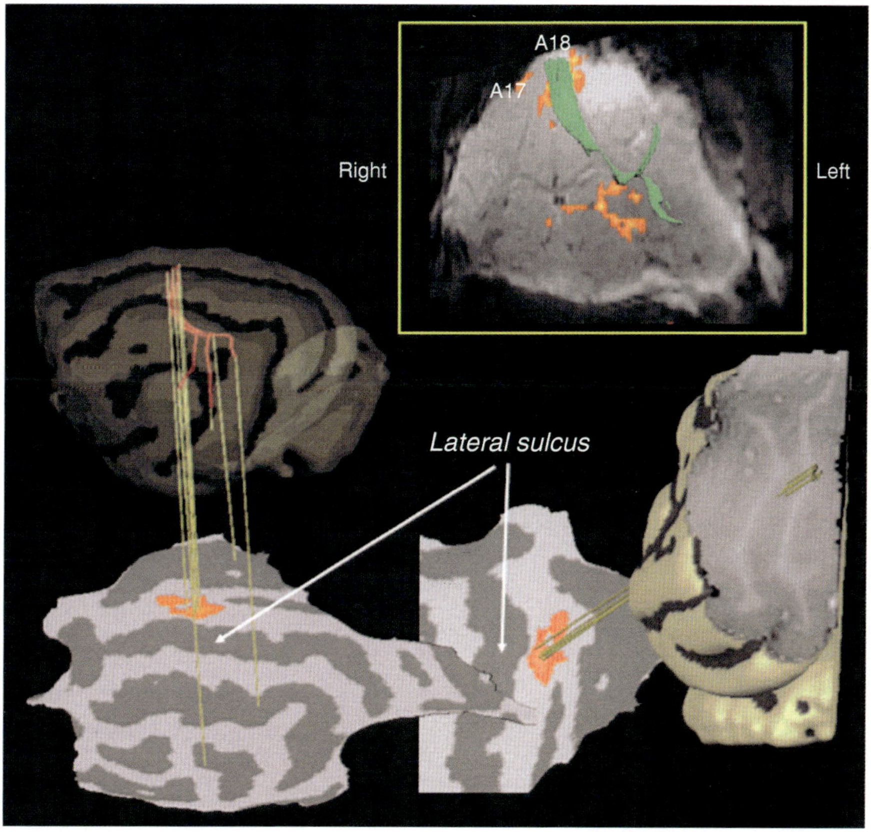

KIM, FIG. 5. Combining fMRI with high-resolution diffusion-tensor imaging (DTI)-based fiber tractography. Here, 3D volume reconstruction of the cat cortex is depicted as a translucent "glass brain" in which the BOLD-based DTI fibers were embedded, thus preserving the true 3D trajectories of the reconstructed fibers. Lower panels: "flattened" representation of cortex. Light- and dark-gray areas code for gyri and sulci, respectively. Corresponding points between reconstructed and flattened cortices were connected through the reconstructed DTI fibers. Right panel: same data as in the left panel, viewed from a different perspective.

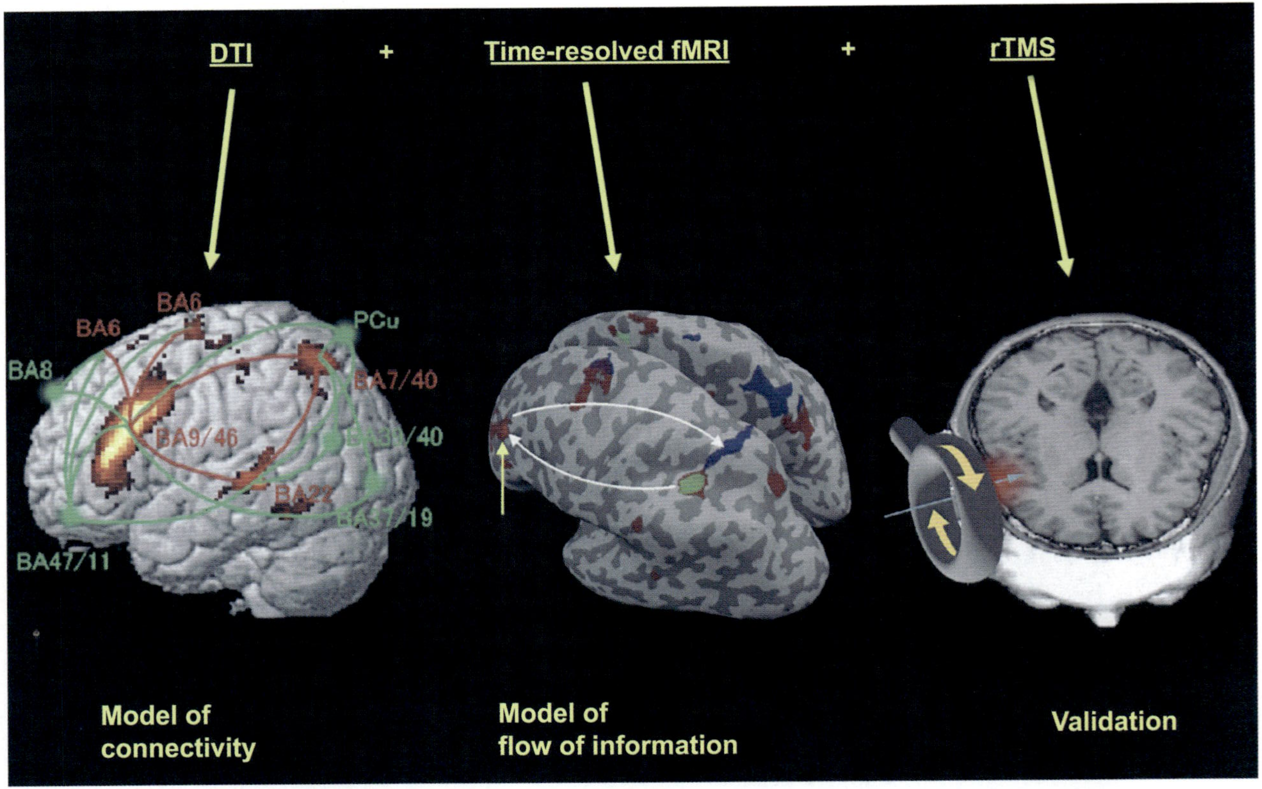

KIM, FIG. 6. Cartoon diagram of the multimodal approach that is increasingly being employed for functional MRI studies. If combined properly, such multimodal studies could potentially provide information on localization (by fMRI), anatomical connectivity (by DTI), flow of information (by effective connectivity), and localized disruption (by rTMS) of cortical functionalities from the same brain.

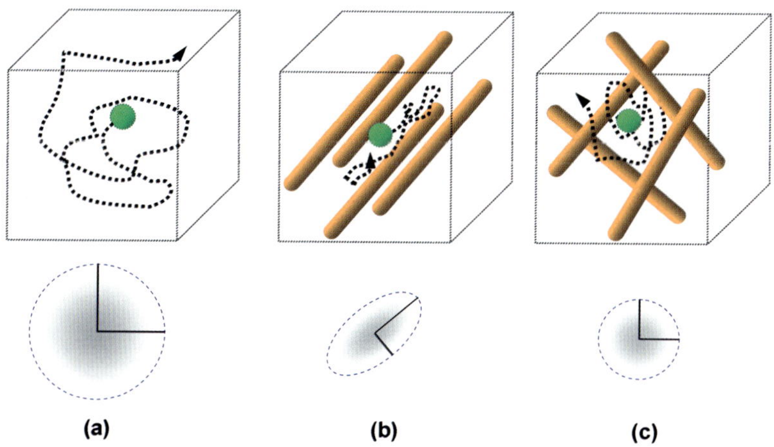

PARK, FIG. 1. Diffusion of the water molecule. Unrestricted isotropic self-diffusion (a), restricted anisotropic diffusion in well-arranged structures (b), and restricted isotropic diffusion in complex structures (c).

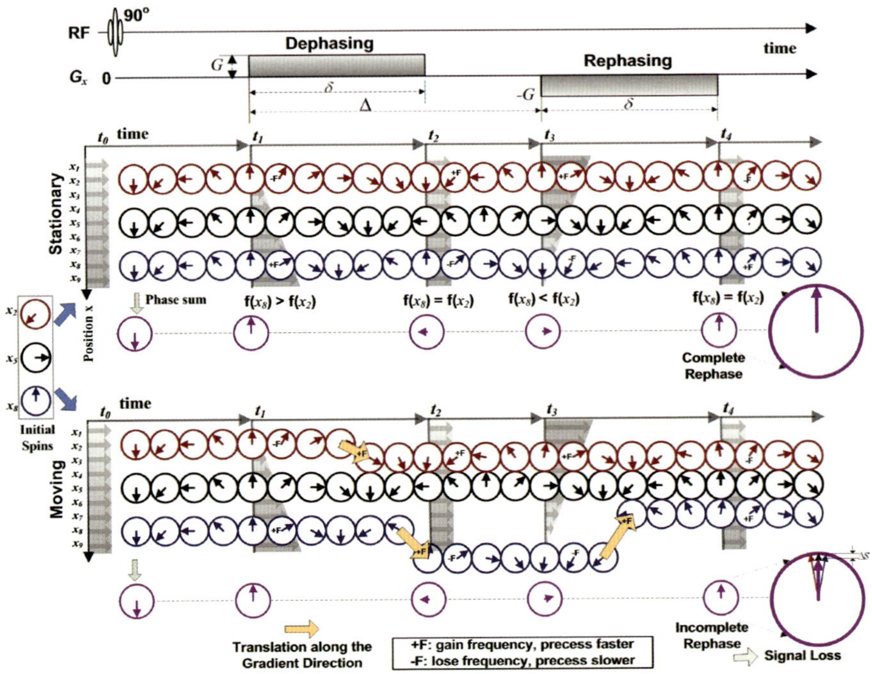

PARK, FIG. 3. Gradient-echo diffusion-weighted imaging (DWI). The phase accumulation of three spins initially located at x_2, x_5, and x_8 in the stationary situation (upper panel) and the moving situation (lower panel) during the gradient echo DWI sequence is explained as time elapses. A 90° excitation radiofrequency pulse renders all spins in phase (t_0), and under the uniform magnetic field, this condition is sustained. When a first gradient magnetic pulse is applied (t_1), the strength of the magnetic field increases along the x, i.e., $B(x_2) < B(x_5) < B(x_8)$, and so does the precession frequency, i.e., $f(x_2) < f(x_5) < f(x_8)$. The phase shifts accumulated due to the frequency difference will be sustained after the application of the uniform magnetic field, i.e., $\mathbf{G} = 0$ (t_2). The second magnetic-field gradient will cause the precession frequency at x_8 to be lower than that at x_2. The passing of the same amount of accumulation time (δ) as for the first gradient magnetic pulse will lead to a complete phase realignment of all spins that are stationary (upper panel). For moving situations, the first spin (initially located at x_2) moves to x_3, which causes the spin to rotate faster than at the previous position rotating with $f(x_2)$. Under the second gradient, the compensating phase accumulates proportionally to the frequency $f(x_3)$, which results in incomplete rephase of the phase accumulation with rotation frequency of both $f(x_2)$ and $f(x_3)$ during the first gradient (lower panel). The spin at x_8 also undergoes similar incomplete rephase at time t_4.

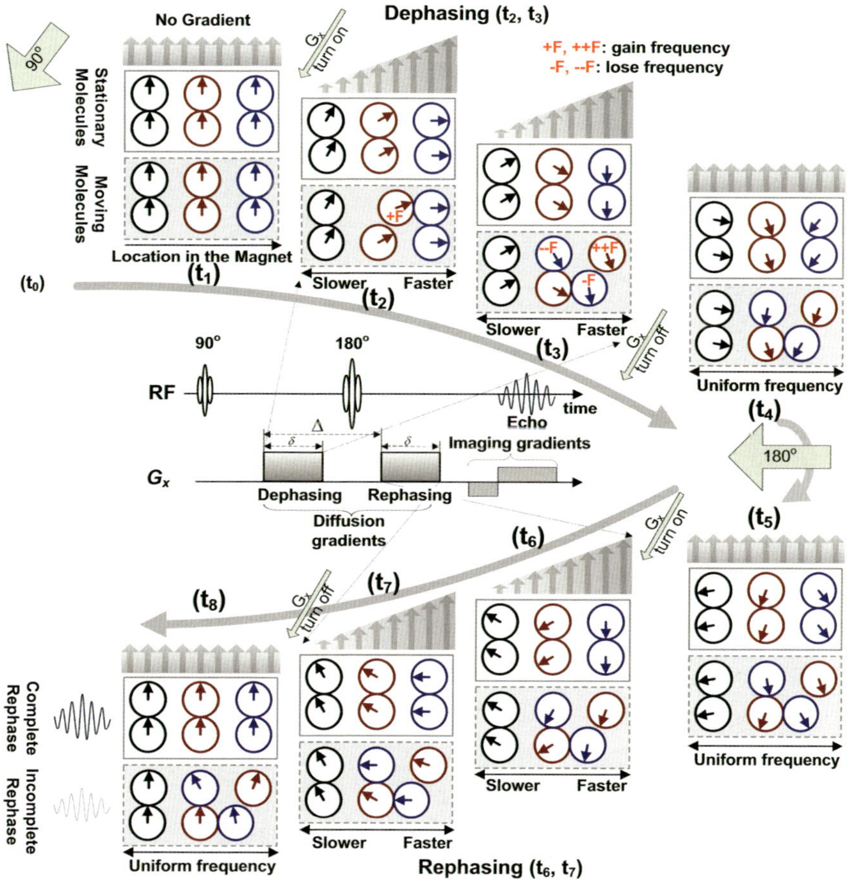

PARK, FIG. 4. Spin-echo diffusion-weighted imaging. The time courses of the phase accumulations of six stationary molecules (upper part) and of six moving molecules (lower part) are explained for the spin-echo sequence. Under the uniform magnetic field with no gradient (t_1), the phase alignment of all individual spins caused by a 90° radiofrequency excitation is sustained. When the first gradient magnetic pulse is on (t_2 and t_3), the location-dependent magnetic fields induce spins to rotate with different frequencies. If a spin moves along the direction of the magnetic-field gradient, the phase of the spin changes. A 180° radiofrequency pulse reverses the phase of all spins (t_4 and t_5). The second gradient (t_6 and t_7) will lead to realignment for a stationary spin (t_8), whereas moving molecules will be rendered out of phase. This out-of-phase condition leads to signal attenuation.

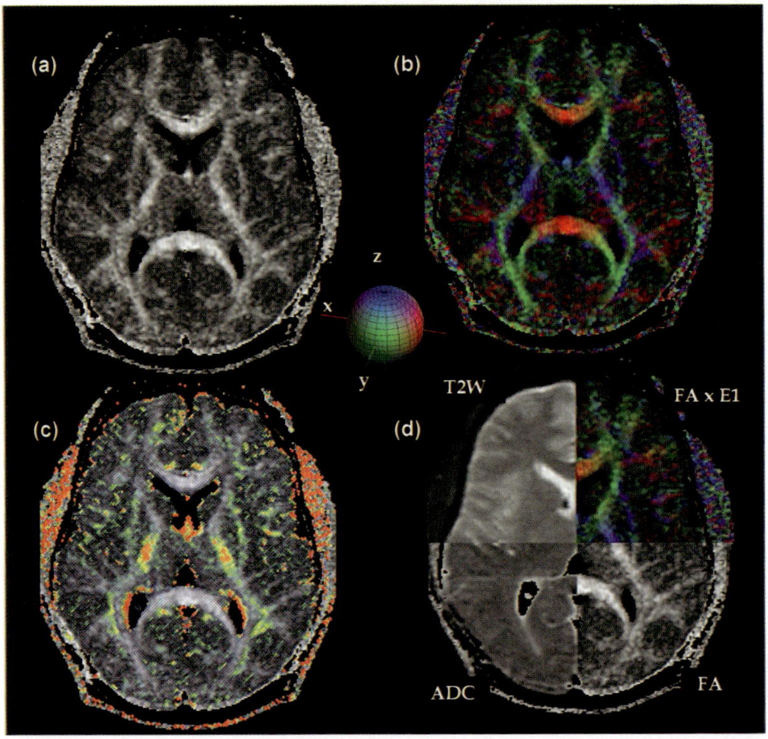

PARK, FIG. 12. Voxel-based visualization methods for the diffusion tensor. An axial slice of a diffusion-tensor image is displayed with the fractional anisotropy (FA) map (a), the color-coded vector map (b), the arrow vector map (c), and a combinatory display with T2, apparent diffusion coefficient (ADC), FA, and color-coded vector map (FA × E1) (d).

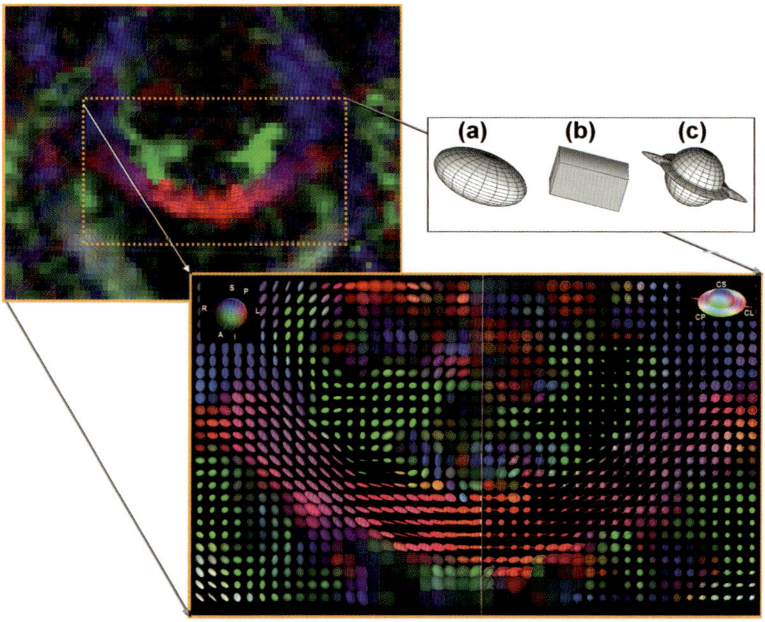

PARK, FIG. 13. Three-dimensional tensor map. A coronal section of the diffusion-tensor image can be displayed with a three-dimensional ellipse (a), a cubic object (b), and a geometric object (c). The zoomed figure shows the ellipsoidal display (left plane) and geometric-object display (right plane). The geometric objects are composed of lines, planes, and spheres according to the portion in the tensor of the linear, planar, and spherical components (c_l, c_p, and c_s).

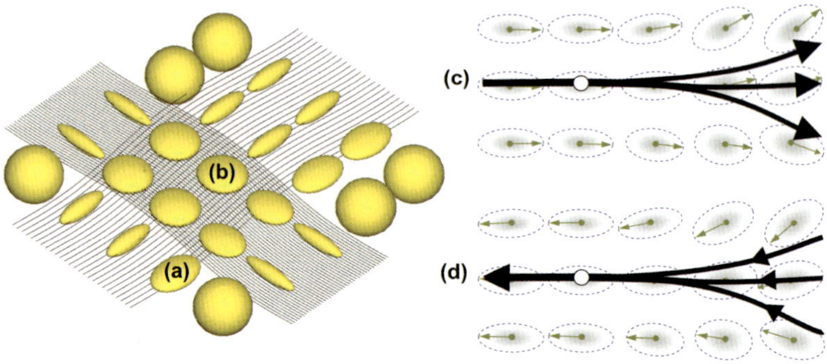

PARK, FIG. 15. Partial volume effects on the diffusion tensor. A tensor at the boundary of the fibers and nonfiber structures (a) and a tensor where two fiber bundles cross have highly augmented ambiguity. At the diverging fibers, the streamline approach cannot explain fiber branching from a single point to several fibers (c). Instead, fiber branching can be explained by the merging of multiple fibers starting from multiple seed points (d).

PARK, FIG. 18. Whole-brain fiber bundles in lateral view. The whole-brain fiber bundles are reconstructed using the fourth-order Runge–Kutta algorithm with seed points in the whole white matter. The diffusion-tensor image is acquired from 3-Tesla GE Echospeed systems using a line-scan diffusion-imaging sequence (LSDI). The fiber tractography was conducted using the DoDTI (*http://neuroimage.yonsei.ac.kr/dodti*).

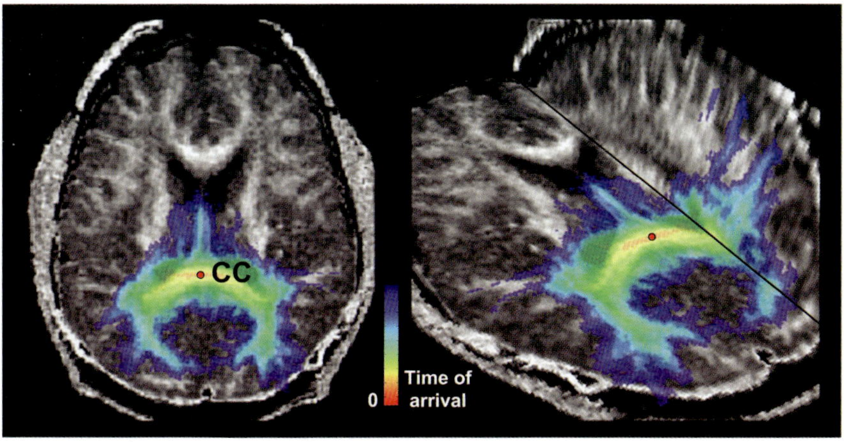

PARK, FIG. 19. Connectivity mapping using fast-marching tractography. A time-of-arrival map is derived using fast-marching tractography from a seed point located in the middle of the corpus callosum. In contrast to streamline approaches, fast-marching tractography has one-to-many mapping that makes it possible to evaluate the relative connectivity between the pairs of points in the brain.

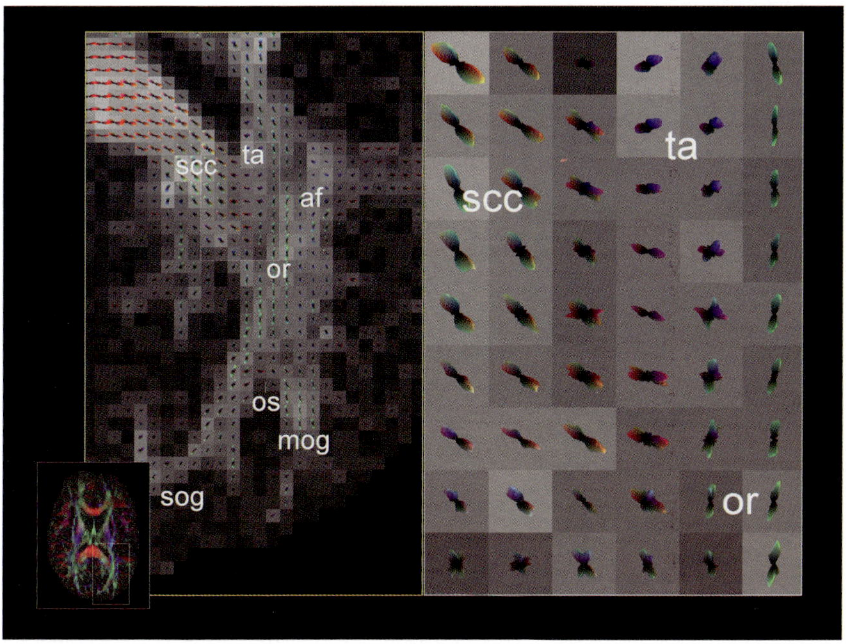

PARK, FIG. 20. Q-ball diffusion image (QBI) of three-way fiber crossing between the splenium of the corpus callosum, the tapetum, and the optic radiation. The QBI data were acquired on a 3-Tesla Siemens Trio with $b = 4000$ s/mm^2, 252 directions, and 2.2-mm isotropic resolution. af, arcuate fasciculus; mog, middle occipital gyrus; or, optic radiation; os, occipital sulcus; scc, splenium of corpus callosum; sog, superior occipital gyrus; ta, tapetum. Courtesy of David S. Tuch, Harvard Medical School. Adapted from Tuch (2004) with permission of Wiley-Liss, Inc., a subsidiary of John Wiley & Sons, Inc. © 2004 Wiley-Liss, Inc.

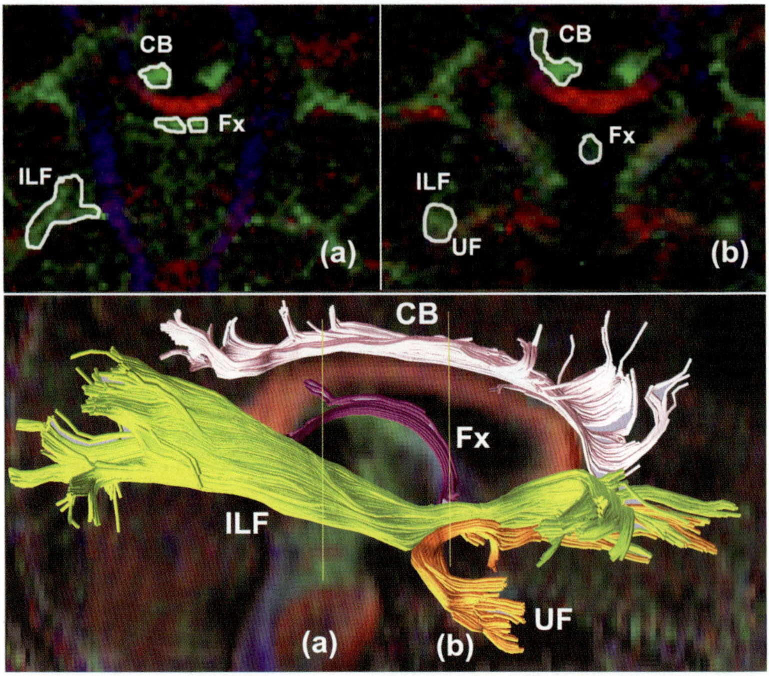

PARK, FIG. 21. Fibers of interest. The cingulum bundle (CB), inferior longitudinal fascicle (ILF), uncinate fascicle (UF), and fornix (Fx) are derived from multiple regions of interest delineated at two separate coronal slices.

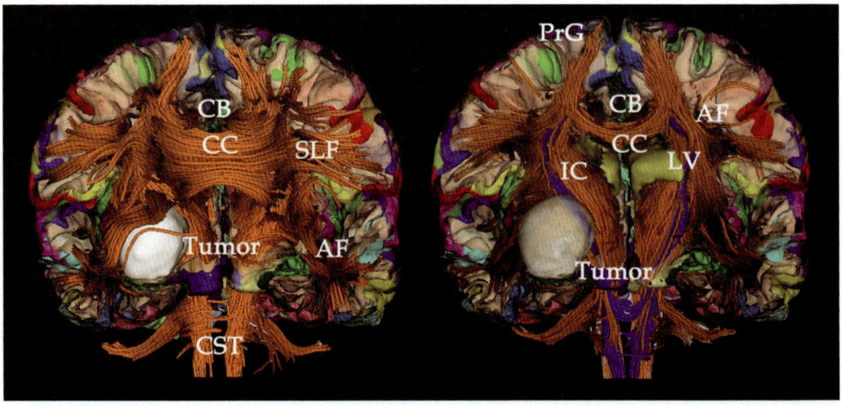

PARK, FIG. 22. Fiber tractography in tumor surgery. Fiber tractography can be used in the surgical planning for a tumor surgery. The left panel shows the tractography of the entire brain displayed with cross-sections of gray- and white-matter surface. The right panels show the fibers (in blue) that originated from the right precentral gyrus (blue color in cortex) and are located closest to the tumor. CC, corpus callosum; SLF, superior longitudinal fasciculus; CB, cingulum bundle; AF, arcuate fibers; IC, internal capsule; CST, cortico-spinal tract; PrG, precentral gyrus; LV, lateral ventricle. Adapted from Park *et al.* (2004) with permission from the American Society of Neuroradiology.

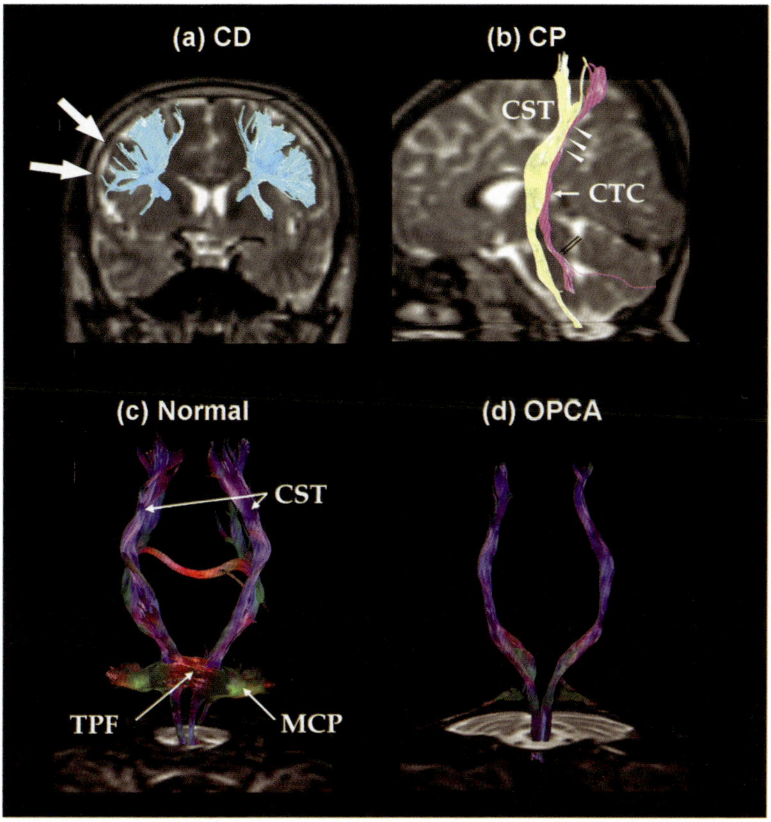

PARK, FIG. 23. Clinical applications of fiber tractography. A patient with cortical dysplasia (CD) shows reduced subcortico–cortical white-matter connectivity (a). In a cerebral palsy (CP) patient, the sensory fibers are reduced but motor fibers are normal (b). In a patient with olivopontocerebellar atrophy (OPCA), transverse pontine fibers (TPF) and the middle cerebellar peduncle (MCP) are reduced or absent (d); compare with the normal subject (c). CST, corticospinal tract; TPF, transverse pontine fibers; CTC, cortico–thalamic connection. Courtesy of Seung-Koo Lee, Department of Diagnostic Radiology, Yonsei University.

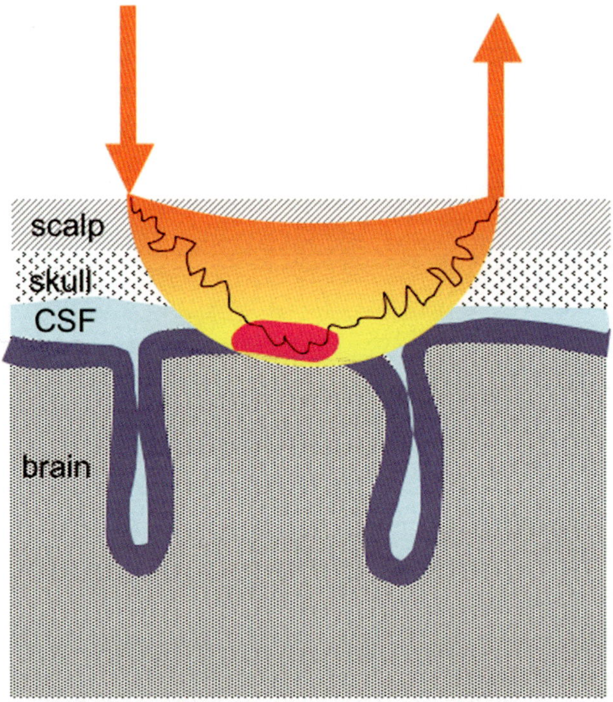

HOSHI, FIG. 1. Schematic representation of near-infrared light traveling through the head. The red area denotes a region of cortical activation. Determination of the path length through this region is required for quantification of changes in cerebral hemoglobin concentration. CSF, cerebrospinal fluid.

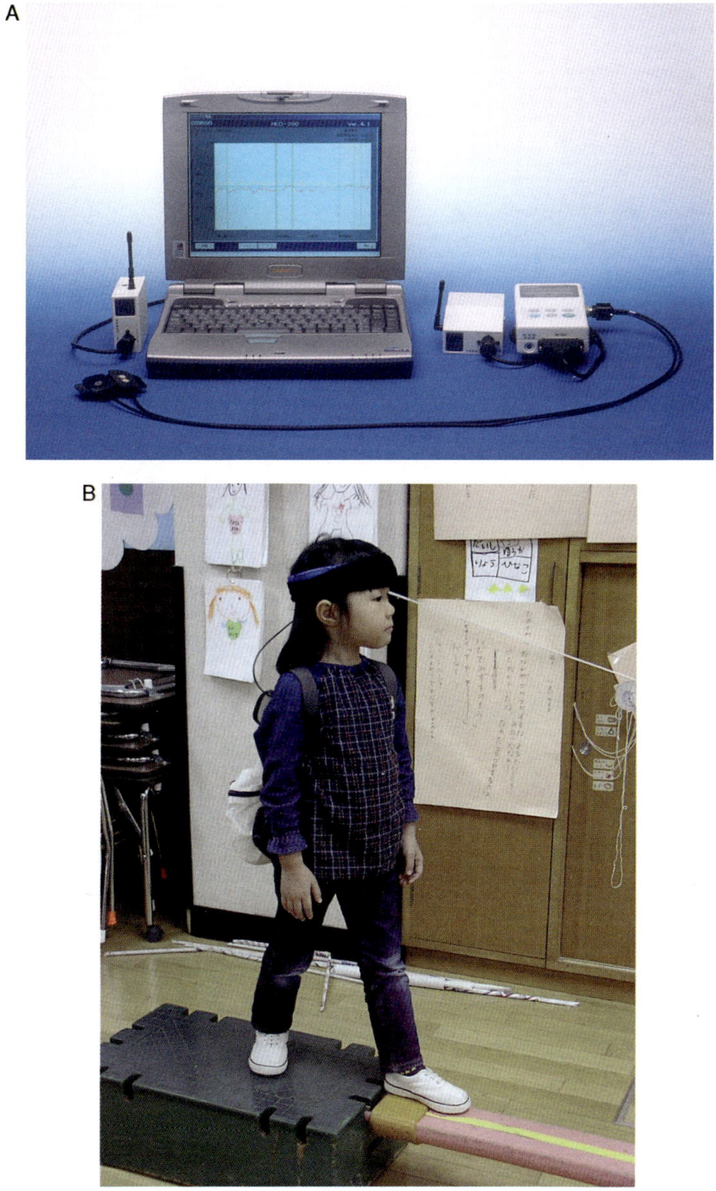

HOSHI, FIG. 9. (A) Wearable near-infrared spectroscopy (NIRS) system. (B) Measurement being performed using the wearable NIRS system.

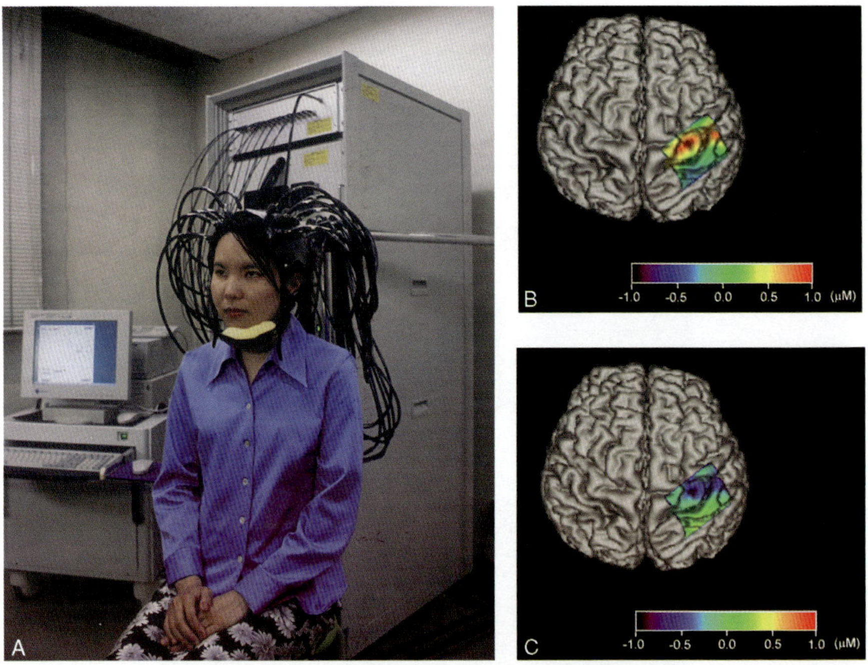

HOSHI, FIG. 10. Multichannel time-resolved optical imaging system (A) and topographical images of oxygenated hemoglobin (B) and deoxygenated hemoglobin (C) during left finger tapping.

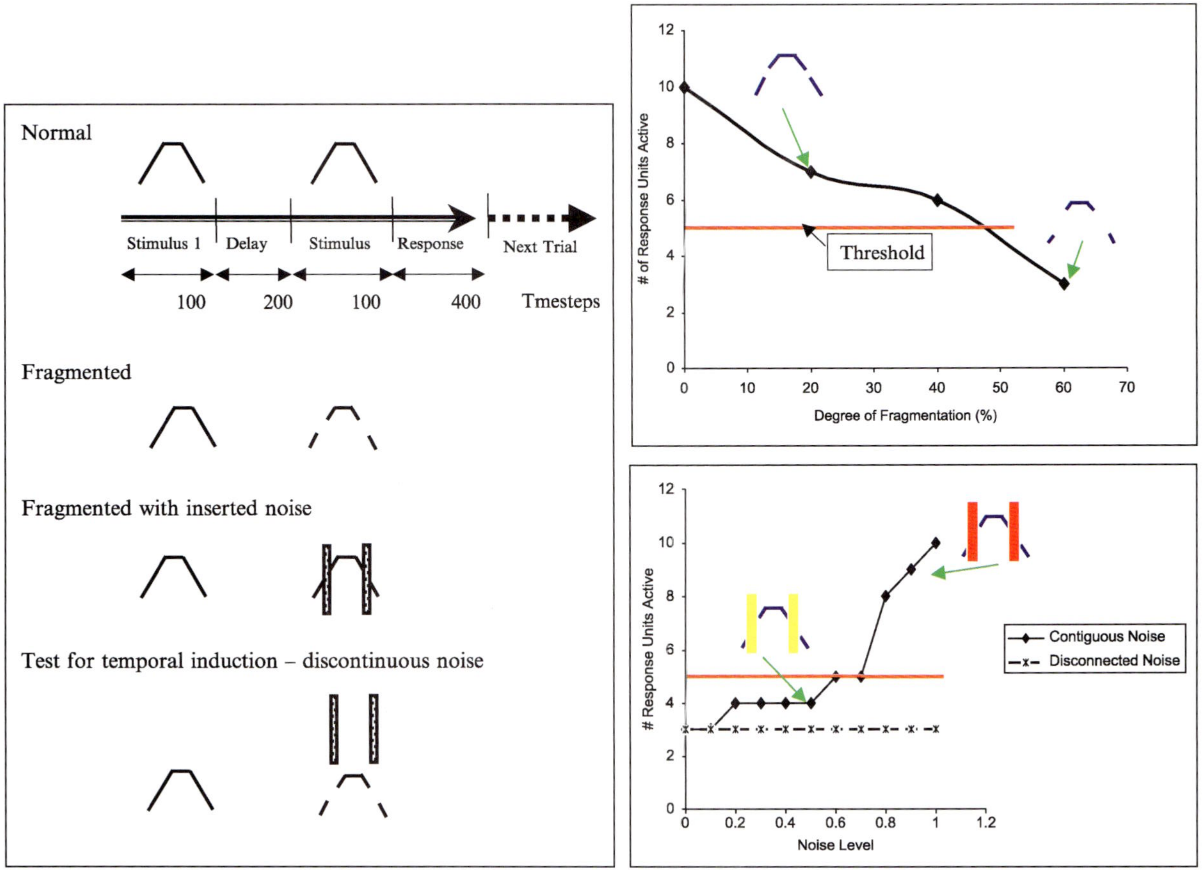

HORWITZ AND GLABUS, FIG. 4. (*Continued*)

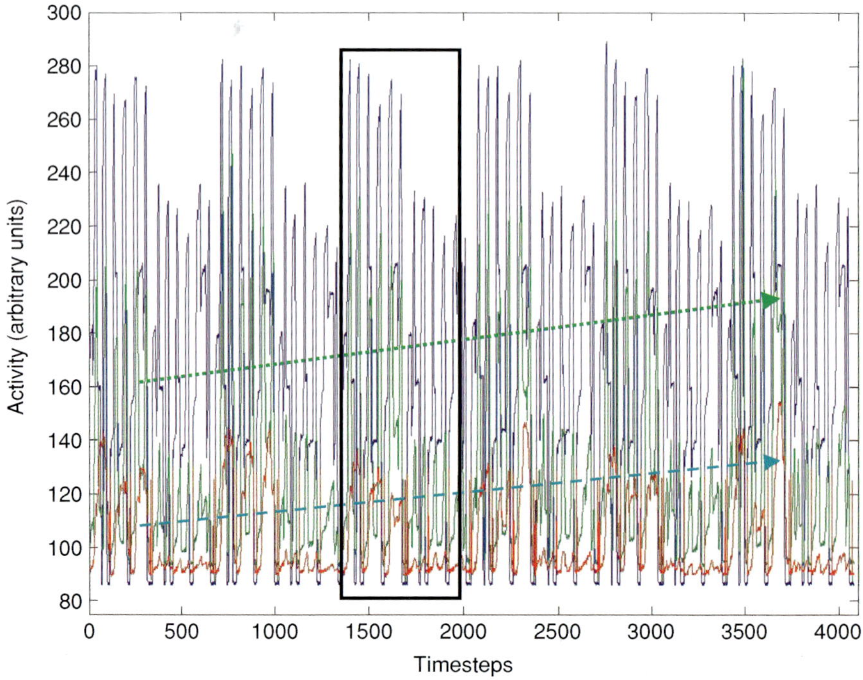

HORWITZ AND GLABUS, FIG. 5. Graphs of the integrated synaptic activity in the V1/V2 (blue), inferior temporal (IT) (green) and D1 (red) modules. As seen in the V1/V2 graph, there were three DMS trials and three control trials for each of six levels of attention (the lowest is on the left; the highest is on the right). The activities in the IT and D1 modules increase with attention, as indicated by the dashed lines. One block, with three DMS and three control trials, is shown in the box. The temporal integration interval was 50 msec. The values of the activity (y-axis) are in arbitrary units. Modified from Horwitz *et al.* (2005).

HORWITZ AND GLABUS, FIG. 4. Left: Illustration of the auditory continuity illusion. Shown at the top is the delayed match-to-sample (DMS) task for a tonal contour (indicated as a frequency-versus-time pattern; each tonal contour is about 350 msec in duration). Fragmented tonal contours, with and without inserted noise, are shown on the bottom. If the noise spans the frequency range of the contour and is loud enough, the fragmented tonal contour is perceived as intact. Right: Performance of the auditory model of Husain *et al.* (2004). Above-threshold activity of 5 or more neurons in the response module (FR) indicates a match between the two stimuli of a DMS trial. The top graph shows that with a short duration gap in the second stimulus, the model (like actual subjects) indicates a match, thus grouping the parts of the tonal contour into a perceptual whole. As the gap widens, a non-match results. If noise is inserted in the gap, and is of weak intensity (yellow), the tonal contour is not considered to continue through the gap (bottom right); if the noise is more intense (red), then perceptual grouping occurs, but only if the band of noise is in the part of frequency space occupied by the tonal contour. See Husain *et al.* (2005) for details.

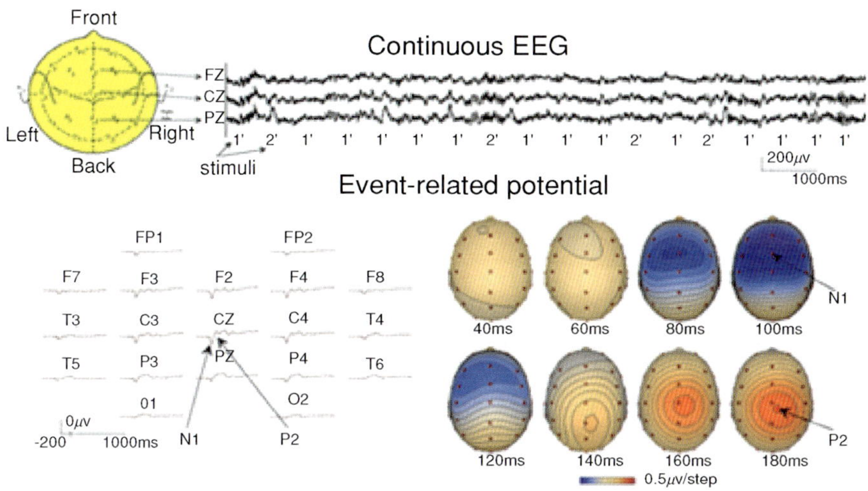

MENON AND CROTTAZ-HERBETTE, FIG. 1. Top: Raw EEG recorded at electrodes FZ, CZ, and PZ during the presentation of two different auditory tones, labeled "1" and "2." The EEG responses following the presentation of the tones are not visible on the raw EEG. Bottom left: Event-related potentials (ERPs) derived by averaging EEG segments following the presentation of the tones (e.g., stimulus 1 on the raw EEG). The N1 and P2 peaks are negative and positive deflections at 100 msec and 200 msec poststimulus, respectively. Bottom right: Topographical maps every 20 ms, from 40 ms to 180 ms poststimulus, presenting the N1 and P2 peaks of the ERPs.

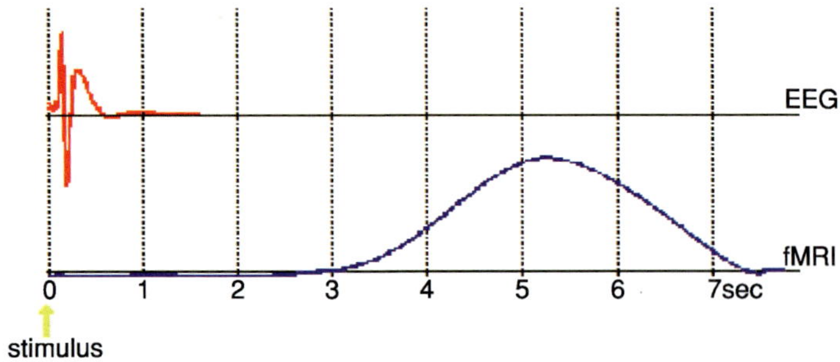

Menon and Crottaz-Herbette, Fig. 2. Relative timing of EEG and fMRI responses following the presentation of a stimulus. The EEG response lasts about 1 second poststimulus, whereas the fMRI (blood oxygen level–dependent) response takes about 2–3 seconds to rise and reaches a maximum at about 5–6 seconds post-stimulus.

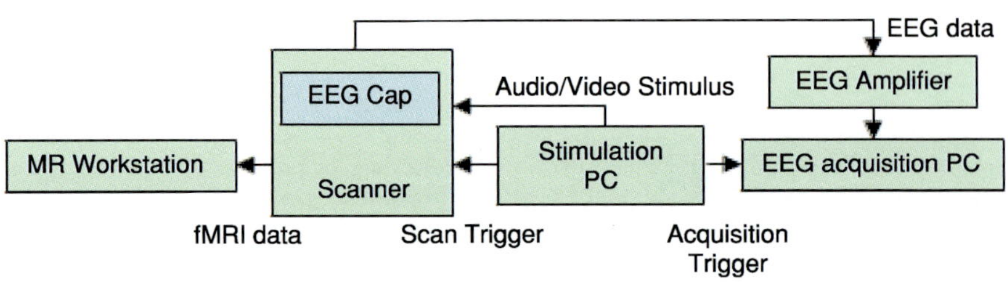

Menon and Crottaz-Herbette, Fig. 3. Schematic diagram showing basic features of a combined EEG–fMRI data acquisition system.

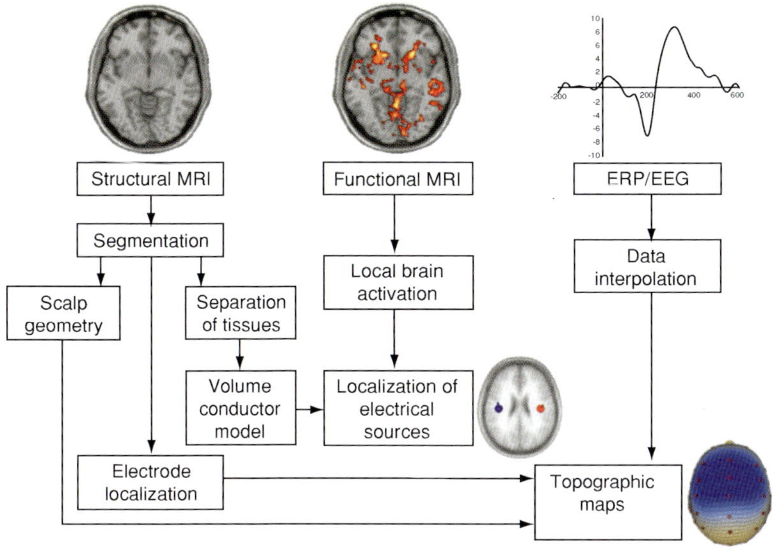

MENON AND CROTTAZ-HERBETTE, FIG. 6. Schematic overview of some of the steps involved in biophysical modeling of EEG data using constraints derived from structural and functional MRI data.

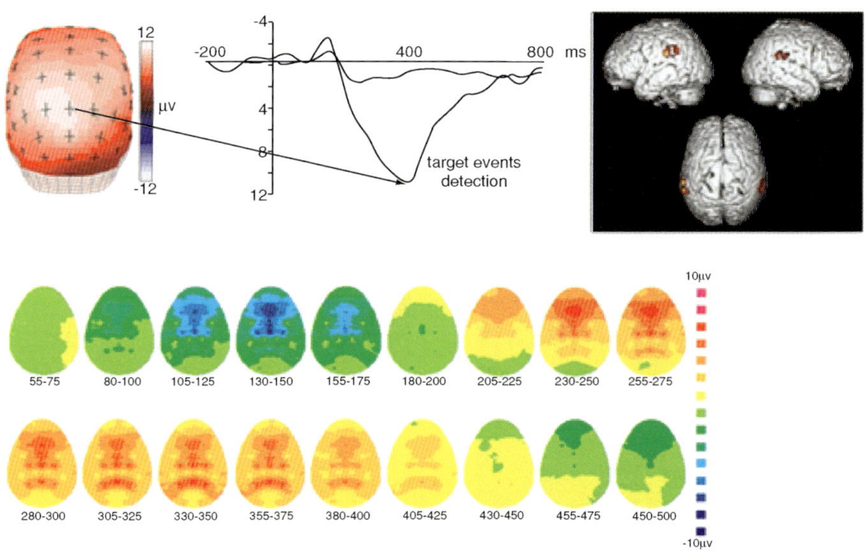

MENON AND CROTTAZ-HERBETTE, FIG. 7. Top left: Topography and time course of event-related potentials (ERPs) from normal adults during an "oddball" target detection task. The P300 component elicited during target detection has maximum amplitude over the parietal cortex approximately 300–400 ms after onset of the deviant stimulus. Top right: Event-related fMRI responses during target detection. Activated regions include the left and right temporo-parietal cortex 1. Bottom: Spatio-temporal dynamics of brain potentials acquired during target detection.

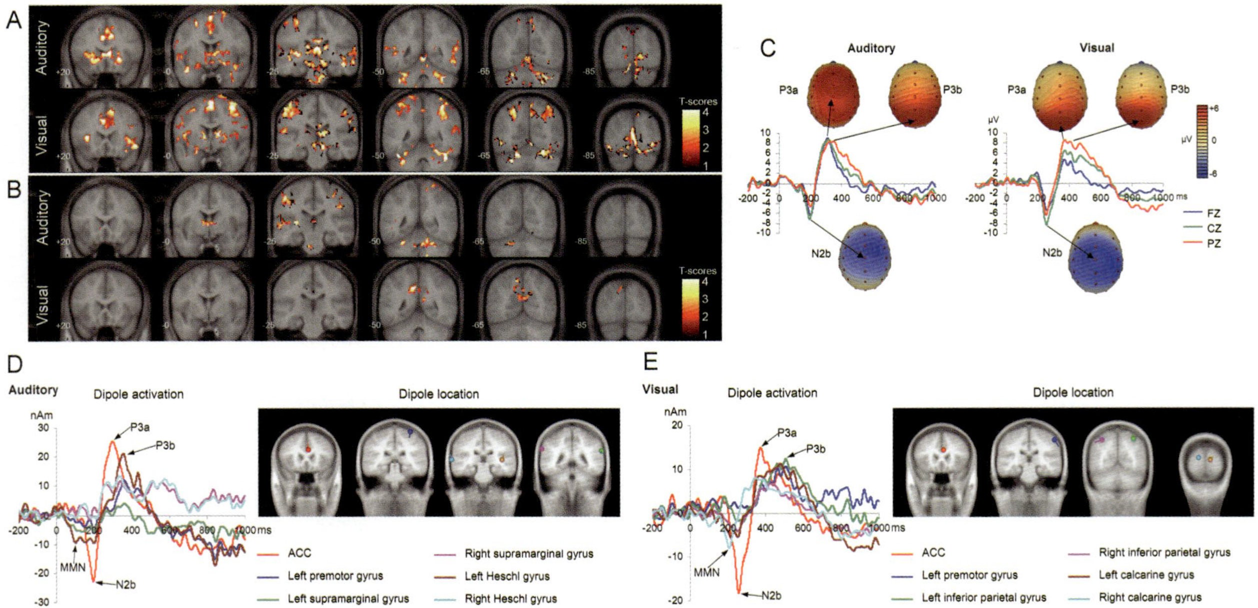

MENON AND CROTTAZ-HERBETTE, FIG. 8. (A) Brain areas that showed significantly greater activation during auditory (top) and visual (bottom) target detection. (B) Brain regions that showed significantly greater connectivity with the anterior cingulate cortex (ACC) during target detection. (C) Difference waveforms (targets minus standards) of the auditory and visual ERPs at electrodes FZ, CZ, and PZ. The topographical maps shown for the N2 and P3 components, however, reveal very little useful information. (D and E) The fMRI-derived dipole model of the ERPs, however, shows a large contribution by the ACC to the N2b–P3a components in both auditory and visual modalities.